Unlocking the Potential of Agri-Food Waste for Innovative Applications and Bio-Based Materials

Unlocking the Potential of Agri-Food Waste for Innovative Applications and Bio-Based Materials

Guest Editors

Silvia Tabasso
Emanuela Calcio Gaudino

MDPI

Basel • Beijing • Wuhan • Barcelona • Belgrade • Novi Sad • Cluj • Manchester

Guest Editors
Silvia Tabasso
Department of Drug Science and Technology
University of Turin
Torino
Italy

Emanuela Calcio Gaudino
Department of Drug Science and Technology
University of Turin
Torino
Italy

Editorial Office
MDPI AG
Grosspeteranlage 5
4052 Basel, Switzerland

This is a reprint of the Special Issue, published open access by the journal *Applied Sciences* (ISSN 2076-3417), freely accessible at: https://www.mdpi.com/journal/applsci/special_issues/7YKNEL7M74.

For citation purposes, cite each article independently as indicated on the article page online and as indicated below:

Lastname, A.A.; Lastname, B.B. Article Title. *Journal Name* **Year**, *Volume Number*, Page Range.

ISBN 978-3-7258-6304-4 (Hbk)
ISBN 978-3-7258-6305-1 (PDF)
https://doi.org/10.3390/books978-3-7258-6305-1

Cover image courtesy of Emanuela Calcio Gaudino

Contents

About the Editors

Silvia Tabasso

Silvia Tabasso is an Associate Professor in Industrial Chemistry at the Department of Drug Science and Technology, University of Turin. She received her Ph.D. in Chemistry from the University of Turin, with a thesis title "Fungal Metabolites: Isolation, Structural Characterization, Bioactivity, and Synthesis". Her research activity has been devoted to the extraction, characterization, synthesis, and possible industrial applications of natural products and derivatives, with the aim of identifying new compounds that can replace those obtained from fossil resources. The common thread linking these different areas of research is sustainability, in terms of (renewable) resources, products, and processes, using green solvents and enabling technologies. Prof. Tabasso is co-author of six book chapters and of 95 scientific publications (h-index 28) in international journals surveyed by ISI and Scopus and with Impact Factors (IFs).

Emanuela Calcio Gaudino

Emanuela Calcio Gaudino received her Ph.D. in Science and Innovative Technologies in 2011 under the supervision of Prof. Cravotto, working on a research project entitled "Innovative Eco-friendly Strategies for Process Intensification in Organic Synthesis." She is currently an Associate Professor of Organic Chemistry at the Department of Drug Science and Technology, University of Turin (Italy). Her research focuses on the synthesis of fine chemicals and bioactive compounds using non-conventional methods and conditions (green solvents and reagents, heterogeneous catalysis, high-energy microenvironments, and enabling technologies). She is also involved in the study of biomass waste as a renewable source of chemicals; in particular, she works on the conversion of biomass into chemicals through non-conventional methods (solvent- and catalyst-free reactions, high-energy microenvironments, etc.). She is the author of 95 papers in peer-reviewed journals (H-index: 29) and 4 book chapters.

Preface

This Reprint compiles the articles published in the Special Issue entitled "Unlocking the Potential of Agri-Food Waste for Innovative Applications and Bio-Based Materials". The Special Issue has been completed, and the present Reprint serves to provide readers with a consolidated collection of the published contributions in a single volume for ease of reference.

Silvia Tabasso and Emanuela Calcio Gaudino
Guest Editors

applied sciences

Editorial

Unlocking the Potential of Agri-Food Waste for Innovative Applications and Bio-Based Materials

Emanuela Calcio Gaudino * and Silvia Tabasso *

Department of Drug Science and Technology, University of Turin, 10125 Turin, Italy
* Correspondence: emanuela.calcio@unito.it (E.C.G.); silvia.tabasso@unito.it (S.T.);
Tel.: +39-011-670-7100 (E.C.G.); +39-011-670-3686 (S.T.)

Abstract: The Special Issue "Unlocking the Potential of Agri-Food Waste for Innovative Applications and Bio-Based Materials" brings together recent advances and emerging strategies for the valorization of agri-food residues. This Editorial provides an overview of the contributions included in the Special Issue, highlighting innovative approaches that convert waste streams into valuable bio-based materials, chemicals, and products. The collected works demonstrate how hydrodynamic, chemical, biological, and catalytic processes can be integrated to achieve sustainable waste management and circular resource recovery. By summarizing the main findings and perspectives, this Editorial emphasizes the growing relevance of agri-food waste valorization within the framework of the circular bioeconomy and encourages further interdisciplinary collaboration to accelerate the transition toward sustainable production systems.

1. Introduction

The twenty-first century is witnessing an unprecedented convergence of two major global challenges: the overproduction of waste and the unsustainable consumption of natural resources. Agriculture and food processing industries, although essential to human survival, are among the largest generators of waste streams. Crop residues, food processing by-products, expired or unsold food, and lignocellulosic biomass are produced in massive amounts each year, with a large fraction underutilized or improperly disposed of. According to the Food and Agriculture Organization (FAO), nearly one-third of the food produced globally for human consumption—approximately 1.3 billion tons per year—is wasted. This waste not only represents an ethical and social issue but also translates into severe environmental consequences, including greenhouse gas emissions, soil and water pollution, and resource depletion [1,2]. In this context, the concept of a circular bioeconomy has emerged as a paradigm shift from the traditional linear "take–make–dispose" model. The valorization of agri-food waste is at the core of this approach, aiming to transform residues into valuable resources and reduce dependence on fossil-based feedstocks. Emerging technologies enable the conversion of agricultural residues into bio-based materials, renewable energy, nutraceuticals, biodegradable composites, or innovative products for agricultural practices [3–5]. These strategies also contribute to achieving the United Nations Sustainable Development Goals (SDGs), particularly SDG 12 ("Responsible Consumption and Production") and SDG 13 ("Climate Action") [6]. This Special Issue, "Unlocking the Potential of Agri-Food Waste for Innovative Applications and Bio-Based Materials", was conceived to capture recent advances in this vibrant field of research. The publications in this Special Issue highlight how different waste streams can be upgraded through thermal, chemical, and biological approaches, offering sustainable alternatives to conventional practices. By presenting either experimental studies or reviews, this Special Issue underlines the transformative power of science and technology in addressing one of the most pressing environmental challenges of our time.

Agri-Food Waste as a Strategic Resource

Agri-food residues encompass a wide variety of materials, from lignocellulosic biomass such as husks, stalks, and shells to protein-rich and lipid-rich by-products of food processing. Their heterogeneous composition makes them suitable for a broad spectrum of valorization routes [7,8]. Among the most prominent technological strategies are

- Thermochemical processes, such as pyrolysis, which can convert biomass into biochar, syngas, and bio-oils with high energy potential [9].
- Biotechnological routes, including fermentation and enzymatic treatments, that enable the extraction or synthesis of biopolymers and bioactive compounds [10].
- Material science approaches, in which natural fibers and biopolymers are integrated into composites or packaging materials to replace petroleum-derived plastics [11].
- Nutraceutical and pharmaceutical valorization, where secondary metabolites from agri-food waste provide bioactive compounds with antimicrobial, antioxidant, or therapeutic functions [12].

The integration of these approaches exemplifies the vision of a circular economy: using what is conventionally regarded as waste to generate high-value products, thereby closing the loop of production and consumption [13].

2. Contributions in This Special Issue

The articles collected in this Special Issue describe the multifaceted opportunities of agri-food waste valorization, spanning applications in energy, antimicrobial compounds, biodegradable materials, and sustainable agricultural inputs.

2.1. *Residues from Oil-Pressing Processes as Biochar for Alternative Fuels*

The authors investigated the pyrolysis of oilseed press cakes from fennel flower, rapeseed, flax, evening primrose, milk thistle, and hemp. Biochar derived from these residues can reach calorific values as high as 27 MJ kg^{-1}—comparable to high-grade fossil fuels—and exhibited reduced emissions of carbon monoxide, nitrogen oxides, and sulfur dioxide [14].

2.2. *Chestnut Burrs as a Source of Antimicrobial Bioactive Compounds*

Chestnut burrs, which are usually discarded as waste, were shown to contain gallic acid, quinic acid, protocatechuic acid, brevifolin carboxylic acid, and ellagic acid. The study demonstrated modest antibacterial activity, particularly against *Enterococcus faecalis*, with a minimum inhibitory concentration (MIC) of 64 μg/mL. Complementary in silico analyses provided insights into molecular targets, supporting potential drug discovery applications [15].

2.3. *The Role of Natural Plant and Animal Fibers in PLA Composites*

The incorporation of flax and sheep wool fibers into polylactic acid (PLA)-based composites improved mechanical properties (flax) and accelerated degradation in soil (wool). Although wool fibers promoted biodegradability, excessive wool content reduced mechanical strength, highlighting the trade-off between durability and degradability in material design [16].

2.4. Harnessing Agri-Food Waste as a Source of Biopolymers for Agriculture

The extraction and synthesis of natural biopolymers from lignin, cellulose, pectin, starch, and chitosan were reviewed. Applications include mulching films, hydrogels, nanocarriers, soil stabilizers, and seed coatings. These sustainable polymers can replace fossil-based plastics in agriculture, embodying a circular economy approach [17].

3. Challenges and Future Perspectives

While the works in this Special Issue clearly demonstrate the potential of agri-food waste valorization, significant challenges remain to be addressed. These include economic feasibility and scale-up as current strategies are often demonstrated at pilot scale, requiring cost-effective processes, robust supply chains, and supportive policies for industrial deployment [18]; the heterogeneity of raw materials, as agri-food waste streams vary widely in composition, complicating standardization and requiring adaptive technologies; regulatory frameworks and safety concerns, as applications in food, feed, or biomedical sectors demand strict compliance with safety standards and clear regulatory pathways [19]; and integration into circular systems, as valorization must be integrated into existing agricultural and industrial systems, optimizing energy, water, and material flows holistically [20]. Future studies will benefit from interdisciplinary research, hybrid technologies combining thermochemical and biological routes, and the integration of digital tools such as life-cycle assessment and artificial intelligence. Societal acceptance and consumer awareness will also be critical to the success of waste-derived products.

4. Conclusions

This Special Issue demonstrates that agri-food waste is not an endpoint but a starting point for innovation. From biochar fuels to antimicrobial extracts and from biodegradable composites to agricultural biopolymers, these contributions underscore the versatility of waste streams as sustainable feedstocks. Valorization of agri-food waste reduces environmental burdens, creates value-added products, and promotes circular economy practices. At the same time, continuous innovation, interdisciplinary collaboration, and supportive policies are essential to overcoming existing barriers.

As Guest Editors, we extend our gratitude to all authors for their contributions, to reviewers for their constructive feedback, and to the editorial staff for their support. It is our hope that this Special Issue inspires further research and collaboration to transform agri-food waste into a springboard for sustainable progress.

Author Contributions: Conceptualization, E.C.G. and S.T.; writing—original draft preparation, E.C.G. and S.T.; writing—review and editing, E.C.G. and S.T. All authors have read and agreed to the published version of the manuscript.

Conflicts of Interest: The authors declare no conflicts of interest.

References

1. FAO. *The State of Food and Agriculture 2023*; Food and Agriculture Organization of the United Nations: Rome, Italy, 2023.
2. Gustavsson, J.; Cederberg, C.; Sonesson, U. *Global Food Losses and Food Waste*; FAO Report: Rome, Italy, 2011.
3. Abass, A.; Gazal, S.; Bonnet, S. Circular Economy Strategies for Agri-Food Production. *Circ. Econ. Sust.* **2025**, *5*, 2467–2493. [CrossRef]
4. Wagh, M.S.; Sowjanya, S.; Nath, P.C.; Chakraborty, A.; Amrit, R.; Mishra, B.; Mishra, A.K.; Mohanta, Y.K. Valorisation of Agro-Industrial Wastes: Circular Bioeconomy and Biorefinery Process—A Sustainable Symphony. *Process Saf. Environ. Prot.* **2024**, *183*, 708–725. [CrossRef]
5. Shaba, E.Y.; Kapoor, R.T.; Sillanpää, M.; Zdarta, J.; Rafatullah, M. Biomass Valorisation: A Sustainable Approach Towards Carbon Neutrality and Circular Economy. In *Biomass Valorization*; Springer: Singapore, 2024. [CrossRef]

6. United Nations. Sustainable Development Goals (SDG 12, SDG 13). Available online: https://sdgs.un.org/goals (accessed on 22 October 2025).
7. Li, Y.; Gupta, R.; Zhang, Q.; You, S. Review of Biochar Production via Crop Residue Pyrolysis: Development and Perspectives. *Bioresour. Technol.* **2023**, *369*, 128423. [CrossRef] [PubMed]
8. Cusenza, M.A.; Longo, S.; Cellura, M.; Guarino, F.; Messineo, A.; Mistretta, M.; Volpe, M. Environmental Assessment of a Waste-to-Energy Practice: The Pyrolysis of Agro-Industrial Biomass Residues. *Sustain. Prod. Consum.* **2021**, *28*, 866–876. [CrossRef]
9. Varkolu, M.; Gundekari, S.; Omvesh; Palla, V.C.S.; Kumar, P.; Bhattacharjee, S.; Vinodkumar, T. Recent Advances in Biochar Production, Characterization, and Environmental Applications. *Catalysts* **2025**, *15*, 243. [CrossRef]
10. Ezeorba, T.P.C.; Okeke, E.S.; Mayel, M.H.; Nwuche, C.O.; Ezike, T.C. Recent Advances in Biotechnological Valorization of Agro-Food Wastes (AFW): Optimizing Integrated Approaches for Sustainable Biorefinery and Circular Bioeconomy. *Bioresour. Technol. Rep.* **2024**, *26*, 101823. [CrossRef]
11. Mishra, R.; Wiener, J.; Militky, J.; Michal, P.; Blanka, T.; Jana, N. Bio-Composites Reinforced with Natural Fibers: Comparative Analysis of Thermal, Static and Dynamic-Mechanical Properties. *Fibers Polym.* **2020**, *21*, 619–627. [CrossRef]
12. Bekavac, N.; Krog, K.; Stanić, A.; Šamec, D.; Šalić, A.; Benković, M.; Jurina, T.; Gajdoš Kljusurić, J.; Valinger, D.; Jurinjak Tušek, A. Valorization of Food Waste: Extracting Bioactive Compounds for Sustainable Health and Environmental Solutions. *Antioxidants* **2025**, *14*, 714. [CrossRef] [PubMed]
13. Nguyen, T.H.; Wang, X.; Utomo, D.; Gage, E.; Xu, B. Circular Bioeconomy and Sustainable Food Systems: What Are the Possible Mechanisms? *Clean. Circ. Bioecon.* **2025**, *11*, 100145. [CrossRef]
14. Saletnik, B.; Czarnota, R.; Maczuga, M.; Saletnik, A.; Bajcar, M.; Zaguła, G.; Puchalski, C. Residues from the Oil Pressing Process as a Substrate for the Production of Alternative Biochar Materials. *Appl. Sci.* **2024**, *14*, 8028. [CrossRef]
15. Trezza, A.; Barletta, R.; Geminiani, M.; Frusciante, L.; Olmastroni, T.; Sannio, F.; Docquier, J.-D.; Santucci, A. Chestnut Burrs as Natural Source of Antimicrobial Bioactive Compounds: A Valorization of Agri-Food Waste. *Appl. Sci.* **2024**, *14*, 6552. [CrossRef]
16. Szczepanik, E.; Szatkowski, P.; Molik, E.; Pielichowska, K. The Effect of Natural Plant and Animal Fibres on PLA Composites Degradation Process. *Appl. Sci.* **2024**, *14*, 5600. [CrossRef]
17. Valle, C.; Voss, M.; Calcio Gaudino, E.; Forte, C.; Cravotto, G.; Tabasso, S. Harnessing Agri-Food Waste as a Source of Biopolymers for Agriculture. *Appl. Sci.* **2024**, *14*, 4089. [CrossRef]
18. Khanna, M.; Zilberman, D.; Hochman, G.; Basso, B. An Economic Perspective of the Circular Bioeconomy in the Food and Agricultural Sector. *Commun. Earth Environ.* **2024**, *5*, 507. [CrossRef]
19. European Commission. *Food Safety and Waste Valorization Regulations*; European Union: Brussels, Belgium, 2024.
20. Romero-Perdomo, F.; González-Curbelo, M.Á. Integrating Multi-Criteria Techniques in Life-Cycle Tools for the Circular Bioeconomy Transition of Agri-Food Waste Biomass: A Systematic Review. *Sustainability* **2023**, *15*, 5026. [CrossRef]

applied sciences

Article

The Impact of Dextrin-Activated Expanded Perlite and Vermiculite Particles on the Performance of Thermal Insulating Rapeseed Oil-Based Polyurethane Foam Composites

Agnė Kairytė * and Aliona Levina

Laboratory of Thermal Insulating Materials and Acoustics, Institute of Building Materials, Faculty of Civil Engineering, Vilnius Gediminas Technical University, Linkmenų st. 28, 08217 Vilnius, Lithuania; levinaaliona@gmail.com
* Correspondence: agne.kairyte@vilniustech.lt

Featured Application: The developed rapeseed oil-based polyurethane foam composites with dextrin-activated expanded vermiculite and perlite fillers could be used as efficient thermal insulating materials for walls, roofs, and floors, reducing heat transfer and energy consumption in buildings.

Abstract: To enhance the performance of polyurethane foams, fillers are often incorporated into the matrix. However, the interaction between the filler and the polyurethane matrix is crucial for achieving the desired property improvements. Therefore, surface modification of the fillers plays a vital role in optimizing this interaction. The current study aims to activate the surface of expanded vermiculite and perlite with dextrin to incorporate additional functional groups on the surface of the fillers via the ball-milling process, thereby improving the reaction with a polymer matrix. Applied surface activation with dextrin resulted in the formation of dextrin-Si-O-Si-dextrin linkages in the fillers, allowing for a maximum improvement of 11% and 9% in water absorption, as well as slightly positive changes in the water contact angle of polyurethane foam with dextrin-activated perlite and vermiculite, respectively, compared to non-activated fillers. It also resulted in noticeable differences in the foaming times and viscosity of the premixes, affecting the structure of rigid polyurethane foam composites. Compared to non-activated perlite and vermiculite filler polyurethane foam composites, the dynamic viscosity of polyurethane foam composites with dextrin-activated perlite and vermiculite reduced maximally 16 and 21 times, respectively. At the same time, the closed cell content increased, resulting in lower thermal conductivity values up to a 7.5 wt.% filler concentration. In addition, a rise in mechanical performance was also achieved. Compressive strength increased by a maximum of 61% and 71%, while tensile strength increased by a maximum of 36% and 20% for polyurethane foam composites with dextrin-activated perlite and vermiculite, respectively.

Keywords: polyurethane foam composites; dextrin; expanded vermiculite; expanded perlite; interface adhesion; surface activation

1. Introduction

Rigid polyurethane (PUR) foams are highly desirable materials, particularly for thermal insulation applications, due to their beneficial characteristics [1]. The formation of PUR foams involves several competing chemical reactions. During urethane formation, the polymer backbone forms. Hydroxyl groups (−OH) from polyol react with –NCO groups from isocyanate to form urethane linkages [2]. When water is used as a chemical blowing agent, it reacts with isocyanate groups to produce unstable carbamic acid, which immediately decomposes to form a urea linkage and release carbon dioxide, acting as a blowing agent and forming the cellular structure. Then, the amine formed during the water–isocyanate reaction can further react with another isocyanate group to form a urea linkage and subsequently with another isocyanate group to form a biuret linkage, which contributes to additional cross-linking and rigidity in the PUR foams [3]. Such materials have a low density, thermal conductivity, and water absorption, as well as sufficient dimensional stability and a relatively high mechanical performance. Currently, the demand for such products in various sectors, including building, automotive, food packaging, furniture, medical, and aviation, has increased [4]. PUR products have become irreplaceable in the modern industry. Its global production exceeds 21.3 million tons annually, making it the sixth-largest category of synthetic materials worldwide [5]. The estimated value of the PUR products market in 2021 was USD 77.9 billion. According to predictions, this value is expected to increase to USD 105.3 billion by 2026 [6].

The popularity of PUR products can be attributed to the increasing cost of energy and growing awareness of the importance of energy conservation. Along with the energy-related aspects, environmental awareness associated with petroleum-based polyurethane PUR foams is of great interest to the scientific and industrial communities. Significant effort has been invested in converting petroleum-based PUR foams into more environmentally friendly alternatives. For instance, polyol is dependent on petroleum-based feedstocks, and its synthesis using biomass such as oils is in high demand. The commonly incorporated vegetable oils for bio polyol synthesis are soybean [7,8], castor [9,10], palm [11,12], rapeseed [13,14], sunflower [15,16], and even waste cooking oils [17]. In addition, Olivito et al. [18] proposed a straightforward and efficient method for obtaining furfural from fructose. Furfural often serves as an intermediate or starting point in multi-step processes and can be utilized in the synthesis of furan-based diols, as indicated by Jang et al. [19]. The synthesis of polyurethane from furan-based diols is similar to the synthesis from conventional polyols, except that the furan-based ones have two primary hydroxyl groups and a furan ring, which enhance the rigidity, stiffness, and thermal stability of the resulting product [20]. Furthermore, studies by Torresi et al. [21] and Skoczinski et al. [22] have shown that furan-based compounds can increase surface water contact angles and reduce the flammability of the resulting products, demonstrating great potential for future use as a main compound in the synthesis of polyurethanes. However, few studies indicated that replacing petroleum-based polyols with bio-polyols can lead to the degradation of several properties. Kurańska et al. [23] determined that adding a rapeseed oil-based polyol slows the reaction time, reduces compressive strength and apparent density, and deteriorates the closed-cell structure. Moreover, the thermal insulating properties also showed an adverse effect with an increase in the amount of rapeseed oil-based polyol. In addition, Tan et al. [24] determined that the thermal conductivity of PUR foams made from biopolyols ages faster than that of conventional ones. It was also noted by Kurańska et al. [25] that water absorption and dimensional instability were increased for PUR foam made from a palm oil-based polyol. The structure was weakened due to the lower miscibility of the bio polyol with petroleum-based polyol and its slower reaction rate.

The properties of PUR foams can be varied, and some drawbacks can be mitigated by incorporating various fillers to develop novel composite PUR foam materials. The most interesting fillers for thermal insulating PUR foams are those capable of improving thermal conductivity, water absorption, and mechanical performance. For instance, Valipour et al. [26] investigated nanoclays as promising fillers for enhancing the microstructural features and thermal insulating properties of PUR foam composites. Authors have concluded that a 22–26% improvement in thermal insulation was achieved. Additionally, unmodified expanded vermiculite (Verm) and perlite (Per) were incorporated into the PUR foam matrix. Wang et al. [27] suggested that Verm did not react with the PUR matrix, thus causing the formation of an irregular structure and weakening the interfacial adhesion. Although Verm and Per are relatively similar, Ai et al. [28] demonstrated that it is possible to obtain thermal insulating PUR foam composites with reduced thermal conductivity while maintaining unchanged mechanical performance. However, none of these studies touched on the impact of water on the resulting foams.

In particular, vermiculite and perlite are lightweight, sound-absorbent materials with excellent thermal insulating properties. Due to their structure, these materials offer intensive volume growth after high-temperature impact, resulting in expansion. Thermal expansion occurs perpendicularly to Verm and Per sheets, thus obtaining a highly porous structure [29]. Thanks to their expansion ability, these materials can be applied to various purposes, such as structural, thermal insulation, acoustic, agricultural, and as a filler for polymer-based products [30,31]. Furthermore, numerous studies have shown that Verm and Per are promising for reducing flammability in various composites [32–34]. However, various fillers lacking proper functional groups on their surface can sometimes lead to the formation of an improper interface between the filler and the polymer matrix. Therefore, research on the functionalization of fillers is currently of great interest to the research community. For instance, Joseph et al. [35], Hu et al. [36], and Li et al. [37] applied silane coupling agents to modify the surface of metal and clay fillers. The results demonstrated outstanding compatibility with dispersing media, enabling the avoidance of particle agglomeration and imparting chemical reactivity to the particles. Miedzińska et al. [38] proposed a more environmentally friendly approach by modifying the ball-milling process to incorporate hydroxyl functional groups into casein, chitosan, and potato protein on the surface of vermiculite. It was determined that proteins and saccharides applied to vermiculite result in PUR foam composites with reduced average cell size, improved mechanical performance, and nearly unchanged resistance to water impact. However, the reaction of the prepared premixes was slowed down with all the surface modifiers used. In addition, the same surface modification method with casein was applied to natural fillers—apricot stones [39]. The same observation about reaction kinetics was noted. Moreover, structural parameters, such as closed cell content, deteriorated, leading to increased thermal conductivity values. Although the suggested surface modification method and materials used are more sustainable, the granted performance characteristics for PUR foam composites are not entirely satisfactory.

Given the bio-based origin and availability of dextrin (Dex), its application as a sole filler surface modifier is promising for producing PUR foam composites with improved microstructural parameters and thermal insulation properties. According to Wang et al. [40], the hydroxyl groups in Dex are more easily exposed and have higher reactivity, making it suitable as a surface modifier for Verm and Per fillers. Therefore, this study aims to develop PUR foam composites with Dex-activated Verm (VermDex) and Dex-activated Per (PerDex) fillers, which could find their application as promising thermal insulation materials with improved microstructure, thermal conductivity, mechanical performance, and resistance to water impact.

2. Materials and Methods

2.1. Raw Materials

Bio-based polyol BioPolyol RD, derived from rapeseed oil, and traditional polyether polyol ROKOPOL RF-152 V were incorporated as primary raw materials for synthesizing rigid polyurethane foam composites. BioPolyol RD was supplied by SIA PolyLabs, Riga,

Latvia. It had a moisture content of <0.2 wt.% and a hydroxyl number of 360 KOH/g. ROKOPOL RF-152 V was purchased from PCC Rokita, Brzeg Dolny, Poland, with a moisture content of 0.07 wt.% and a hydroxyl number of 444 KOH/g. A polymeric 4,4-diphenylmethane diisocyanate, Lupranat M20S, was incorporated as a hardener and purchased from BASF, Ludwigshafen, Germany. Water was used as a chemical blowing agent. Polycat 9 from Air Products and Chemicals, Inc., Allentown, PA, USA, was incorporated for the blowing and gelling reactions. A silicone surfactant ST-52 was obtained from Shijiazhuang Chuanghong Technology Co., Ltd., Shijiazhuang, China. Expanded vermiculite (Verm) and perlite (Per) with an initial moisture content of 0.01% were used to reinforce the inorganic fillers. These fillers were purchased from JSC Biosprendimų laboratorija, Elektrėnai, Lithuania. Dextrin from JSC Suvena, Panevėžys, Lithuania, was used as a fillers' surface activator.

2.2. Preparation of Dextrin Modified Fillers and Polyurethane Foam Composites

Verm and Per fillers were separately ball-milled with dextrin (Dex) (Verm:Dex and Per:Dex weight ratio of 1:1) using a ball-milling machine with 6 ceramic balls for 1 h at 2500 rpm. The resulting VermDex and PerDex fillers were dried for 48 h at 105 °C in a ventilated oven, and the moisture contents were determined to be 6.3 wt.% and 7.5 wt.%, respectively. The water amount in PUR foam compositions presented in Table 1 was recalculated based on the moisture contents of Verm, Per, VermDex, and PerDex fillers. Isocyanate index was calculated according to the methodology provided by Ivdre et al. [41].

Table 1. Composition of polyurethane foam composites with fillers.

Component	Amount of Component, pbw				
	Control PUR	PUR-Verm	PUR-Per	PUR-VermDex	PUR-PerDex
Component A:					
BioPolyol RD	60	60	60	60	60
ROKOPOL RF-152 V	60	60	60	60	60
Water	3	3	3	3	3
Polycat 9	0.7	0.7	0.7	0.7	0.7
ST-52	3.6	3.6	3.6	3.6	3.6
Filler, wt.%	0	2.5; 5; 7.5; 10	2.5; 5; 7.5; 10	2.5; 5; 7.5; 10	2.5; 5; 7.5; 10
Component B:					
Isocyanate	Isocyanate index for all compositions—125				

Component A, consisting of the required polyols, water, catalyst, surfactant, and fillers, was thoroughly mixed for 60 s at 1800 rpm. Then, the calculated amount of Component B (isocyanate) was added, and the mixture was further mixed for 10 s at 1800 rpm and poured into 300 × 300 mm-sized molds. The resulting polyurethane foam composites were left to harden for 24 h at a temperature of 23 ± 5 °C and 50 ± 5% relative air humidity. After hardening, the demolded samples were further conditioned at a temperature of 23 ± 5 °C and 50 ± 5% relative air humidity before being cut into samples and tested.

2.3. Testing Methods

The moisture content of all particles was determined using a KERN DAB moisture analyzer (KERN & Sohn GmbH, Balingen, Germany).

The granulometry determination of all particles was conducted with a set of sieves consisting of the bottom (0 mm), 0.063 mm, 0.10 mm, 0.20 mm, 0.50 mm, 0.63 mm, and 1.25 mm sieves.

Fourier transform infrared (FTIR) spectroscopy was conducted for all particles using Lumos II (Bruker Optics GmbH, Ettlingen, Germany). Thirty scans at a 600–4000 cm^{-1} range with a scan interval of 4 cm^{-1} were taken for each particle type.

The dynamic viscosity test was conducted for polyol and filler premixes using an SV-10 viscometer (A&D Company Ltd., Tokyo, Japan) at a velocity of 6 rpm with an accuracy of 0.01 mPa·s.

The characteristic foaming times were determined using a stopwatch with an accuracy of 0.5 s according to the requirements of EN 14315-1 [42].

The structural analysis of the Verm, Per, VermDex, and PerDex particles, control PUR foam, and PUR foam composites was performed using an optical microscope Smart 5MP PRO (Delta Optical, Gdańsk, Poland) with a maximum magnification of up to 300 times. The average cell size of the resulting PUR foam composites was obtained using optical images using ImageJ software, v. 1.51j8.

The closed-cell content was determined using the EN ISO 4590 [43] method 2 requirements for samples with a size of 100 mm × 30 mm × 30 mm.

The thermal conductivity determination was performed at an average temperature of 10 °C on samples with a size of 300 mm × 300 mm × 50 mm, based on EN 12667 [44] requirements, using a heat flow meter FOX 304 (TA Instruments, Newcastle, DE, USA) with active edge insulation. The direction of the heat flow was upward during the test. The difference between cold and hot plates was 20 °C.

The short-term water absorption by partial immersion for control rigid polyurethane foam composites and rigid polyurethane foam composites filled with Verm, Per, VermDex, and PerDex particles was determined based on the ISO 29767 [45] method A requirements for samples measuring 200 mm × 200 mm × 50 mm. Excess water in the samples was drained for 10 min using a drainage stand at 45° after the test.

The surface hydrophobicity of control rigid polyurethane foam and rigid polyurethane foam composites filled with different particles was determined by the water contact angle method. The test was implemented using a goniometer with an optical system (Ossila Ltd., Sheffield, UK) to capture the angle of water drop on the tested surfaces. Water drops of 1 μL were deposited using a micrometer syringe. The results were taken 10 s after water droplet deposition. Measurements were done at least three times on each sample.

The tensile and compressive strengths were determined in accordance with EN 1607 [46] and ISO 29469 [47] requirements. The samples used for the tests were 50 mm × 50 mm × 50 mm in size. A Universal Testing Machine, H10KS Hounsfield (Tinius Olsen Ltd., Surrey, UK), will monitor the behavior of the samples during compression and tension.

Experimental data processing and reliability evaluations were performed using mathematical and statistical methods with the software STATISTICA v.8.

3. Results and Discussion

3.1. Characterization of Fillers

The results show that the size of the non-activated Per and Verm fillers (Figure 1b,d) was up to 1.25 mm. The highest percentage of particles was observed at 0.2 mm. Per filler consisted of snowflake-like particles (Figure 1a). In contrast, Verm particles were larger in size and non-homogeneous in shape (Figure 1c). In addition, Figure 1f,h demonstrate Dex-activated Per and Verm fillers (PerDex and VermDex, respectively). It can be seen that the fillers' particle size distribution ranged up to 0.63 mm, with the highest accumulation of particles observed on the 0.1 mm sized sieve. The smaller particles in both PerDex and VermDex fillers were obtained through the ball-milling process, as confirmed by Sakr et al. [48], which alters their surface, size, and chemical reactivity. PerDex filler also has a snowflake-like structure with the addition of round-shaped particles (Figure 1e), while VermDex filler is more like sand-like particles with brownish inserts of Dex particles (Figure 1g). In order to check the efficacy of Dex adsorption on the surface of Per and Verm fillers, FTIR spectra of each of the filler types are obtained, and the results are presented in Figure 2.

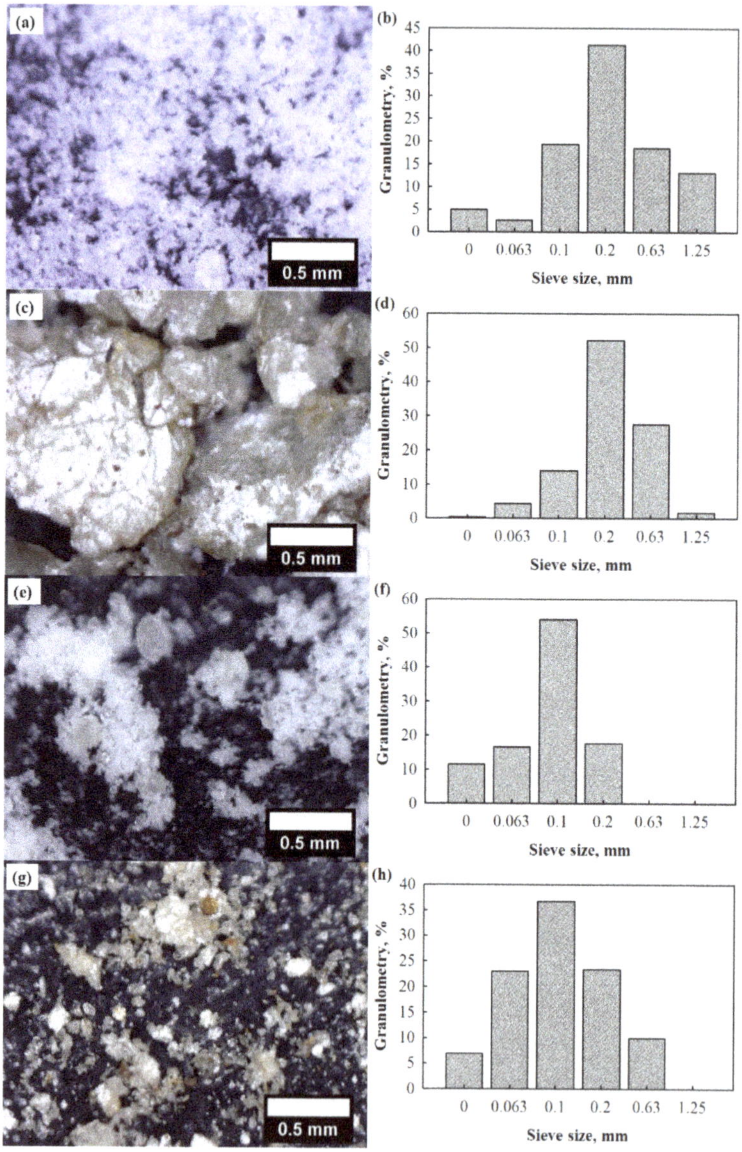

Figure 1. Optical images and granulometry of fillers: (**a**,**b**) Per; (**c**,**d**) Verm; (**e**,**f**) PerDex, and (**g**,**h**) VermDex.

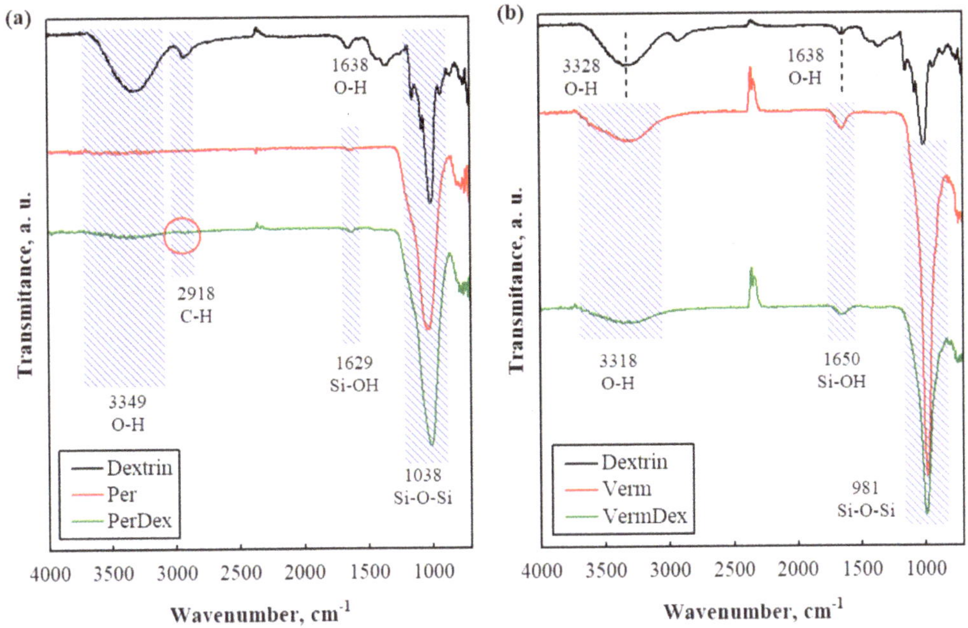

Figure 2. FTIR spectra of fillers: (**a**) Dex, Per, and PerDex (red circle indicate C-H stretching) and (**b**) Dex, Verm, and VermDex.

FTIR analysis was employed to assess the presence of functional groups in the fillers before and after activation. Figure 2 presents FTIR spectra of the Per and PerDex, Verm, and VermDex fillers. For the Per filler (Figure 2a), the band located in the 1669–1577 cm^{-1} range, with a peak at 1629 cm^{-1}, is assigned to the vibration of hydroxyl groups. A large band observed at 1038 cm^{-1}, within the range of 1280–864 cm^{-1}, corresponds to the stretching vibrations of Si-O-Si [49]. Further, the peak at 789 cm^{-1} is assigned to the Si-O stretching vibration of Si-O-Al [50].

After the activation of the Per with Dex, a new band is noticed at the range of 3359–3082 cm^{-1} with a peak intensity at 3349 cm^{-1}, which is associated with the presence of water molecules in the material structure, while a less pronounced band at 2918 cm^{-1} due to the C-H stretching of the -CH_2 group in the case of Dex absorbed Per. This indicates that hydroxyl groups of Dex only partially react with silanols in Per filler.

For Verm filler (Figure 2b), the broad band between 3726 cm^{-1} and 3016 cm^{-1} is related to interlayer water molecules' −OH stretching vibration. The peak of 3416 cm^{-1} is attributed to the silanol groups in Verm and VermDex [51] and −OH vibrations. The spectrum of Verm and VermDex exhibited a weak band at 1648 cm^{-1}, corresponding to Si-OH vibrations. In addition, a strong band in the range of 1210–827 cm^{-1}, with a peak at 981 cm^{-1}, is associated with the Si-O stretching vibration [52].

Contrary to the case of Per and PerDex fillers, after activation with Dex, a hydroxyl stretching frequency, which should be visible at ~2918 cm^{-1} in VermDex filler, has no intensity, thus indicating that the −OH groups of Dex are fully reacted with the silanol groups of Verm. In this case, the formation of a Dex–O–Si–O–Dex linkage could be assumed. Similar observations were made in the study of Salih et al. [53]. The authors researched the chitosan adsorption on vermiculite. However, they noticed only a partial reaction of chitosan with silanols in Verm filler.

The proposed Verm and Per particles activation with the Dex scheme is presented in Figure 3. The formation of O-H groups from Dex on the surface of the analyzed fillers will contribute to the reaction with N-C-O groups from isocyanate.

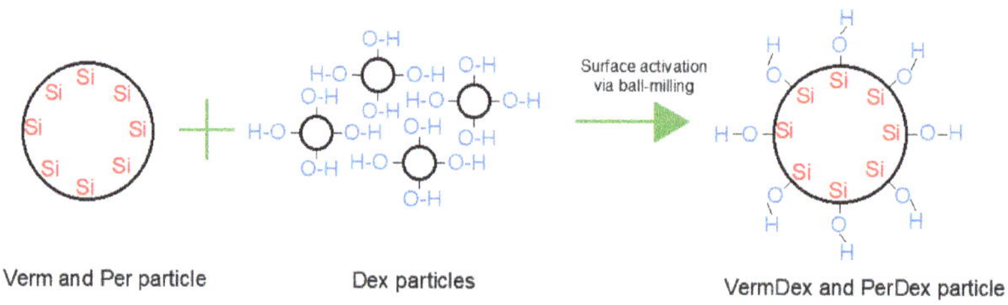

Figure 3. Proposed Verm and Per particles activation with Dex.

As hydroxyl groups are important for the synthesis of PUR foam, Dex-activated fillers will react with N-C-O groups with isocyanate, thus assumably improving the interfacial adhesion between VermDex, PerDex fillers, and the polymeric PUR foam composites matrix.

3.2. *Reaction Kinetics of Premixes*

It is known that adding various fillers generally increases the viscosity of the PUR foam mixtures, which can negatively affect the mixing of reactants, potentially slowing down the overall reaction rate [54]. The results of the main characteristic foaming times and dynamic viscosity are presented in Table 2.

Table 2. Reaction kinetics of control and fillers modified premixes.

Premix	Parameter			
	Cream Time, s	Gel Time, s	Tack-Free Time, s	Dynamic Viscosity, mPa·s
Control PUR	20 ± 2	76 ± 2	102 ± 4	521 ± 10
PUR/Per-2.5	29 ± 2	114 ± 4	156 ± 3	1456 ± 15
PUR/Per-5	22 ± 2	110 ± 2	159 ± 2	3875 ± 20
PUR/Per-7.5	19 ± 3	104 ± 2	162 ± 3	28,699 ± 25
PUR/Per-10	16 ± 2	96 ± 3	164 ± 3	50,987 ± 18
PUR/Verm-2.5	28 ± 1	107 ± 2	140 ± 2	2350 ± 14
PUR/Verm-5	24 ± 2	100 ± 3	165 ± 2	7172 ± 10
PUR/Verm-7.5	21 ± 2	94 ± 3	197 ± 3	34,569 ± 15
PUR/Verm-10	18 ± 2	140 ± 3	237 ± 3	59,230 ± 19
PUR/PerDex-2.5	31 ± 3	112 ± 2	160 ± 3	988 ± 21
PUR/PerDex-5	26 ± 2	110 ± 1	174 ± 2	1798 ± 18
PUR/PerDex-7.5	20 ± 3	110 ± 2	181 ± 3	2230 ± 16
PUR/PerDex-10	16 ± 2	109 ± 2	189 ± 4	3042 ± 19
PUR/VermDex-2.5	30 ± 2	99 ± 2	179 ± 3	652 ± 5
PUR/VermDex-5	28 ± 3	98 ± 3	184 ± 2	1153 ± 15
PUR/VermDex-7.5	23 ± 2	95 ± 2	189 ± 4	1690 ± 12
PUR/VermDex-10	20 ± 3	94 ± 2	195 ± 2	2742 ± 20

Unsurprisingly, the premixes of PUR/Per and PUR/Verm foam composites negatively impacted cream, gel, and tack-free times—for instance, the addition of 2.5 wt.% Per and Verm increased cream time by 45% and 40% and gel time by 50% and 41%, respectively, compared to the control PUR foam premix. However, further addition of fillers improved the parameter, and at 10 wt.% Per and Verm concentration became almost the same as for the control premix. Similar observations can be made for PUR/PerDex and PUR/VermDex premixes, indicating that PerDex and VermDex fillers do not highly impact the cream and gel times, which are comparable to the premixes of PUR/Per and PUR/Verm foam composites. The obtained results are in excellent agreement with the ones obtained by Barczewski et al. [29], who indicated that fillers similar to Verm and Per do not typically accelerate the initial stages of foam formation. However, the same conclusion cannot be made for tack-free time. The maximum tack-free times were observed for all PUR foam composites at 10 wt.% fillers concentrations, i.e., by 61% for PUR/Per, 132% for PUR/Verm, 85% for PUR/PerDex, and 91% for PUR/VermDex compared to the premix of control PUR foam. It means that all fillers interfere with the nucleation and growth of the bubbles, potentially altering the foaming kinetics. Therefore, a slower or less efficient foaming process can indirectly lead to a longer time for the surface to become tack-free. Cetin et al. [30] and Sari et al. [55] showed that expanded Verm and Per can be used for oil absorption. Fillers with high oil absorption tend to increase viscosity more significantly, as can be seen from the results obtained in Table 2. In general, all premixes show the same behavior, i.e., a constant increase in viscosity with adding fillers. The parameters of PUR/Per and PUR/Verm foam composites were up to 97 and 113 times higher for 10 wt.% of Per and Verm addition compared to control PUR foam. It is unsurprising as few studies [56,57] indicate a deterioration in the rheological characteristics of highly filled mixtures. Interesting observations can be made for PUR/PerDex and PUR/VermDex foam composites. Compared to PUR/Per and PUR/Verm foam composites, the dynamic viscosity of PUR foam composite systems with Dex-activated Per and Verm fillers was reduced 16 and 21 times, respectively, at the highest concentration of fillers. It can be assumed that incorporating hydroxyl groups on the surface of fillers can interact with polar components in the premix surrounding more easily, thus creating a repulsive force between filler particles, preventing them from agglomeration.

3.3. Microstructure and Thermal Conductivity

It is common to find that incorporating various fillers can affect the density of the foam due to the interference of the solid particles with the polymer and foaming kinetics. The density results are summarized in Table 3.

Table 3. Structural parameters of PUR foam composites.

PUR Foam Composite	Parameter			
	Apparent Density, kg/m^3	Closed Cell Content, vol.%	Cell Size, mm	Wall Thickness, μm
Control PUR	43.7 ± 4	94.4 ± 0.52	1.32 ± 0.12	177 ± 5
PUR/Per-2.5	44.5 ± 3	94.9 ± 0.63	0.812 ± 0.11	100 ± 4
PUR/Per-5	47.4 ± 5	95.4 ± 0.55	0.618 ± 0.12	88.6 ± 6
PUR/Per-7.5	47.6 ± 2	97.3 ± 0.42	0.725 ± 0.10	74.5 ± 6
PUR/Per-10	47.4 ± 3	94.6 ± 0.44	0.837 ± 0.16	65.5 ± 2
PUR/Verm-2.5	45.7 ± 5	94.0 ± 0.22	0.882 ± 0.14	76.7 ± 6
PUR/Verm-5	44.9 ± 4	91.4 ± 0.26	0.703 ± 0.19	71.6 ± 4
PUR/Verm-7.5	46.8 ± 3	91.6 ± 0.38	0.754 ± 0.13	65.3 ± 3
PUR/Verm-10	52.3 ± 2	91.2 ± 0.44	0.772 ± 0.12	62.7 ± 5
PUR/PerDex-2.5	49.4 ± 4	97.6 ± 0.62	1.12 ± 0.22	138 ± 6
PUR/PerDex-5	52.3 ± 5	97.3 ± 0.51	0.780 ± 0.14	155 ± 4
PUR/PerDex-7.5	56.4 ± 5	97.9 ± 0.38	0.623 ± 0.16	168 ± 3
PUR/PerDex-10	63.7 ± 4	97.0 ± 0.49	0.592 ± 0.12	185 ± 6
PUR/VermDex-2.5	48.0 ± 3	96.1 ± 0.29	1.01 ± 0.23	113 ± 4
PUR/VermDex-5	51.0 ± 2	98.0 ± 0.49	0.795 ± 0.14	115 ± 3
PUR/VermDex-7.5	56.4 ± 3	94.4 ± 0.64	0.832 ± 0.16	119 ± 4
PUR/VermDex-10	59.8 ± 2	94.2 ± 0.58	0.865 ± 0.17	119 ± 3

They confirm that the parameter increase depends on the increase in filler concentration and agree with the results reported in the study by Gama et al. [58]. In this case, the apparent density of PUR/Per foam composites increased by a maximum of 9%, whereas for PUR/Verm foam composites, the parameter reached its highest value at 10 wt.% Verm addition. It is also worth noting that Dex-activated Per and Verm particles caused an even greater increase in the apparent density of PUR foam composites. For instance, the highest increase was 46% for PUR/PerDex foam composites and 37% for PUR/VermDex foam composites. The apparent densities of Dex-activated and non-activated particles-filled PUR foam composites could be attributed to differences in the bulk densities of the fillers.

To evaluate the impact of non-activated and Dex-activated Per and Verm particles on the thermal conductivity of PUR foam composites, graphs were presented based on the concentration of each filler added (Figure 4). It can be seen from Figure 4a–d that the difference in thermal conductivity between the PUR/Per, PUR/Verm, PUR/PerDex, and PUR/VermDex foam composites was insignificant. However, a 3–4% reduction in the parameter can be observed, i.e., from 0.0295 W/(m·K) to 0.0283 W/(m·K) for 2.5 wt.% filler addition (Figure 4a), to 0.0284 W/(m·K) for 5 wt.% willer addition (Figure 4b) and to 0.0285 W/(m·K) for 7.5 wt.% filler addition compared to control PUR foam. The cell size results in Table 2 can explain this.

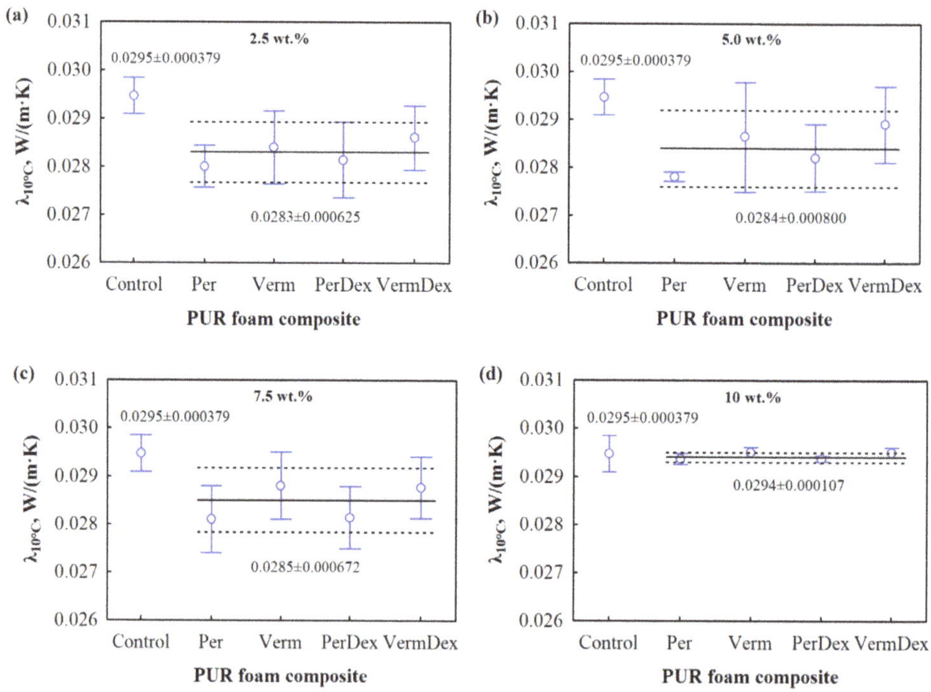

Figure 4. Comparison of thermal conductivity of PUR foam composites with different fillers at (**a**) 2.5 wt.%; (**b**) 5 wt.%; (**c**) 7.5 wt.%, and (**d**) 10 wt.%.

A closer examination of the results in Table 3 and Figure 5 revealed that all fillers (Per, Verm, PerDex, VermDex) and their concentrations (2.5–10 wt.%) reduced average cell size by a minimum of 15% and by a maximum of 55% compared to control PUR foam indicating that the filler particles act as nucleating agents. Similar observations were made by Lee et al. [59], who investigated PUR foam composites with silica aerogels and silica nanoparticles. Even though the closed cell content (Table 3) of all PUR foam composites is higher than that of a control PUR foam, Soloveva et al. [60] determined that the cell size causes the highest impact on thermal conductivity. They explained that in larger cells, the contribution of radiative heat transfer to the thermal conductivity value increases. Figure 4d shows that the further addition of all fillers, i.e., 10 wt.%, yields a thermal conductivity value almost identical to that of the control PUR foam. The excessive addition of fillers led to even worse results, with the cell size becoming larger and non-uniform.

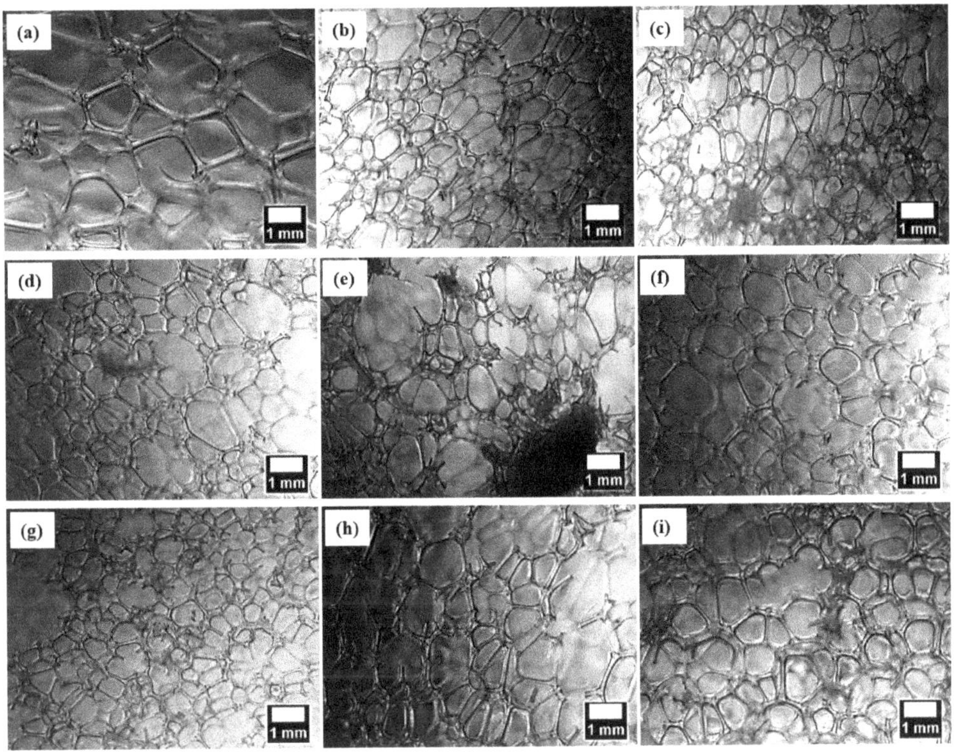

Figure 5. Optical imaging of PUR foam composites' microstructure at x80 magnification: (**a**) control PUR foam; (**b**) PUR/Per-5; (**c**) PUR/Per-10; (**d**) PUR/Verm-5; (**e**) PUR/Verm-10; (**f**) PUR/PerDex-5; (**g**) PUR/PerDex-10; (**h**) PUR/VermDex-5; and (**i**) PUR/VermDex-10.

The microstructure of all PUR foam composites at 10 wt.% filler concentrations (Figure 5c,e,g,i) was also destroyed and exhibited interconnected cells, unlike the control PUR foam, which displayed the typical cellular structure. This could be attributed to the worsened thermal conductivity values of the PUR foam composites. Therefore, the maximum addition of all fillers based on thermal conductivity is 7.5 wt.%, which yields the highest thermal insulation performance in PUR foam composites.

However, the statistical analysis of the thermal conductivity of PUR foam composites with 2.5 wt.% different fillers showed (Figure 4a) that there are no differences between the PUR/Per-2.5, PUR/Verm-2.5, PUR/PerDex-2.5, and PUR/VermDex-2.5 composites. The F-criterion is 0.48, $p = 0.70$, and $R^2 = 0.154$. Further, the statistical analysis of the thermal conductivity of PUR foam composites with 5 wt.% different fillers showed (Figure 4b) that there are also no differences between the PUR/Per-5, PUR/Verm-5, PUR/PerDex-5, and PUR/VermDex-5 foam composites. The F-criterion is 1.14, $p = 0.39$, and $R^2 = 0.299$. Moreover, the statistical analysis of the thermal conductivity of PUR foam composites with 7.5 wt.% different fillers showed (Figure 4c) that there are no differences between the

PUR/Per-7.5, PUR/Verm-7.5, PUR/PerDex-7.5, and PUR/VermDex-7.5 foam composites. The F-criterion is 0.98, $p = 0.45$, and $R^2 = 0.269$. In addition, the statistical analysis of the thermal conductivity of PUR foam composites with 10 wt.% different fillers showed (Figure 4d) that there are no differences between the PUR/Per-10, PUR/Verm-10, PUR/PerDex-10, and PUR/VermDex-10 foam composites. The F-criterion is 1.94, $p = 0.20$, and $R^2 = 0.421$. After repeated comparisons of the PUR/Per, PUR/Verm, PUR/PerDex, and PUR/VermDex foam composites with the control PUR, F-criterion is 0.39, $p = 0.81$, and $R^2 = 0.135$, indicating a statistically insignificant difference between the mean values.

3.4. Water Resistance Performance

Figure 6 presents water absorption behavior for control PUR foam and PUR foam composites with each type of filler at concentrations ranging from 2.5% to 10%. Water absorption at each concentration of filler is presented separately to highlight the impact of each filler type on the parameter more clearly.

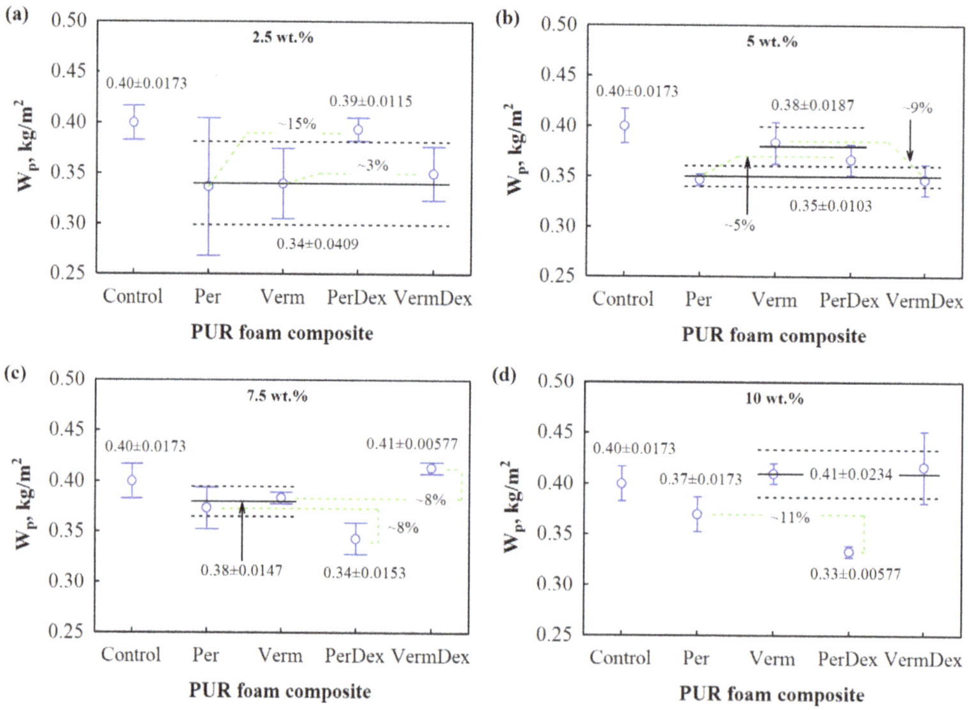

Figure 6. Comparison of short-term water absorption by partial immersion of PUR foam composites with different fillers at (**a**) 2.5 wt.%; (**b**) 5 wt.%; (**c**) 7.5 wt.%; and (**d**) 10 wt.%.

Considering the concentration of each filler type, a reduction in water absorption is noted for Per and Verm fillers. The parameter is reduced by 15% at 2.5 wt.% Per filler concentration (Figure 6a). The further addition of Per increases water absorption, but it remains 7.5% lower than control PUR foam (Figure 6b–d). It can be explained by the fact that Per does not absorb much water. On the contrary, it holds water within the tiny crevices that pockmark its surface [61]. Additionally, the obtained water absorption results correlate with the closed-cell content for PUR/Verm composites, as shown in Table 3.

A similar but slightly more intense tendency is observed for Verm filler-modified PUR foam composites. The incorporation of 2.5 wt.%, 5 wt.%, and 7.5 wt.% Verm by 15%, 5%, and 5%, respectively, decreases water absorption compared to the control PUR foam (Figure 6a–c). However, a further 10 wt.% Verm deteriorates the parameter by 2.5% compared to control PUR foam (Figure 6d). It is well known that expanded Verm absorbs water like a sponge whether it is used in polymer composites, cement-based composites, or mortars [62,63]. Its flakes expand when it sops up water. This means that structural damage occurs. For instance, Table 3 shows that incorporating Verm in any concentration reduces the number of closed cells, which may also contribute to the deterioration of moisture-related properties.

Dex-activated Per (PerDex) fillers show a decreasing tendency in water absorption values, even at a maximum addition of 10 wt.% PerDex (Figure 6d), PUR foam composites exhibit ~18% lower water absorption than control PUR foam and ~11% compared to untreated Per filler. However, Dex-activated Verm (VermDex) filler positively impacts the parameter at only up to 5 wt.% addition (Figure 6b,c), i.e., maximum by ~13% compared to control PUR foam and by ~8% compared to unmodified Verm. Further incorporation of this filler up to 10 wt.% deteriorates the resistance to water by a maximum of 5% compared to control PUR foam and by ~3% compared to unmodified Verm (Figure 6c,d). As indicated in FTIR spectra of VermDex (Figure 2), Verm activation with Dex allows the formation of Dex–Si–O–Si–Dex linkages, which inhibit the filler's participation in further reaction with isocyanate to form a strong interface between filler particles and polymer matrix. On the contrary, PerDex filler has hydroxyl groups, which only partially react with silanols in Per. Therefore, participation in the reaction during PUR foam composites synthesis is not entirely restricted, thus allowing the formation of a more reliable interface between particles and PUR foam composites. Therefore, the FTIR spectra of control PUR foam and PUR foam composites with Dex-activated fillers are presented in Figure 7.

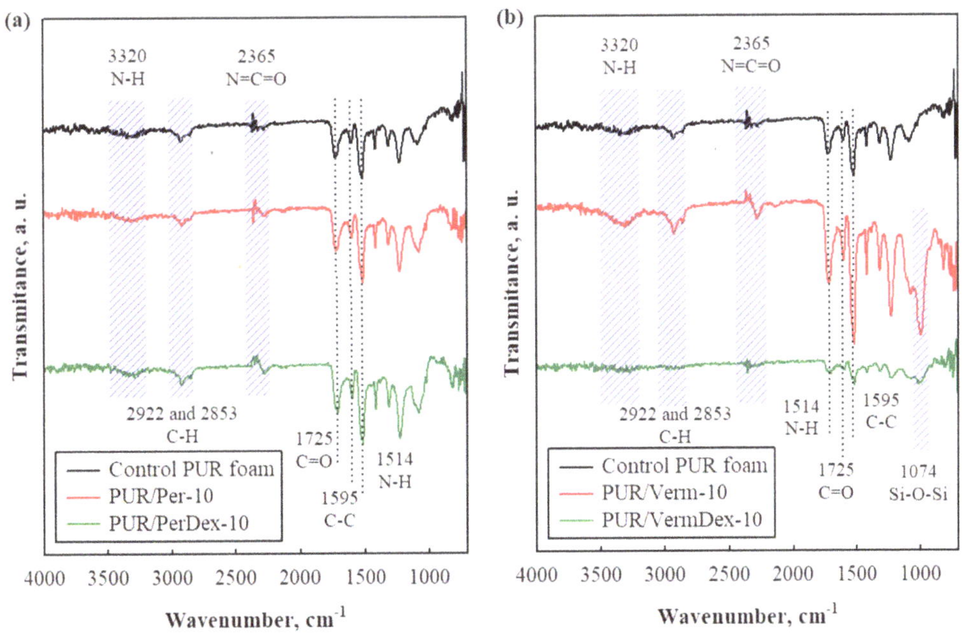

Figure 7. FTIR spectra of PUR foam composites with (**a**) Per and PerDex and (**b**) Verm and VermDex.

It can be seen that all PUR foam composites have similar patterns. The intensity at 3320 cm^{-1} is assigned to the stretching band of N-H. In addition, asymmetrical and symmetrical vibrations of C-H take place at wavenumbers of 2922 cm^{-1} and 2853 cm^{-1}, respectively. Furthermore, unreacted N=C=O groups are observed at 2365 cm^{-1}, indicating the excess of isocyanate used in the synthesis of the control PUR foam and PUR foam composites. However, Figure 7b shows that the addition of VermDex reduced the N=C=O intensity to a minimum, resulting in the formation of complete urethane linkages. The band at 1725 cm^{-1} indicate the stretching vibrations of C=O, while intensities at 1595 cm^{-1} and 1514 cm^{-1} are attributed to C-C and N-H bending, respectively. Interestingly, the FTIR spectra of PUR/VermDex-10 foam composites (Figure 7b) are characterized by an additional band at 1074 cm^{-1}, which is attributed to Si-O-Si.

It is considered that a chemical reaction has occurred between VermDex particles and the PUR matrix. However, no changes in the FTIR spectra of PUR/PerDex-10 (Figure 7a) are visible. The possible reaction of Dex-activated fillers and polyol with isocyanate is graphically presented in Figure 8. The graphical representation shows that Dex-activated fillers have hydroxyl groups incorporated on their surface, which, together with the hydroxyls from polyol, successfully react with an excess of isocyanate, thus forming urethane linkages with the incorporated fillers in PUR chains.

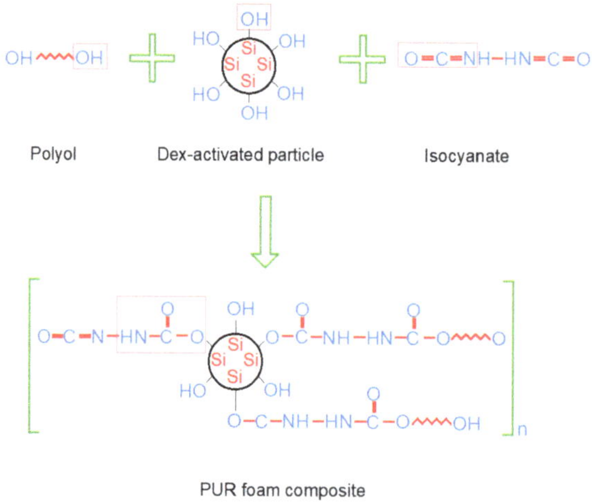

Figure 8. Graphical representation of PUR foam composites production.

The statistical analysis of short-term water absorption of PUR foam composites with 2.5 wt.% different fillers showed (Figure 6a) that there are no differences between the PUR/Per-2.5, PUR/Verm-2.5, and PUR/VermDex-2.5 foam composites. The F-criterion is 0.066, $p = 0.94$, and $R^2 = 0.0216$. However, a significant difference occurs between the aforementioned PUR foam composites and control PUR as well as PUR/Perdex-2.5. When comparing the PUR/PerDex-2.5 foam composite with the control PUR, the analysis showed that there is no difference between the average values of short-term water absorption. The F-criterion statistic is 0.31, $p = 0.61$, and $R^2 = 0.0714$. Further, the statistical analysis of the short-term water absorption of PUR foam composites with 5 wt.% different fillers showed (Figure 6b) that there are no differences between the PUR/Per-5 and PUR/VermDex-5 foam composites. The F-criterion is 0.0, $p = 1.0$, and $R^2 = 0$. Comparing PUR/Verm-5 and PUR-PerDex-5, the statistical analysis showed a statistically insignificant difference in the results. The F-criterion is 1.25, $p = 0.33$, and $R^2 = 0.238$. However, significant differences were observed between PUR/Per-5 and PUR/VermDex-5 and PUR/Verm-5 and PUR/PerDex-5 foam composites. In addition, all PUR foam composites with 5 wt.% fillers had significant difference in short-term water absorption compared to control PUR. Moreover, the statistical analysis of the short-term water absorption of PUR foam composites with 7.5 wt.% different fillers showed (Figure 6c) that there are no differences between the PUR/Per-7.5 and

PUR/Verm-7.5 foam composites. The F-criterion is 0.64, $p = 0.47$, and $R^2 = 0.138$. Statistically significant differences were observed between PUR/Per-7.5 and PUR/Verm-7.5 and control PUR, PUR/PerDex-7.5, and PUR/VermDex-7.5. Lastly, the statistical analysis of the short-term water absorption of PUR foam composites with 10 wt.% different fillers showed (Figure 6d) that there were no differences between the PUR/Verm-10 and PUR/VermDex-10 foam composites. The F-criterion statistic is 0.10, $p = 0.78$, and $R^2 = 0.0244$. When comparing the control PUR and PUR/Verm-10 and PUR/Vermdex-10 foam composites, the analysis showed that there was no difference between the mean values of short-term water absorption. The F-criterion is 0.39, $p = 0.69$, and $R^2 = 0.114$. However, data processing showed that significant differences were obtained between control PUR, PUR/Verm-10, and PUR/VermDex-10 and PUR/Per-10 and PUR/PerDex-10.

Materials' surface properties play a crucial role in various processes, including hydrophobicity. To determine the hydrophobic character of the PUR foam composites' surface, water contact angle measurements were performed by the sessile drop method after 10 s (Figure 9).

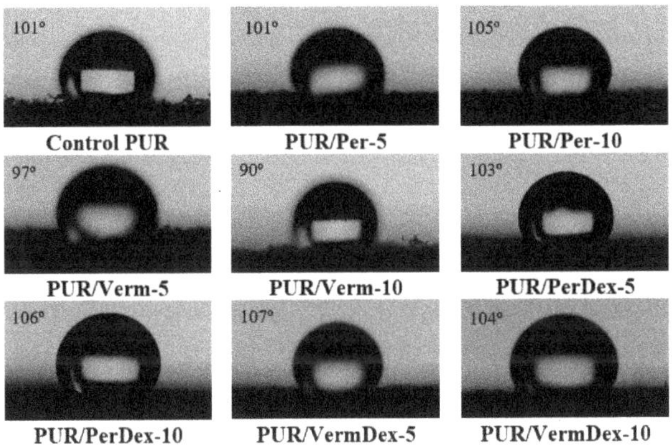

Figure 9. The water contact angle of PUR foam composites with different fillers at 5 wt.% and 10 wt.% concentrations.

According to the results given in Figure 9, as the Per filler concentration in PUR foam composites increased, the hydrophobic character of the composites increased only slightly, i.e., from 101° to 105° compared to the control PUR foam composite. Similarly, regarding water absorption results, the Verm filler-modified PUR foam composites still exhibited a hydrophobic character but with the lowest water contact angle among all fillers analyzed at a 10 wt.% concentration. Cetin et al. [30] demonstrated that Verm exhibits oleophilic behavior in flexible polyurethane foams, and its diesel contact angle decreases significantly with an increase in Verm concentration.

The surface activation of PerDex and VermDex fillers highly improved the hydrophobic character of the resulting surfaces of PUR foam composites. The water contact angle of PerDex filler-modified PUR foam composites increased to 103° and 106° at 5 wt.% and 10 wt.% PerDex, respectively, while for VermDex filler-modified PUR foam composites, it increased to 107° and 104° at 5 wt.% and 10 wt.% VermDex, respectively. The obtained tendency for water contact angle correlates with the water absorption results.

3.5. Mechanical Performance

The mechanical properties of thermal insulating PUR foams are crucial for their performance and longevity in various applications. While their primary function is thermal insulation, their ability to withstand mechanical stresses significantly impacts their effectiveness and suitability for specific uses. Therefore, Figures 10 and 11 present the compressive and tensile strengths of control PUR foam and modified PUR foam composites.

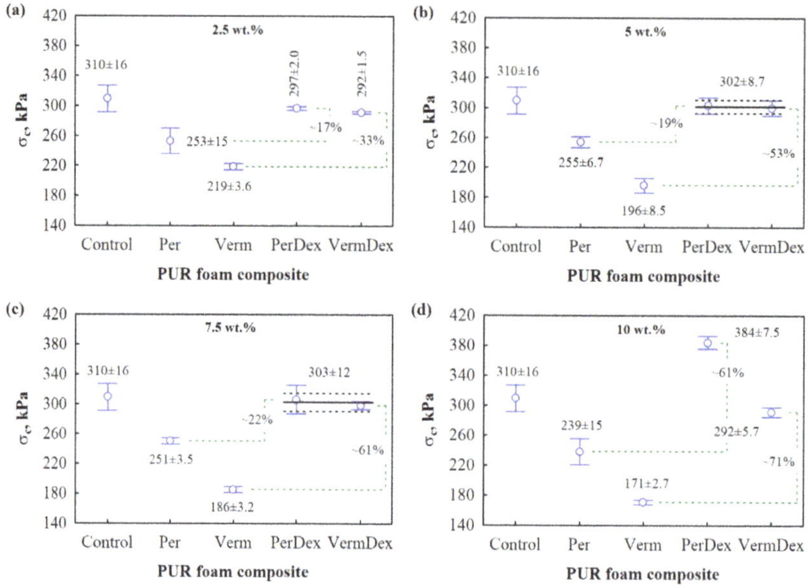

Figure 10. Compressive strength of PUR foam composites with different fillers at (**a**) 2.5 wt.%; (**b**) 5 wt.%; (**c**) 7.5 wt.%; and (**d**) 10 wt.%.

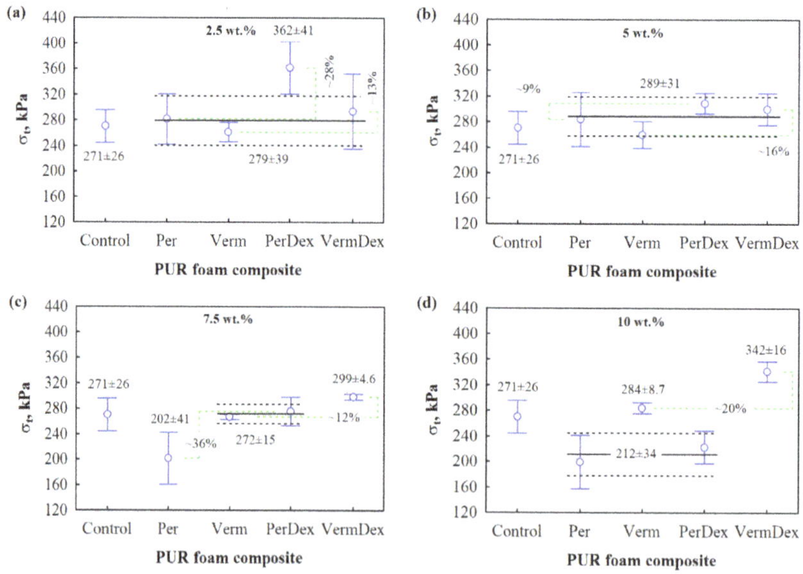

Figure 11. Tensile strength of PUR foam composites with different fillers at (**a**) 2.5 wt.%; (**b**) 5 wt.%; (**c**) 7.5 wt.%; and (**d**) 10 wt.%.

It can be seen that PUR/Per and PUR/Verm foam composites with 2.5–10 wt.% filler exhibit lower compressive strength values compared to control PUR foam. For instance, 2.5 wt.% and 5 wt.% Per filler (Figure 10a,b) reduces the parameter by 18%, while the same concentration of Verm filler—by 30% and 37%, respectively. With the further addition of 7.5 wt.% and 10 wt.% Per filler (Figure 10c,d), the compressive strength also deteriorates by 19% and 23%, respectively. A similar tendency of the same Verm filler concentrations can also be observed; for example, the parameter reduces by 40% and 45%, respectively. However, Xia and Wang [64] determined that the addition of Verm at concentrations up to 10 wt.% increases the compressive strength of PUR/melamine phenol phosphate composite foams, indicating that the positive results are attributed to fillers' insertion into the cell walls, which then strengthens the cellular matrix. The difference in the results obtained may be due to the size of the fillers used and the varying elasticity of the cellular morphology of PUR foam composites. Thinner cell walls and struts in PUR/Per and PUR/Verm foam composites (Table 3) provide fracture induction. As a result, less energy is absorbed, leading to cell buckling and irreversible damage [65]. Moreover, a decrease in compressive strength was observed in PUR foam composites filled with expandable graphite [66], which causes defects in the analyzed products.

The activation of Per and Verm fillers has a significant impact on the compressive strength of all PUR foam composites. Compared to PUR/Per and PUR/Verm foam composites, the maximum increase of 61% and 71% is shown by PUR/PerDex-10 and PUR/VermDex-10, respectively. However, only PUR/PerDex-10 has 24% greater compressive strength than control PUR foam. The positive interaction between Dex-activated fillers and PUR foam can be explained by the FTIR results, which show that hydroxyl groups from Dex form the linkages with silicates in Verm and Per, leading to a more stable and covalent attachment of Dex and anchoring points for attaching polymeric matrix. Therefore, an enhanced interfacial inorganic–organic interaction is observed. Similar conclusions have been drawn by Miedzińska et al. [38], who demonstrated that significant mechanical performance can be achieved by activating inorganic fillers with potato protein, chitosan, and casein.

The statistical analysis of compressive strength of PUR foam composites with 2.5 wt.% different fillers showed (Figure 10a) that there are differences between the averages of all PUR foam composites compared to the control PUR. The F-criterion is 42.0, p = 0, and R^2 = 0.944. Further, the statistical analysis of the compressive strength of PUR foam composites with 5 wt.% different fillers showed (Figure 10b) that there are no differences between the PUR/PerDex-5 and PUR/VermDex-5 foam composites. The F-criterion statistic is 0.15, p = 0.72, and R^2 = 0.356. When comparing the PUR/PerDex-5 and PUR/VermDex-5 foam composites with the control PUR, the analysis showed that there is no difference between the average values. The F-criterion statistic is 0.51, p = 0.63, and R^2 = 0.145. Moreover, the statistical analysis of compressive strength of PUR foam composites with 7.5 wt.% different fillers showed (Figure 10c) that there were no differences between the PUR/PerDex-7.5 and PUR/VermDex-7.5 foam composites. The F-criterion is 0.57, p = 0.49, and R^2 = 0.125. When comparing the PUR/PerDex-7.5 and PUR/VermDex-7.5 foam composites with the control PUR, the analysis showed that there is no difference between the average values. The F-criterion is 0.51, p = 0.63, and R^2 = 0.146. However, the significant differences are observed between control PUR, PUR/PerDex-7.5, and PUR/VermDex-7.5 and PUR/Per-7.5 and PUR/Verm-7.5. Lastly, the statistical analysis of the compressive strength of PUR foam composites with 10 wt.% different fillers showed (Figure 10d) that there are significant differences between compressive strength of all PUR foam composites compared to the control PUR. The F-criterion is 164, p = 0, and R^2 = 0.985.

Furthermore, Figure 11 presents the tensile strength results of control PUR foam and modified PUR foam composites.

For PUR/Per and PUR/Verm foam composites with filler concentrations of 2.5–5 wt.%, no significant changes in tensile strength are observed, indicating that these fillers do not deteriorate the parameter at lower concentrations. However, PUR/Per-7.5 and PUR/Per-10 foam composites showed a 25% and 22% decrease in tensile strength compared to the control PUR foam, while PUR/Verm-7.5 remained unchanged, and PUR/Verm-10 had a 5% higher tensile strength compared to the control PUR foam. It seems that Per filler has a higher negative impact on tensile properties. Additionally, both fillers were found to be inefficient under compression and tension loads.

However, different tendencies can be seen for PUR foam composites with Dex-activated Per and Verm particles. The highest tensile strength for PUR/PerDex foam composites is observed at a 2.5 wt.% PerDex concentration. The parameter increased by 34% compared to the control PUR foam, although further addition of PerDex filler deteriorated the parameter. However, compared to the control PUR foam, the tensile strength remains higher, up to 7.5 wt.% PerDex concentration. Smaller and more uniform cell sizes and wall thickness (Table 3) enhance tensile strength. A finer cellular structure provides a larger surface area for stress distribution, preventing localized failure. Such an effect is related to improved interfacial adhesion associated with the PerDex surface. Hejna et al. [67] reached a similar conclusion when modifying flexible polyurethane foam with potassium peroxide-oxidized tire rubber. Even though the cells of PUR/Per-10 foam composite are smaller, the destruction of microstructure at higher loading of PerDex filler takes place as indicated, as can be seen in Figure 5g. Therefore, a 22% reduction in tensile strength (Figure 11d) at 10 wt.% PerDex addition can be seen. Moreover, a slightly different conclusion can be made for PUR/VermDex foam composites. The tensile strength of these PUR foam composites at any loading of VermDex filler increased, and the highest parameter was 26% higher than that of the control PUR foam composite. The presence of Dex on the Verm surface might be creating stronger interactions with PUR chains. This enhances interaction and allows for more efficient stress transfer from the PUR matrix to stronger VermDex particles when the foam composite is under tensile stress.

The statistical analysis of the tensile strength of PUR foam composites with 2.5 wt.% different fillers showed (Figure 11a) that there are no differences between the PUR/Per-2.5, PUR/Verm-2.5, and PUR/VermDex-2.5 foam composites. The F-criterion is 0.49, p = 0.64, and R^2 = 0.139. When comparing the PUR/Per-2.5, PUR/Verm-2.5, and PUR/VermDex-2.5 foam composites with the control PUR, the analysis showed that there is no difference between the average values of tensile strength. The F-criterion is 0.42, p = 0.74, and R^2 = 0.136. Further, the statistical analysis of the tensile strength of PUR foam composites with 5 wt.% different fillers showed (Figure 11b) that there are no differences between the PUR/Per-5, PUR/Verm-5, PUR/PerDex-5, and PUR/VermDex-5 foam composites. The F-criterion statistic is 1.81, p = 0.22, and R^2 = 0.404. When comparing the control PUR and PUR/Per-5, PUR/Verm-5, PUR/PerDex-5, and PUR/VermDex-5 foam composites, the statistical analysis showed a statistically insignificant difference in the results. The F-criterion statistic is 1.65, p = 0.24, and R^2 = 0.398. Moreover, the statistical analysis of the tensile strength of PUR foam composites with 7.5 wt.% different fillers showed (Figure 11c) that there are no differences between the PUR/Verm-7.5 and PUR/PerDex-7.5 foam composites. The F-criterion is 0.45, p = 0.54, and R^2 = 0.102. Comparing the PUR/Verm-7.5 and PUR/PerDex-7.5 foam composites with the control PUR, the analysis showed that there is no difference between the average values of tensile strength. The F-criterion is 0.15, p = 0.86, and R^2 = 0.0474. However, a significant difference is observed between PUR/Per-7.5 and PUR/VermDex-7.5 and PUR/Verm-7.5 and PUR/PerDex-7.5. Lastly, the statistical analysis of the tensile strength of PUR foam composites with 10 wt.% different fillers showed (Figure 11d) that there are no differences between PUR/Per-10 and PUR/PerDex-10 foam composites. The F-criterion is 0.69, p = 0.45, and R^2 = 0.148. Comparing the control PUR and PUR/Verm-10 foam composite, the analysis showed that there is no difference between the average values of tensile strength. The F-test is 0.73, $p = 0.44$, and $R^2 = 0.154$. However, a significant difference is obtained between control PUR, PUR/Verm-10, and PUR/VermDex-10 and PUR/Per-10 and PUR/Verm-10.

4. Conclusions

The presented study aimed to research the impact of non-activated and Dex-activated Per and Verm fillers on the processing times, microstructure, thermal conductivity, water absorption, and strength properties of PUR foam composites. Applied activation resulted in the development of Per and Verm fillers' surfaces containing hydroxyl groups.

Dex-activated Per and Verm fillers were introduced into a PUR matrix. The introduction of non-activated and Dex-activated fillers resulted in an elongation of processing times compared to control PUR foam premixes. Such an effect was primarily due to the increased viscosity of the resulting premixes caused by the introduction of solid particles. However, the surface activation with Dex caused a positive effect due to the availability of hydroxyl groups. Changes in processing times were reflected in the cellular structure of PUR foam composites.

For PUR/PerDex and PUR/VermDex foam composites, a typical increase in apparent density, closed cell content, wall thickness, and reduction in average cell size were observed. At the same time, Dex activation catalyzed the polymerization reaction compared to non-activated Per and Verm fillers. The quality of cellular structure directly affected the compressive and tensile strengths. Therefore, superior compressive strength was observed for PUR/PerDex and tensile strength for PUR/VermDex foam composites at 10 wt.% PerDex and VermDex loadings due to the changed surface interaction abilities of the Dex-activated fillers.

Regarding thermal conductivity, the maximum reduction in the parameter was achieved at 2.5 wt.% non-activated and Dex-activated fillers loadings. Considering water resistance, changes in the surface of Dex-activated Per and Verm fillers were reflected in the water absorption and contact angle properties of the PUR foam composites. Generally, the presented results indicate that the Dex activation of Per and Verm surface results in the formation of Dex–Si–O–Si–Dex linkages, which inhibit the filler's participation in further reaction with isocyanate to form the strong interface between filler particles and polymer matrix. Such an effect resulted in reduced water absorption and increased water contact angles.

Author Contributions: Conceptualization, A.K.; methodology, A.K.; software, A.L.; validation, A.K. and A.L.; formal analysis, A.K.; investigation, A.K. and A.L.; resources, A.K.; data curation, A.K. and A.L.; writing—original draft preparation, A.K. and A.L.; writing—review and editing, A.K. and A.L.; visualization, A.L.; supervision, A.K.; funding acquisition, A.K. All authors have read and agreed to the published version of the manuscript.

Funding: This research received no external funding.

Institutional Review Board Statement: Not applicable.

Informed Consent Statement: Not applicable.

Data Availability Statement: All the data obtained is presented within the article.

Conflicts of Interest: The authors declare no conflicts of interest.

Abbreviations

The following abbreviations are used in this manuscript:

PUR	Rigid polyurethane foam
Dex	Dextrin
Verm	Expanded vermiculite
Per	Expanded perlite
VermDex	Dextrin-activated expanded vermiculite
PerDex	Dextrin-activated expanded perlite

References

1. Cheng, C.; Zhu, Z.; Wang, M.; Li, J.; Sun, H.; Jiao, R.; Li, A. Enhanced flame retardancy and thermal insulation of flexible polyurethane foam decorated with DMMP-loaded CuO hollow spheres. *Colloids Surf. A Physicochem. Eng. Asp.* **2025**, *718*, 136961. [CrossRef]
2. Mendija, L.C.C.; Dingcong, R.G., Jr.; Alfeche, F.L.A.M.; Al-Moameri, H.H.; Dumancas, G.G.; Tan, N.P.B.; Malaluan, R.M.; Alguno, A.C.; Lubguban, A.A. Elucidating the Impact of Polyol Functional Moieties on Exothermic Poly(urethane-urea) Polymerization: A Thermo-Kinetic Simulation Approach. *Sustainability* **2024**, *16*, 4587. [CrossRef]
3. Mouren, A.; Avérous, L. Sustainable cycloaliphatic polyurethanes: From synthesis to applications. *Chem. Soc. Rev.* **2023**, *52*, 277–317. [CrossRef]
4. Malucelli, G. Flame retardant surface treatments for rigid polyurethane foams used in the building sector: Current state-of-the-art and perspectives. *Constr. Build. Mater.* **2025**, *472*, 140947. [CrossRef]
5. Chen, M.; Yuan, Y.; Wang, W.; Xu, L. Recent Advances in Flame-Retardant Flexible Polyurethane Foams. *Fire* **2025**, *8*, 90. [CrossRef]
6. Polyurethane Market. Available online: https://www.marketsandmarkets.com (accessed on 21 March 2025).
7. Zhang, Y.; Li, P.; Sun, S. Biopolyol synthesis via ring-opening of epoxidized soybean oil with glycerol utilizing a novel sulfonic-functionalized ionic liquid as the catalyst at low temperature. *Ind. Crops Prod.* **2024**, *222*, 119546. [CrossRef]
8. Li, G.-Z.; Wang, Q.; Zhu, C.; Zhang, S.; Wang, F.; Tao, L.; Jiang, Y.; Zhang, Q.; Wang, W.; Han, R. Synthesis and Performance of Epoxy-Terminated Hyperbranched Polymers Based on Epoxidized Soybean Oil. *Molecules* **2025**, *30*, 583. [CrossRef]
9. Gurgel, D.; Bresolin, D.; Sayer, C.; Filho, L.C.; de Araújo, P.H.H. Flexible polyurethane foams produced from industrial residues and castor oil. *Ind. Crops Prod.* **2021**, *164*, 113377. [CrossRef]
10. Ciastowicz, Ż.; Pamuła, R.; Bobak, Ł.; Białowiec, A. Characterization of Vegetable Oils for Direct Use in Polyurethane-Based Adhesives: Physicochemical and Compatibility Assessment. *Materials* **2025**, *18*, 918. [CrossRef]
11. Mekarat, S.; Suwan, A.; Chaisit, T.; Thongseenuch, S.; Saetung, A.; Saetung, N. Synthesis of palm oil-based bio-polyol by thiol-ene reaction: Preliminary study of its potential as cationic waterborne polyurethane for coating application. *Prog. Org. Coat.* **2024**, *194*, 108586. [CrossRef]
12. Palaczek, K.; Kurańska, M.; Auguścik-Królikowska, M.; Prociak, A.; Ryszkowska, J. Open-cell polyurethane foams of very low density modified with various palm oil-based bio-polyols in accordance with cleaner production. *J. Clean. Prod.* **2021**, *290*, 125875. [CrossRef]
13. Abolins, A.; Eihe, D.; Pomilovskis, R.; Fridrihsone, A.; Kirpluks, M. Rapeseed oil as feedstock for the polymeric materials via Michael addition reaction. *Ind. Crops Prod.* **2023**, *204*, 117367. [CrossRef]
14. Aydoğmuş, E.; Kamişli, F. New commercial polyurethane synthesized with biopolyol obtained from canola oil: Optimization, characterization, and thermophysical properties. *J. Mol. Struct.* **2022**, *1256*, 132495. [CrossRef]
15. Czifrák, K.; Lakatos, C.; Cserháti, C.; Vecsei, G.; Zsuga, M.; Kéki, S. Bio-Based Polyurethane Networks Containing Sunflower Oil Based Polyols. *Int. J. Mol. Sci.* **2024**, *25*, 7300. [CrossRef]
16. Suthar, V.; Asare, M.A.; de Souza, F.M.; Gupta, R.K. Effect of Graphene Oxide and Reduced Graphene Oxide on the Properties of Sunflower Oil-Based Polyurethane Films. *Polymers* **2022**, *14*, 4974. [CrossRef]
17. Polaczek, K.; Kurańska, M.; Prociak, A. Open-cell bio-polyurethane foams based on bio-polyols from used cooking oil. *J. Clean. Prod.* **2022**, *359*, 132107. [CrossRef]
18. Olivito, F.; Jagdale, P.; Oza, G. Direct Production of Furfural from Fructose Catalyzed by Iron(III) Sulfate Using a Simple Distillation Apparatus. *ACS Sustain. Chem. Eng.* **2023**, *11*, 17595–17599. [CrossRef]
19. Jang, J.Y.; Byun, Y.; You, S.; Kim, S.; Lee, D.-M.; Kim, S.-W.; Son, S.U. Polyurethanes Synthesized Using Biomass-derived Furan Diols as Sustainable Triboelectric Materials. *Chem. Commun.* **2024**, *60*, 9741–9744. [CrossRef]

20. Chang, H.; Gilcher, E.B.; Huber, G.W.; Dumesic, J.A. Synthesis of performance-advantaged polyurethanes and polyesters from biomass-derived monomers by aldol-condensation of 5-hydroxymethyl furfural and hydrogenation. *Green Chem.* **2021**, *23*, 4355–4364. [CrossRef]
21. Torresi, S.; Calvo-Correas, T.; Basasoro, S.; Guaresti, O.; Alonso-Varona, A.; Gabilondo, N.; Eceiza, A. Furan-containing biobased polyurethane nanofibers: A new versatile and green support clickable via Diels-Alder reaction. *React. Funct. Polym.* **2022**, *178*, 105353. [CrossRef]
22. Skoczinski, P.; Cangahuala, M.K.E.; Maniar, D.; Albach, R.W.; Bittner, N.; Loos, K. Biocatalytic Synthesis of Furan-Based Oligomer Diols with Enhanced End-Group Fidelity. *ACS Sustain. Chem. Eng.* **2020**, *8*, 1068–1086. [CrossRef]
23. Kurańska, M.; Pinto, J.A.; Salach, K.; Barreiro, M.F.; Prociak, A. Synthesis of thermal insulating polyurethane foams from lignin and rapeseed based polyols: A comparative study. *Ind. Crops Prod.* **2020**, *143*, 111882. [CrossRef]
24. Tan, S.; Abraham, T.; Ference, D.; MacOsko, C.W. Rigid polyurethane foams from a soybean oil-based polyol. *Polymer* **2011**, *52*, 2840–2846. [CrossRef]
25. Kurańska, M.; Marcovich, N.E.; Prociak, A.; Malewska, E.; Kulpa, K. Open cell semi-rigid polyurethane foams synthesized using palmoil-based bio-polyol. *Ind. Crops Prod.* **2017**, *102*, 88–96.
26. Valipour, F.; Dehghan, S.F.; Hajizadeh, R. The effect of nano- and microfillers on thermal properties of Polyurethane foam. *Int. J. Environ. Sci. Technol.* **2022**, *19*, 541–552. [CrossRef]
27. Wang, H.; Li, T.-T.; Ren, H.; Peng, H.; Huang, S.-Y.; Lin, Q.; Lin, J.-H.; Lou, C.-W. Expanded Vermiculite-Filled Polyurethane Foam-Core Bionic Composites: Preparation and Thermal, Compression, and Dynamic Cushion Properties. *Polymers* **2019**, *11*, 1028. [CrossRef]
28. Ai, M.; Cao, L.; Zhao, X.; Xiang, Z.; Guo, X. Preparation and Characterization of Polyurethane Rigid Foam/Expanded Perlite Thermal Insulation Composites. *Adv. Mater. Res.* **2010**, *96*, 141–144. [CrossRef]
29. Barczewski, M.; Kurańska, M.; Sałasińska, K.; Aniśko, J.; Szulc, J.; Szafraniak-Wiza, I.; Prociak, A.; Polaczek, K.; Uram, K.; Surmacz, K.; et al. Comprehensive Analysis of the Influence of Expanded Vermiculite on the Foaming Process and Selected Properties of Composite Rigid Polyurethane Foams. *Polymers* **2022**, *14*, 4967. [CrossRef]
30. Cetin, M.S.; Toprakci, O.; Taskin, O.S.; Aksu, A.; Toprakci, H.A.K. Expanded vermiculite-filled flexible polymer composites. *J. Elastomers Plast.* **2022**, *54*, 145–168. [CrossRef]
31. Jaworska, B.; Stańczak, D.; Łukowski, P. The Influence of Waste Perlite Powder on Selected Mechanical Properties of Polymer–Cement Composites. *Buildings* **2024**, *14*, 181. [CrossRef]
32. Szadkowski, B.; Marzec, A.; Rybiński, P.; Żukowski, W.; Zaborski, M. Characterization of Ethylene–propylene Composites Filled with Perlite and Vermiculite Minerals: Mechanical, Barrier, and Flammability Properties. *Materials* **2020**, *13*, 585. [CrossRef]
33. Erdem, R. Uses of Vermiculite and Perlite in FireRetardant Textile Applications. *Fash. Technol.* **2023**, *8*, 000691.
34. Lanzón, M.; Castellón, F.J.; Ayala, M. Effect of the expanded perlite dose on the fire performance of gypsum plasters. *Constr. Build. Mater.* **2022**, *346*, 128494. [CrossRef]
35. Joseph, E.J.; Akshayraj, V.R.; Panneerselvam, K. Surface modification of tungsten fillers for application in polymer matrix composites. *Mater. Today Proc.* **2021**, *45*, 7930–7933. [CrossRef]
36. Hu, Y.; Liu, Y.; Zheng, S.; Kang, W. Progress in Application of Silane Coupling Agent for Clay Modification to Flame Retardant Polymer. *Molecules* **2024**, *29*, 4143. [CrossRef]
37. Li, H.; Song, S.-I.; Jang, K.-S. The Role of Surface Treatment and Coupling Agents for Adhesion between Stainless Steel (SUS) and Polyamide (PA) of Heterojunction Bilayer Composites. *Polymers* **2024**, *16*, 896. [CrossRef]
38. Miedzińska, K.; Członka, S.; Strąkowska, A.; Strzelec, K. Vermiculite Filler Modified with Casein, Chitosan, and Potato Protein as a Flame Retardant for Polyurethane Foams. *Int. J. Mol. Sci.* **2021**, *22*, 10825. [CrossRef]
39. Członka, S.; Kairytė, A.; Miedzińska, K.; Strąkowska, A. Casein/Apricot Filler in the Production of Flame-Retardant Polyurethane Composites. *Materials* **2021**, *14*, 3620. [CrossRef]
40. Wang, W.; Wang, D.; Xia, B.; Li, T.; Wang, Y.; Zhang, X.; Bai, H.; Chen, M.; Dong, W. Rigid polyurethane foams based on dextrin and glycerol. *Ind. Crops Prod.* **2022**, *177*, 114479. [CrossRef]
41. Ivdre, A.; Abolins, A.; Sevastyanova, I.; Kirpluks, M.; Cabulis, U.; Merijs-Meri, R. Rigid Polyurethane Foams with Various Isocyanate Indices Based on Polyols from Rapeseed Oil and Waste PET. *Polymers* **2020**, *12*, 738. [CrossRef]
42. *EN 14315-1*; Thermal Insulating Products for Buildings—In-Situ Formed Sprayed Rigid Polyurethane (PUR) and Polyisocyanurate (PIR) Foam Products—Part 1: Specification for the Rigid Foam Spray System before Installation. European Standardization Committee: Brussels, Belgium, 2013.
43. *ISO 4590*; Rigid Cellular Plastics—Determination of the Volume Percentage of Open Cells and of Closed Cells. International Standardization Organization: Geneva, Switzerland, 2016.
44. *EN 12667*; Thermal Performance of Building Materials and Products—Determination of Thermal Resistance by Means of Guarded Hot Plate and Heat Flow Meter Methods—Products of High and Medium Thermal Resistance. European Standardization Committee: Brussels, Belgium, 2001.

45. *ISO 29767*; Thermal Insulating Products for Building Applications—Determination of Short-Term Water Absorption by Partial Immersion. International Standardization Organization: Geneva, Switzerland, 2019.
46. *EN 1607*; Thermal Insulating Products for Building Applications—Determination of Tensile Strength Perpendicular to Faces. European Standardization Committee: Brussels, Belgium, 2013.
47. *ISO 29469*; Thermal Insulating Products for Building Applications—Determination of Compression Behaviour. International Standardization Organization: Geneva, Switzerland, 2022.
48. Sakr, H.; Ammar, A.; Zaki, H.; Salama, M.A.; Ali, M. Impact of ball milling on physicochemical, structural, and functional properties of *Moringa oleifera* L. leaf powders. *J. Food Meas. Charact.* **2023**, *18*, 320–330. [CrossRef]
49. Saufi, H.; El Alouani, M.; Alehyen, S.; El Achouri, M.; Aride, J.; Taibi, M. Photocatalytic Degradation of Methylene Blue from Aqueous Medium onto Perlite-Based Geopolymer. *Int. J. Chem. Eng.* **2020**, *2020*, 9498349. [CrossRef]
50. Lee, W.K.W.; van Deventer, J.S.J. Use of Infrared Spectroscopy to Study Geopolymerization of Heterogeneous Amorphous Aluminosilicates. *Langmuir* **2003**, *19*, 8726–8734. [CrossRef]
51. Tran, L.; Wu, P.; Zhu, Y.; Yang, L.; Zhu, N. Highly enhanced adsorption for the removal of Hg(II) from aqueous solution by Mercaptoethylamine/Mercaptopropyltrimethoxysilane functionalized vermiculites. *J. Colloid Interface Sci.* **2015**, *445*, 348–356. [CrossRef]
52. Duman, O.; Tunç, S.; Polat, T.G. Determination of adsorptive properties of expanded vermiculite for the removal of C.I. Basic Red 9 from aqueous solution: Kinetic, isotherm and thermodynamic studies. *Appl. Clay Sci.* **2015**, *109*, 22–32. [CrossRef]
53. Salih, S.S.; Kadhom, M.; Shihab, M.A.; Ghosh, T.K. Competitive Adsorption of Pb(II) and Phenol Onto Modified Chitosan/Vermiculite Adsorbents. *J. Polym. Environ.* **2022**, *30*, 4238–4251. [CrossRef]
54. Mamiński, M.Ł.; Więcław-Midor, A.; Parzuchowski, P.G. The Effect of Silica-Filler on Polyurethane Adhesives Based on Renewable Resource for Wood Bonding. *Polymers* **2020**, *12*, 2177. [CrossRef]
55. Sari, Y.; Utama, A.H.; Irawan, A.; Andiana, P.; Faradila, R.; Manab, A.; Radiati, L.E. Optimizing Dextrin Levels for Enhanced Physicochemical Properties and Microstructure of Sumbawa Mare's Milk Powder. *Adv. Anim. Vet. Sci.* **2025**, *13*, 431–439. [CrossRef]
56. Arai, Y.; Tomai, T.; Seong, G.; Ito, R.; Yoko, A.; Adschiri, T. Unified formulation of particle-size-dependent viscosity for Newtonian dispersions of micro- and nanoparticles. *J. Mol. Liq.* **2024**, *411*, 125659. [CrossRef]
57. Dorokhov, A.S.; Goncharova, Y.A.; Sviridov, A.S.; Tuzhilin, S.P.; Alekhina, R.A. Measurement of the Dynamic Viscosity of Filled Composites on the Basis of Thermosetting Cast Polyurethane. *Theor. Found. Chem. Eng.* **2023**, *57*, 1116–1119. [CrossRef]
58. Gama, N.V.; Amaral, C.; Silva, T.; Vicente, R.; Coutinho, J.A.P.; Barros-Timmons, A.; Ferreira, A. Thermal Energy Storage and Mechanical Performance of Crude Glycerol Polyurethane Composite Foams Containing Phase Change Materials and Expandable Graphite. *Materials* **2018**, *11*, 1896. [CrossRef]
59. Lee, D.I.; Ha, Y.H.; Jeon, H.; Kim, S.H. Preparation and Properties of Polyurethane Composite Foams with Silica-Based Fillers. *Appl. Sci.* **2022**, *12*, 7418. [CrossRef]
60. Soloveva, O.V.; Solovev, S.A.; Vankov, Y.V.; Shakurova, R.Z. Experimental Studies of the Effective Thermal Conductivity of Polyurethane Foams with Different Morphologies. *Processes* **2022**, *10*, 2257. [CrossRef]
61. Vijayaraghavan, K.; Raja, F.D. Experimental characterisation and evaluation of perlite as a sorbent for heavy metal ions in single and quaternary solutions. *J. Water Process Eng.* **2014**, *4*, 179–184. [CrossRef]
62. Marcos, C. Effect of Water Immersion on Raw and Expanded Ugandan Vermiculite. *Minerals* **2022**, *12*, 23. [CrossRef]
63. Assis Neto, P.C.D.; Sales, L.P.B.; Oliveira, P.K.S.; Silva, I.C.D.; Barros, I.M.D.S.; Nóbrega, A.F.D.; Carneiro, A.M.P. Expanded Vermiculite: A Short Review about Its Production, Characteristics, and Effects on the Properties of Lightweight Mortars. *Buildings* **2023**, *13*, 823. [CrossRef]
64. Xia, Z.; Wang, Z. Rigid polyurethane/expanded vermiculite/melamine phenylphosphate composite foams with good flame retardant and mechanical properties. *e-Polymers* **2019**, *19*, 563–573. [CrossRef]
65. Członka, S.; Bertino, M.F.; Strzelec, K.; Strąkowska, A.; Masłowski, M. Rigid polyurethane foams reinforced with solid waste generated in leather industry. *Polym. Test.* **2018**, *69*, 225–237. [CrossRef]
66. de Souza, F.M.; Choi, J.; Bhoyate, S.; Kahol, P.K.; Gupta, R.K. Expendable Graphite as an Efficient Flame-Retardant for Novel Partial Bio-Based Rigid Polyurethane Foams. *J. Carbon Res.* **2020**, *6*, 27. [CrossRef]
67. Hejna, A.; Olszewski, A.; Zedler, Ł.; Kosmela, P.; Formela, K. The Impact of Ground Tire Rubber Oxidation with H_2O_2 and $KMnO_4$ on the Structure and Performance of Flexible Polyurethane/Ground Tire Rubber Composite Foams. *Materials* **2021**, *14*, 499. [CrossRef]

applied sciences

MDPI

Article

The Potential of Environmental Non-*Saccharomyces* Yeast to Valorise Apple Pomace During Low-Temperature Fermentation

Wiktoria Liszkowska [1,*], Ilona Motyl [1], Katarzyna Pielech-Przybylska [2], Julia Dzierżanowska [3], Sara Motyl [4] and Joanna Berlowska [1]

1 Department of Environmental Biotechnology, Lodz University of Technology, 171/173 Wolczanska Street, 90-530 Lodz, Poland; ilona.motyl@p.lodz.pl (I.M.); joanna.berlowska@p.lodz.pl (J.B.)
2 Institute of Fermentation Technology and Microbiology, Lodz University of Technology, 171/173 Wolczanska Street, 90-530 Lodz, Poland; katarzyna.pielech-przybylska@p.lodz.pl
3 Faculty of Biotechnology and Food Sciences, Lodz University of Technology, 171/173 Wolczanska Street, 90-057 Lodz, Poland; 230419@edu.p.lodz.pl
4 School of Mathematics & Sciences, Lindsey Wilson College, 210 Lindsey Wilson Street, Columbia, KY 42728, USA; sara.motyl@lindsey.edu
* Correspondence: wiktoria.liszkowska@dokt.p.lodz.pl

Abstract: It is estimated that approximately 25% of waste remains after the apple juice pressing process. Combining this waste biomass with valuable compounds creates the potential for reuse. To create a cost-efficient ecological process without any expensive steps, the aim of this research was to examine the potential of using non-*Saccharomyces* yeasts (*Kazachstania barnettii* D1, *Hanseniaspora uvarum* D9, and *Wickerhamomyces anomalus* D11) for the low-temperature valorisation of apple pomace. The scope encompassed characteristics of apple pomace and the evaluation of yeast growth and metabolic activity, including carbohydrate consumption, enzymatic activity, and the biosynthesis of volatile organic compounds. Moreover, the effect of inoculum size on biomass increases and the productivity of metabolites during the fermentation of apple pomace were evaluated. To investigate the potential intensification of the process, the experiment was performed on hydrolysed and untreated apple pomace. The obtained results indicate that yeast growth was satisfactory regardless of the inoculum size in both fermentation media. Various activities of peptidases, esterases, phosphatases, and glucosidases were observed. The yeast isolates presented metabolic activity during the process which was confirmed by the production of ethanol and acetic acid. Moreover, a significant amount of volatile organic compounds, especially esters, were synthesised, which have a positive impact on the sensory profile of fermented apple pomace. In general, the hydrolysis of apple pomace did not result in better yeast activity and the formation of aroma compounds.

Keywords: volatile organic compounds; unconventional yeast; low-temperature fermentation; waste valorisation; apple pomace

1. Introduction

The growing global population has increasing demands for food, water, and energy. Increased production and consumption inevitably lead to the production of various types of waste, including agricultural, food, wood, municipal solid, and sewage sludge wastes. These wastes are directly connected to the core needs of food, energy, and water, which are vital for the sustenance and growth of human societies [1–3]. Their diverse composition and characteristics make them more challenging to manage uniformly. The improper management of waste biomass can lead to environmental issues, such as greenhouse gas emissions (particularly methane from decomposing organic matter) and the pollution of soil and water, as well as inefficient use of potentially valuable resources. Adequately processing and utilising waste biomass often requires specific technologies and infrastructures, such as facilities for composting, anaerobic digestion, or biomass-to-energy conversion. However, waste biomass can also serve as a renewable material capable of supplying sustainable energy and a range of bio-based materials.

In the food industry, the apple juice sector generates notable amounts of waste by-products. According to the World Apple and Pear Association (WAPA), China is the leading producer of apples at about 36.09 million tons. Turkey, for the first time, secured the second position, increasing production from 4.44 to 4.66 million tons. The United States occupies the third place with 4.51 million tons. The fourth major global apple producer is Poland with 4.50 million tons [4]. The large-scale production of apples and, subsequently, apple juice results in significant quantities of waste biomass (approximately 25%), primarily in the form of pomace—the solid residue left after the extraction of juice, including the skins, pulp, seeds, and stems. Managing this by-product efficiently presents a substantial challenge as it constitutes a large portion of organic waste in the food processing industry. The most popular method of apple pomace disposal is to transfer it directly to the soil in a landfill. Due to the fact that this waste material contains more than 70% of water, carbohydrates, and organic acids, which lead to enhanced microbial fermentation, it may cause a serious pollution problem because the excessive growth of microorganisms may decrease the available nitrogen in the soil and truly affect the C/N ratio [5]. Nonetheless, the ready availability of apple pomace and its beneficial components for human health offer possibilities for economically viable post-processing opportunities [4,6].

Apple pomace contains several notable compounds, including dietary fibres, polyphenols, and bioactive compounds such as phenolic acids (chlorogenic acid), flavonoids (catechins and epicatechins), and dihydrochalcone (phlorizin) [7]. The valorisation of apple pomace using microorganisms, particularly yeast, is attracting growing interest from both researchers and the food industry. Due to the relatively high amounts of carbohydrates in apple pomace, yeasts seem to be a good choice for reducing this type of by-product. Currently, many techniques of apple pomace management are available. Apart from the biofuel industry, where products are mainly fermented by microorganisms, existing techniques are mostly focused on the extraction of various compounds from the raw material [8,9]. Such processes usually require expensive and specialised equipment, and they can generate wastes. However, the microbiological way of processing side-streams is the promising solution within the circular economy assumptions. The potential of apple pomace as a substrate for microbial growth is well known [5]. After fermentation, numerous compounds such as pectin, dietary fibre, xyloglucan, natural antioxidant, and aromatic compounds are extracted [10].

In recent years, the potential of non-*Saccharomyces* yeast has been gaining recognition. Historically linked with spontaneous fermentation processes, such as in the production of lambic beers, these wild yeasts are now being explored for their ability to diversify flavours and aromas in fermented products, thereby enhancing their organoleptic qualities [11,12]. However, research and applications in this area remain quite limited. In contrast to baker's yeast, many non-*Saccharomyces* yeast strains have been reported to metabolise various types of carbohydrates. *Kluyveromyces lactis*, for example, utilises lactose. Spencer and coworkers highlighted the ability *of Yarrowia lipolytica* and *Candida maltose* to assimilate hydrocarbons, fatty acids, and lipids [13]. Pentoses such as xylose and arabinose can also be assimilated by certain yeast strains, for instance, *Pichia stipitis* and *Candida shehatae* [14]. Some non-*Saccharomyces* yeasts present the abilities to utilise more complex carbon sources than *S. cerevisiae* and to survive in harsher conditions [15].

There is growing awareness of the advantages of conducting bioprocesses at a reduced temperature. Low-temperature fermentation in industrial processes is typically defined within the range of 10–18 °C for fruit-based fermentations. In contrast, high-temperature fermentation is carried out at the temperature above 25 °C, which is close to the optimal temperature for yeast activity [16–18]. Using lower temperatures offers a great number of benefits, including achieving energy savings by eliminating the need to use additional heat during the process, thus reducing costs [19]. At low temperatures, slower microbial metabolism leads to more controlled fermentation, reducing unwanted reactions and by-products like volatile acidity, higher alcohols, and aldehydes while preserving the balance of acids and aroma compounds [20,21]. Microorganisms adapted to low temperatures exhibit high growth rates, enzymatic activity, and catalytic efficiency at temperatures much lower than their mesophilic counterparts. Thus, their use can reduce the risk of microbial contamination [22]. Furthermore, the fermentation process not only leads to the formation of specific products with unique flavour and aroma characteristics but is also utilised in the production of natural food additives, probiotics, and functional foods. Incorporating microorganisms in fermentation results in a final product that offers many health benefits for humans [23].

A particularly desirable group of compounds which can be formed successfully by yeasts are volatile organic compounds (VOCs). These compounds are characterised by a low molecular weight and usually present a distinctive odour. They are generally produced by microorganisms during the fermentation process as by-products of carbon metabolism [24,25]. In the natural environment, VOCs produced by yeast serve multiple purposes, including for communication between cells and as a defence mechanism [26]. Certain yeast strains, such as *Wickerhamomyces anomalus*, *Metschnikowia pulcherrima*, *Aureobasidium pullulans*, and *Saccharomyces cerevisiae*, produce VOCs that can potentially suppress the growth and development of pathogens [27]. In industry, they can contribute to the sensory profiles of fermented products. The production of volatiles by yeast has attracted great interest, especially due to its potential to enrich the diversity of aromas in fermented foods and beverages. The production of volatile compounds is connected to the parameters of the fermentation process, the composition of the medium, the pH, temperature, and strain selection. Thus, the amount and profile of VOCs produced by yeast can differ from species to species [26]. The most abundant VOCs are typically esters, alcohols, acids, aldehydes, ketones, and lactones, as well as sulphur and nitrogen compounds [25].

Apple pomace was selected as a raw material because Poland is one of the leading apple producers in the world. During the processing of apples, enormous amounts of waste biomass are generated. However, apple pomace contains numerous valuable compounds that can be reused in an environmentally friendly way, including via microbiological fermentation. The purpose of this study was to present the potential benefits of low-temperature yeast fermentation for the valorisation of apple pomace. Although the process is longer than normal due to the high cost of maintaining elevated temperatures in large industrial settings, activity at low temperature was a key criterion for the selection of the yeasts; thus, the fermentation activity of unconventional yeast strains cultured in apple pomace at a low temperature of 15 °C was investigated. A temperature of 15 °C is within the range of temperatures considered as low. It is also a temperature that can be achieved in industrial production facilities without the use of temperature control devices. This study presents an analysis of the production of ethanol and acetic acid, the utilisation of carbohydrates in the fermentation medium, enzymatic activity determined using API ZYM tests, and the profiles of VOCs. To evaluate the potential for intensifying the process, all experiments were conducted using both untreated apple pomace and pomace hydrolysed by pectinolytic and cellulolytic enzymes.

2. Materials and Methods

2.1. Yeast Strains and Cultivation

Three environmental non-*Saccharomyces* yeast isolates were used in the following experiments. All three yeasts were isolated from plant material from the environment, adopted to grow well at 15 °C and initially characterised at the Lodz University of Technology's Department of Environmental Biotechnology [24]. The following RNA sequences were deposited in the NCBI database:

- *Kazachstania barnettii* D1 from pickled beetroot (accession no.: OQ275007.1);
- *Hanseniaspora uvarum* D9 from grape (accession no.: OQ275152.1);
- *Wickerhamomyces anomalus* D11 from wheat flour (accession no.: OQ275025.1).

The strains were cultured in 10 mL of wort broth (Merck, Darmstadt, Germany) in test tubes at 15 °C for 7 days. Yeast growth was assessed by measuring optical density using a densitometer (DEN-1B, Grant Instruments (Cambridge) Ltd., Shepreth, UK). The low temperature was implemented at the activation stage to maintain constant activity of tested strains during the whole process.

2.2. Apple Pomace Preparation

Apples of the "Ligol" variety (previously characterised by Yoon and coworkers [25]) were washed and dried. Apple pomace was prepared in the laboratory by pressing fruits using a slow-speed juicer (Gallet model HS 703, Prague, Czech Republic). The pomace was subsequently passed through the machine again to remove the rest of the juice. The pressed apple mass was suspended in a ratio of 3:1 in water, and the final suspension was pasteurised at a temperature of 78 °C for 20 min.

2.3. Dry Mass Content

Content of dry mass was analysed by comparing the weight of a sample before and after drying at 135 °C for 2 h. The ratio between mass after drying to mass before drying indicates the percentage of dry mass content [28,29].

2.4. Enzymatic Hydrolysis of Prepared Apple Pomace

In order to fully evaluate the potential of the composed fermentation medium, the prepared suspension was parallelly subjected to enzymatic hydrolysis. The pectinolytic enzymes used were Viscozyme® (Novozymes, Bagsvžrd, Denmark), indicated as "V", and Rohapect 10 L (AKE, Pabianice, Poland), indicated as "R". The cellulolytic enzyme was Rohament® CL (AB Enzymes GmbH, Darmstadt, Germany), indicated as "R1".

Two different volumes were tested—0.25 mL and 0.5 mL of each enzyme per 1 g of dry mass of apple pomace. Both doses were used according to the manufacturer's instructions. The enzyme combinations are given in Table 1. Each variant was evaluated using the percentage of liquidity to identify the most suitable combination for further experimentation. It was calculated as the difference in mass of liquid fraction after and before hydrolysis divided by the total mass.

Table 1. Examined combinations of enzymes.

Sample Name	Enzyme Used for Hydrolysis	Amount of Enzyme per 1 g of Dry Mass [mL]
0.25_V	Viscozyme®	0.25
0.5_V	Viscozyme®	0.5
0.25_R	Rohapect 10 L	0.25

Table 1. *Cont.*

Sample Name	Enzyme Used for Hydrolysis	Amount of Enzyme per 1 g of Dry Mass [mL]
0.5_R	Rohapect 10 L	0.5
0.25_VR1	Viscozyme® + Rohament® CL	0.25 + 0.25
0.5_VR1	Viscozyme® + Rohament® CL	0.5 + 0.5
0.25_RR1	Rohapect 10 L + Rohament® CL	0.25 + 0.25
0.5_RR1	Rohapect 10 L + Rohament® CL	0.5 + 0.5

2.5. Carbohydrate Content Analysis

To assess the concentrations of different carbohydrates in the tested samples, assays were performed for arabinose, raffinose, galacturonic acid, glucose, fructose, and sucrose. Each test was performed four times at different stages of the experiment. First, the pure apple pomace was characterised. Then, the apple pomace was examined after hydrolysis to select the appropriate enzyme combination. Finally, the samples of both hydrolysed and untreated apple pomace were tested after yeast fermentation.

To measure the carbohydrate concentration in g/L, assay kits from Megazyme (Bray, Ireland) were used. These kits can be used for the measurement and analysis of various carbohydrates in various materials, including foods, feeds, beverages, and plant products. The tests were conducted according to the manufacturer's instructions. All tests were performed on 96-well plates. For the analysis, a spectrophotometer (Thermo Scientific Multiskan GO, Thermo Fisher Scientific, Waltham, MA, USA) was used.

2.6. Liquid-State Fermentation of Apple Pomace and Yeast Growth

After activating the yeast strains, the yeast suspensions were standardised to 10^7 CFU/mL, and the untreated and hydrolysed apple pomace samples (60 mL of each) were inoculated with 1%, 5%, or 10% inoculum. Fermentation was performed at 15 °C for 5 days.

In order to confirm the ability of the analysed yeast strains to use apple pomace as a fermentation medium, the growth of the selected yeast strains on the untreated and hydrolysed apple pomace was examined before and after 5 days of fermentation. Culturing was carried out on sterile wort broth with gentamycin using the pour plate method at 15 °C for 5 days. After this period, the colonies were counted, and the average CFU/mL was calculated for each strain.

2.7. Evaluation of Enzymatic Activity of Yeasts: API ZYM

The aim of the analysis was to evaluate the potential of utilising carbohydrates released from apple pomace. The enzymatic activity of the tested yeast isolates at their stationary growth phase was evaluated by API ZYM (bioMérieux, Marcy-l'Étoile, France) according to the manufacturer's instructions. API ZYM is a semi-quantitative method for the investigation of enzymatic activity. This technique is applicable to all types of samples (microorganisms, cell suspensions, tissues, biological fluids, etc.). It enables the systematic and rapid study of 19 enzymatic reactions using very small sample quantities. The examined yeasts were subjected to an analysis of enzymatic activity at the temperature of 15 °C.

2.8. Quantitative Analysis of Ethanol and Acetic Acid Produced During Fermentation Using High-Performance Liquid Chromatography (HPLC)

Fermentation activity of examined yeasts was evaluated by measuring the concentrations of ethanol and acetic acid produced during the fermentation of apple pomace according to the method used by Balcerek and coworkers [30]. They were determined by high-performance liquid chromatography, Agilent, model 1260 Infinity (Agilent Technologies, Santa Clara, CA, USA) equipped with a refractometric detector which operated at 55 °C. An amount of 20 µL of sample was injected, and the compounds were separated on a Hi-Plex H column (7.7 × 300 mm, 8 µm) at 60 °C (Agilent Technologies, Santa Clara, CA, USA). The liquid phase was 5 mM H_2SO_4 with a flow rate of 0.7 mL/min.

A mass spectrophotometer was used for a quantitative analysis of ethanol, working in selected ion monitoring (SIM) mode.

2.9. Quantitative Analysis of Volatile Organic Compounds (VOCs) Produced During Fermentation Using SHS-GC-MS

Biosynthesis of volatile organic compounds (VOCs) was examined by gas chromatography (GC) and mass spectrometry (MS) using the Static Headspace (SHS) technique according to the method described by Pielech-Przybylska and coworkers [31] with several modifications. The samples were stored in falcon tubes in a freezer at −25 °C. First, a qualitative analysis was conducted to identify VOCs in the samples. During analysis, the mass spectrometer was operated in full scan mode (a scan range of m/z 27–350 was acquired). The VOCs were identified by comparing the obtained mass spectrum to the mass spectrum of a standard substance and the mass spectrum from a mass spectrum library (NIST/EPA/NIH Mass Spectra Library, Version 2.0 g). A total of 29 compounds were identified. For the quantitative measurements, selected ion monitoring (SIM) mode was used. Examples of GC-MS ion chromatograms after qualitative analysis are in the Supplementary Materials ("Supplementary reports S1–S6").

Before qualitative and quantitative analyses, 8 mL of each sample was transferred to 20 mL vials, which were then sealed with an aluminium cap. The prepared vials were placed on the feeder plate of a Headspace Sampler. During the analysis, the autosampler transferred the samples to a heater, where they were thermostated for 20 min at 50 °C. During thermostating, the vial was shaken (71 shakes/min with an acceleration of 260 cm/s^2). Next, the surface phase was collected and transferred to the inlet of the GC apparatus. The time of filing the loop was 0.2 min. The time of equilibration was 0.5 min. The dosing time was 0.7 min. The loop and the transfer line temperatures were 60 °C and 70 °C, respectively. Sample dosing was conducted in a split stream system (qualitative analysis: 2:1 split; quantitative analysis: 25:1 split). Separation of the compounds was performed on a capillary column (Rxi-5ms, Restek) with dimensions of 60 m (length) × 250 µm (inside diameter) × 0.25 µm (thickness of the stationary phase), while applying the following programmed temperature increase: 30 °C (10 min); increasing by 5 °C/min. to 80 °C (2 min); and increasing by 10 °C/min to 240 °C (3.5 min). The flow of carrier gas (helium) was 1 mL/min. GC inlet temperature: 250 °C; transfer line to MS temperature: 250 °C; ion source temperature: 230 °C; quadrupole temperature: 150 °C.

2.10. Statistical Analysis

The two-way analysis of variance (ANOVA) test was used to examine differences, considered statistically significant when $p < 0.0001$. The variables were the amount of inoculum and strain type. Statistical analysis was performed using XLSTAT software (Addinsoft, version 2022.2.1, Seattle, WA, USA).

3. Results

3.1. Enzymatic Hydrolysis of Apple Pomace

To ensure effective microbial growth, it is essential to provide carbohydrates in the fermentation medium as a carbon source. In the case of yeast, the most common are glucose and fructose. Data in the literature indicate that apple pomace contains 2.5–22.7% glucose and 11.5–49.8% fructose, respectively [32]. Moreover, apple pomace contains polymer fractions, including the soluble fraction pectin and the insoluble fractions cellulose, hemicellulose, and lignin [5,33,34]. Thus, conducting the process on a large biomass load presents technological problems. Firstly, pectins exhibit gelling properties that make it difficult to maintain a homogeneous fermentation environment. Secondly, insoluble fibre fractions are carbohydrates that are difficult for yeast to digest [35]. Therefore, the enzymatic hydrolysis of apple pomace was performed. In addition to better liquefaction of the fermentation substrate, hydrolysis leads to an increase in the concentration of carbohydrates assimilated by yeast. The addition of cellulolytic enzymes during biomass processing increases the release of glucose from lignocellulosic biomass by improving the access to cellulose [36]. The dry matter content of the analysed pomace was approximately 19–21%. Commercially available multienzyme preparations were used in the context of the applied nature of the research conducted. Two commercial preparations with pectinolytic properties and one commercial cellulolytic preparation were used for hydrolysis. They allowed for the experiment to be scaled up from the laboratory scale to the industrial scale. To choose the most suitable enzyme for use in the hydrolysis of the apple pomace prior to yeast fermentation, the degree of liquidity was calculated in the pomace treated with selected combinations of enzymes. The data for all the samples are presented in Figure 1.

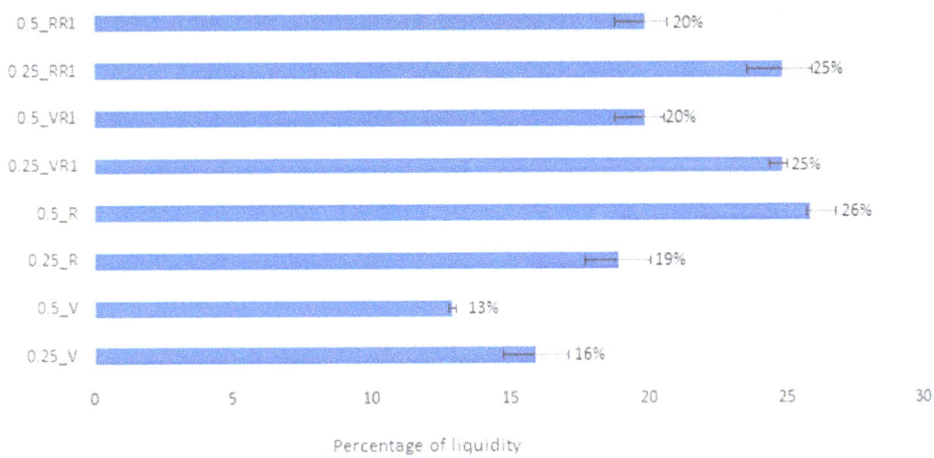

Figure 1. Degree of liquidity of apple pomace treated with various combinations of hydrolytic enzymes. Abbreviations are as defined in Table 1.

The apple pomace sample treated with 0.5 mL of Rohapect 10 L (0.5_R), 0.25 mL of Viscozyme® together with 0.25 mL of Rohament® CL (0.25_VR1), and 0.25 mL of Rohapect 10 L together with 0.25 mL of Rohament® CL (0.25_RR1) showed the highest percentage of liquidity. The initial mass of liquidity of each sample was 18 g. Comparing the overall mass of liquid in samples before and after hydrolysis shows that treatment with enzymes improved the degree of liquidity in the apple pomace. It was found that the most suitable sample contained 0.25 mL per 1 g of dry mass of Viscozyme® and 0.25 mL per 1 g of dry mass of Rohament® CL (indicated as 0.25_VR1) as it expressed a high percentage of overall liquidity. The obtained results prove that enzymatic hydrolysis could improve the liquidity of apple pomace by breaking down structural polysaccharides. These findings align with previous studies on fruit by-product processing [37]. The combination of pectinases and cellulases proved to enhance substrate liquefaction and increased the content of sugars in the fermentation medium [38]. These findings highlight the importance of optimising enzyme selection and dosage, which is crucial for the efficiency of liquefaction and minimising processing challenges related to high viscosity and insoluble fibre fractions in apple pomace [39].

Yeast growth and fermentative abilities are strictly dependent on the amounts of carbohydrates present in the medium. Therefore, to further assess the properties of the prepared apple pomace used as a medium, the concentrations of selected carbohydrates were measured. Enzymatic assay kits by Megazyme were used to analyse raffinose, galacturonic acid, arabinose, glucose, fructose, and sucrose. The results are presented in Table 2. The combination of enzymes selected for the further experiments was highlighted.

Table 2. Carbohydrate content in examined apple pomace before and after hydrolysis. Abbreviations: "Untreated AP" refers to apple pomace untreated with enzymes. "Hydrolysed AP" refers to apple pomace treated with various combinations of enzymes; the rest are as defined Table 1.

		Concentration [g/L]					
		Arabinose	Galacturonic Acid	Raffinose	Glucose	Fructose	Sucrose
Untreated AP		0.10 ± 0.03	0.06 ± 0.01	0.07 ± 0.02	2.22 ± 0.39	6.66 ± 1.08	7.36 ± 0.11
Hydrolysed AP	0.25_V	1.42 ± 0.27	3.57 ± 0.40	1.51 ± 0.23	5.67 ± 2.73	10.91 ± 5.67	4.84 ± 1.70
	0.5_V	1.70 ± 0.09	3.16 ± 0.23	1.45 ± 0.15	4.52 ± 0.61	7.68 ± 0.60	3.66 ± 1.45
	0.25_R	1.39 ± 0.19	2.91 ± 0.25	1.02 ± 0.23	3.12 ± 0.23	7.22 ± 0.47	5.22 ± 1.56
	0.5_R	1.59 ± 0.39	2.90 ± 0.18	1.37 ± 0.49	3.98 ± 0.56	7.97 ± 0.59	2.74 ± 1.21
	0.25_VR1	1.62 ± 0.21	3.39 ± 0.33	1.19 ± 0.34	5.50 ± 0.83	7.50 ± 1.47	3.72 ± 1.26
	0.5_VR1	1.79 ± 0.11	3.42 ± 0.52	1.31 ± 0.15	6.51 ± 1.66	7.19 ± 0.82	3.69 ± 0.98
	0.25_RR1	1.82 ± 0.77	2.80 ± 0.19	1.07 ± 0.58	5.55 ± 2.08	7.10 ± 1.38	2.98 ± 1.15
	0.5_RR1	1.58 ± 0.12	2.89 ± 0.08	1.62 ± 0.23	4.28 ± 0.56	6.00 ± 0.40	4.15 ± 1.05

From the data collected in Table 2, it can be seen that the concentrations of useful carbohydrates in the examined samples, especially glucose and fructose, had increased. The selected combination of enzymes, 0.25 mL of Viscozyme® and 0.25 mL of Rohament® CL (0.25_VR1), improved the content and availability of these carbohydrates. A decrease in the sucrose concentration was also observed in all samples with a simultaneous increase in glucose and fructose concentrations. Therefore, glucose may come not only from the enzymatic hydrolysis of cellulose, but also from the decomposition of sucrose. Analysing the hydrolysis of the pectin fraction, the predictably largest increase in concentration occurred for galacturonic acid, which is the main structural monomer of pectins. A slightly smaller increase occurred in the arabinose and raffinose contents. They may be present in the side chains of the main polygalacturonic chain, so their concentrations may be lower [40]. Other studies proved that the enzymatic hydrolysis of apple pomace using combinations of pectinases and cellulases significantly enhances the concentration of fermentable sugars in the fermentation medium and improves substrate availability for yeast fermentation [38].

3.2. Apple Pomace and Its Hydrolysate as a Culture Medium for Yeast

The growth of the selected yeast strains at 15 °C was calculated by counting the values of CFU/mL before and after 5 days of the fermentation process. The objective of this experiment was to evaluate the effect of the relation between inoculum size and biomass increase during fermentation of apple pomace. In the literature, various inoculation rates were used, for example, 2–20% in winemaking and 5–10% in brewery [41–43]. In the case of small inoculum doses (e.g., 1%), the nutrient limitation of the medium appears later. Most often, the initial number of yeasts is in the range of 10^5–10^6 CFU/mL of the fermentation medium, and this amount determines their optimal activity in technological processes [44]. This is confirmed by scientific studies which show that the initial amount of 10^6 CFU/mL determines the growth of yeast to the level of 10^8 CFU/mL [45,46]. The results are presented in Figure 2.

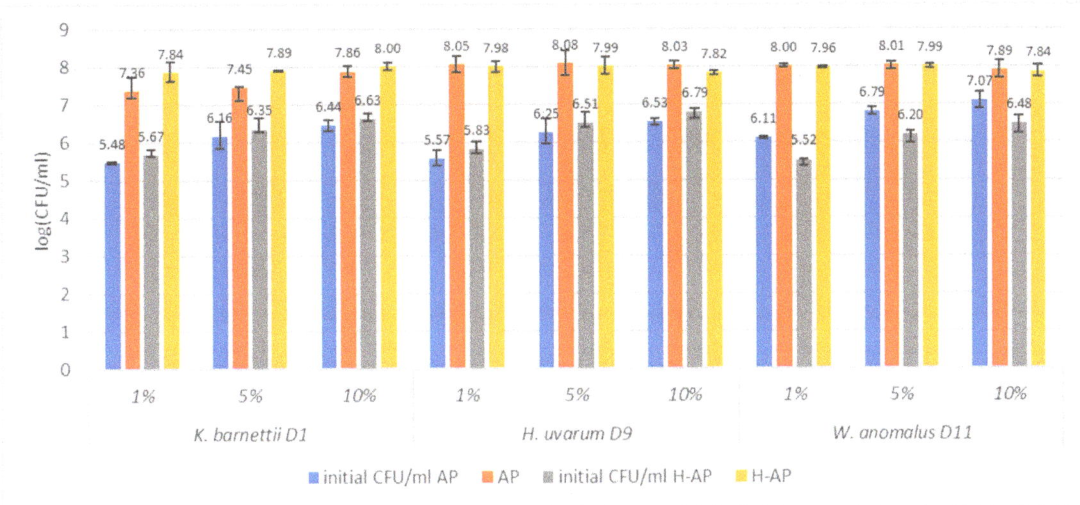

Figure 2. CFU/mL values of examined yeast strains before and after fermentation of untreated and hydrolysed apple pomace. Abbreviations: 1%, 5%, and 10% refer to samples with various inoculation levels; "AP" refers to apple pomace untreated with enzymes; "H-AP" refers to hydrolysed apple pomace treated with 0.25 mL of Viscozyme® and 0.25 mL of Rohament® CL enzymes.

The measured yeast growth confirms that all of the selected yeast strains were able to grow at the tested temperature of 15 °C and remained active throughout fermentation. It can be seen that the cell growth of the examined yeast strains was around 10^8 CFU/mL. Slightly higher results were obtained for *Kazachstania barnettii* D1 when using hydrolysed apple pomace as a medium. The highest final CFU/mL value was detected at 10% for both untreated and hydrolysed apple pomace. For the *Hanseniaspora uvarum* D9 and *Wickerhamomyces anomalus* D11 yeast strains, the difference between the CFU/mL values at the starting point and after cultivation was the largest in the samples with 5% inoculation. A similar result for the total number of cells in CFU/mL after 5 days of cultivation at 15 °C was observed for the *Wickerhamomyces anomalus* D11 yeast strain for the sample with 1% inoculum. All of the examined yeast strains performed well and yielded results suggesting that their adaptation to a low temperature before conducting the main process was successful. In general, the number of cells remained similar regardless of the level of inoculation. The numbers of yeast cells expressed in CFU/mL after 5 days of cultivation did not differ significantly whether 1%, 5%, or 10% of inoculum was used.

An important fact influencing the above process is the size of the yeast culture and thermal treatment of apple pomace. Moreover, before the fermentation process, the medium was subjected to thermal treatment to sterilise the environment through the thermal inactivation of the natural microflora of the pomace. Under model conditions, the initial dose of inoculum did not significantly affect the growth of yeast biomass. However, the trend described in the scientific literature has been maintained—as the inoculum increases, yeast propagation in the environment is faster, and microorganisms are more active and produce more repeatable experimental results [44]. On the other hand, the sterilisation of waste materials is not practiced in the industry as biomass valorisation is carried out on large amounts of plant material. Large amounts of inoculum allow for the control of the non-sterile environment and may inhibit naturally occurring microorganisms. Neelakantam and coworkers confirmed the correlation of the inoculum size and growth inhibition of undesired lactic acid bacteria in corn mash. By using higher inoculation rates, it is possible to inhibit the metabolism of contaminating *Lactobacilli* [47]. Another study was conducted by Chen and coworkers, who analysed the effect of inoculum size on protein productivity during the fermentation process. They proved that a lower inoculation level leads to slower propagation of the environment by yeast, which determines decreased protein productivity. However, using too much yeast is also not beneficial because there is an overconsumption of carbohydrates during the adaptation phase, and it results in protein having lower productivity [48].

The concentrations of the selected carbohydrates in the untreated and hydrolysed apple pomace samples before and after yeast fermentation were examined using enzymatic assays for raffinose, galacturonic acid, arabinose, glucose, fructose, and sucrose. The data are shown in Figure 3.

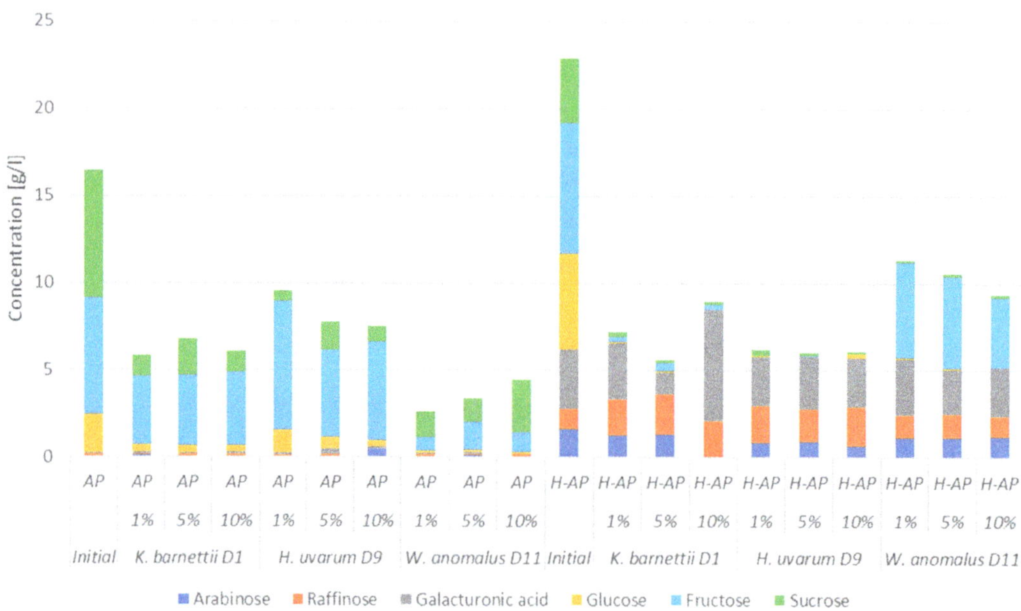

Figure 3. The carbohydrate contents in apple pomace samples before and after fermentation. Abbreviations: "Initial" refers to the carbohydrate content before the fermentation process; 1%, 5%, and 10% refer to samples with various inoculation levels; "AP" refers to yeast cultured with apple pomace untreated with enzymes; "H-AP" refers to yeast cultured with hydrolysed apple pomace treated with 0.25 mL of Viscozyme® and 0.25 mL of Rohament® CL enzymes.

The content of arabinose in the pure apple pomace was found to be quite low. It was significantly improved by treatment with the selected enzymes, rising by almost 17 times. However, after fermentation, no notable decrease in the arabinose concentration was observed, meaning that this carbohydrate was not preferably utilised by the examined yeast strains. The most significant changes in concentration were detected in the case of *Hanseniaspora uvarum* D9 after fermentation of untreated apple pomace with 10% inoculum.

The increase in arabinose concentration may indicate the ability of this yeast to hydrolyse pectin, releasing sugar molecules [49]. The untreated apple pomace did not yield a high concentration of raffinose. The content of raffinose after enzymatic treatment was moderately improved. However, samples collected after 5 days of fermentation, whether using untreated or hydrolysed apple pomace, yielded very similar results to those measured before the process. This suggests that raffinose was not metabolised as a carbon source by the examined yeast strains since the concentrations before and after hydrolysis remained much the same. The content of galacturonic acid in the pure apple pomace was very low. After hydrolysis, the measured concentration was significantly higher due to the pectinolytic activity of the used enzymes. However, the results before and after fermentation did not differ significantly. For all examined strains, regardless of whether hydrolysis was conducted or not, the collected samples yielded results similar to those obtained before yeast fermentation.

The concentration of glucose in the untreated apple pomace was quite high. Enzymatic treatment significantly improved the level of glucose. The sample collected after fermentation yielded a significantly lower concentration of glucose in all cases regardless of the inoculation level. This was to be expected since glucose is a preferred carbon source for yeast fermentation [50]. These results show that during fermentation, the glucose was almost completely utilised by the examined yeast strains. Fructose was the most abundant carbohydrate in the tested samples. This was expected as fruits such as apples are naturally high in fructose. The apple pomace that underwent enzymatic treatment showed an even higher concentration due to the enzymatic activity of the used commercial preparations. An especially low concentration of fructose was observed for *Hanseniaspora uvarum* D9, isolated from grapes, and *Kazachstania barnettii* D1 after the fermentation of hydrolysed apple pomace. In the samples collected after the fermentation of untreated apple pomace, this trend was repeated for the samples with *Wickerhamomyces anomalus* D11, and only trace amounts of fructose were detected after fermentation. Overall, the samples collected after fermentation showed only a slight decrease with 10% inoculum. No major discrepancies between the percentage of inoculum used and the detected carbohydrate concentrations were observed. The utilisation of sucrose during fermentation was generally at a similar level for all tested yeast isolates, suggesting that the tested strains demonstrated invertase enzymatic activity.

Overall, the results for the carbohydrate content suggest that glucose and fructose were preferentially utilised by the yeasts for fermentation. The levels of these carbohydrates were quite high, even in the untreated apple pomace, and the concentrations were further improved by treatment with pectinolytic and cellulolytic enzymes (Viscozyme® and Rohament® CL). However, no close connection between the amount of inoculum used for fermentation and the amounts of carbohydrates utilised during fermentation was detected. The results for the arabinose assay suggest that this carbohydrate was metabolised by *Hanseniaspora uvarum* yeast, although not to such a high extent as glucose or fructose. The results for the fructose assay confirm that this carbohydrate was metabolised by the tested strains [51,52].

3.3. Enzymatic Profile of Yeast During Low-Temperature Fermentation

The enzymatic activity of a yeast strain refers to its ability to utilise enzymes to catalyse various chemical reactions, especially during fermentation. Yeasts use enzymes to transform substances like carbohydrates into primarily ethanol and other compounds. Yeast can produce various enzymes, including alcohol dehydrogenase (responsible for converting ethanol into acetaldehyde and simultaneously regenerating NADH to NAD+), aldehyde dehydrogenase (responsible for converting aldehydes into acetic acid), or glycolytic enzymes (responsible for converting glucose into pyruvate and producing energy in the form of ATP) [53–55]. A basic assessment of the enzymatic activity of yeast was performed using API ZYM.

The enzymatic activity of yeast is closely related to its assimilation profile. Enzymatic activity plays key role in the process as enzymes are responsible for converting more complex compounds into simple compounds. These simple compounds are then used by the yeast as energy substances and precursors for the synthesis of various metabolites, such as alcohols, acids, and esters, which can influence the aroma and flavour of final fermented food products and beverages. As a result, studies of yeast enzymatic activity can enable food producers to control and modify the aromatic and sensory profiles of fermented products, which is important for developing products with unique flavours [56]. The results for the tested strains at 15 °C are summarised in Table 3.

Table 3. Enzymatic activity of tested yeast strains at 15 °C, measured by API ZYM tests. Numbers in columns indicate intensity of colouration on 5-point scale (from 0 to 5): 0—no reaction; 1—weak reaction; 2—moderate reaction; 3—strong reaction; 4—very strong reaction; 5—extremely strong reaction.

	K. barnettii D1	*H. uvarum* D9	*W. anomalus* D11
Alkaline phosphatase	0	0	5
Esterase (C 4)	0	0	4
Leucine arylamidase	5	5	5
Valine arylamidase	5	5	5
Acid phosphatase	5	5	5
Naphthol-AS-BI-phosphohydrolase	5	5	5
α-Glucosidase	0	0	5
ß-Glucosidase	0	3	5

In order to conduct fermentation efficiently, knowledge of the basic enzymatic and assimilation activities of the yeast is needed. In the present study, the broadest spectrum was presented by *Wickerhamomyces anomalus* D11. In the tested fermentation environment (apple pomace with a pH of around 4), some of these enzymes may exhibit only minimal activity or remain inactive. In yeast strains, the activities of peptidases, esterases, phosphatases, and glucosidases were observed. Both peptidases and phosphatases play critical roles in the process of precursor formation, which can influence the subsequent formation of volatile organic compounds responsible for unique tastes and aromas. Their activity affects various biochemical pathways, impacting the availability and transformation of precursors into aroma compounds. C4 esterases work by cleaving α-(1,4)-glycosidic bonds, releasing acetyl residues from pectins, which leads to the depolymerization of the chain and the formation of pectic acid and acetic acid [57]. The organic acid is then used in the biosynthesis of volatile organic compounds. α-Glucosidase hydrolyses oligosaccharides by cutting α-(1,4)-glycosidic bonds, releasing glucose molecules. ß-Glucosidase hydrolyses cellobiose and cello-oligosaccharides, resulting in the production of glucose through further bioconversion. Considering the results of enzymatic activity in this study and information about assimilation profiles from previous research on these yeasts, *Hanseniaspora uvarum* D9 and *Wickerhamomyces anomalus* D11 present cellulolytic activity in the process of hydrolysis cellobiose, an intermediate product in the formation of glucose, through hydrolysis by ß-glucosidase [58]. As a consequence, the glucose released during enzymatic processes, as well as the glucose naturally present in the fermentation environment, can be efficiently utilised by microorganisms. *Wickerhamomyces anomalus* demonstrates esterase activity, which leads to the use of carbohydrates available in the fermentation medium. This allows for the subsequent biosynthesis of aroma compounds. The highest activity in all analysed yeasts was determined for leucine, valine arylamidase, and acid phosphatase. Taking into account the composition of apple pomace—which includes large amounts of simple sugars and insoluble fibre fractions—the tested strains show enzymatic activity that allows them to effectively hydrolyse available compounds. In the case of the yeast *K. barnettii* D1, the hydrolysis of apple pomace before fermentation may improve the process because this strain at a low temperature does not have the enzymatic activity of ß-glucosidase that allows for hydrolysing more complex fractions as effectively as the *H. uvarum* D9 and *W. anomalus* D11 yeast strains [59].

3.4. Fermentation Activity of Yeast

The fermentation activity of the examined yeast isolates was assessed in terms of their ability to produce ethanol during alcoholic fermentation, which can be oxidised into acetic acid. Under aerobic conditions, yeast can oxidise ethanol to acetaldehyde and then to acetic acid [60]. The reaction involves enzymes such as alcohol dehydrogenase and aldehyde dehydrogenase, which catalyse the relevant chemical reactions. As a result, acetic acid is produced. Known for their preservative properties, organic acids can be used for natural food preservation and as ingredients in various beverages. They also contribute flavour and aroma [52]. The concentrations of ethyl alcohol and acetic acid produced during fermentation are presented in Table 4.

Table 4. Acetic acid and ethanol contents after yeast fermentation at 15 °C in untreated and hydrolysed apple pomace. Abbreviations: "Hydrolysed AP" refers to apple pomace treated with 0.25 mL of Viscozyme® and 0.25 mL of Rohament® CL enzymes; "Untreated AP" refers to apple pomace untreated with enzymes.

	Inoculum [%]	Acetic Acid [g/L]		Ethanol [g/L]	
		Untreated AP	**Hydrolysed AP**	**Untreated AP**	**Hydrolysed AP**
K. barnettii D1	1%	0.18 ± 0.01	0.24 ± 0.01	1.19 ± 0.04	1.06 ± 0.03
	5%	0.20 ± 0.01	0.18 ± 0.01	0.92 ± 0.03	1.06 ± 0.04
	10%	0.06 ± 0.00	0.21 ± 0.01	0.89 ± 0.03	0.93 ± 0.03
H. uvarum D9	1%	0.54 ± 0.02	0.20 ± 0.01	5.79 ± 0.21	3.48 ± 0.12
	5%	0.51 ± 0.02	0.13 ± 0.00	0.79 ± 0.03	1.81 ± 0.05
	10%	0.44 ± 0.02	0.12 ± 0.00	2.02 ± 0.07	1.04 ± 0.04
W. anomalus D11	1%	0.04 ± 0.00	0.03 ± 0.01	0.59 ± 0.02	2.88 ± 0.10
	5%	0.10 ± 0.00	0.02 ± 0.00	0.64 ± 0.02	3.71 ± 0.13
	10%	0.04 ± 0.00	0.04 ± 0.00	0.48 ± 0.02	6.05 ± 0.16

After the fermentation of apple pomace treated with pectinolytic and cellulolytic enzymes, the observed concentrations of acetic acid were higher than those after the fermentation of untreated apple pomace. For *H. uvarum* D9 and *W. anomalus* D11, the concentration of acetic acid did not exceed 0.2 g/L. The highest concentration was found in the sample collected after fermentation by *K. barnettii* D1 with 1% and 10% inoculum, while *W. anomalus* D11 with 5% inoculum yielded a very low concentration in comparison. Acetic acid was formed in all samples fermented with untreated apple pomace. *Hanseniaspora uvarum* D9 produced the highest concentration, especially with 1% and 5% inoculum, while 10% inoculum give a lower concentration. *Wickerhamomyces anomalus* D11 produced the smallest amount of acetic acid, with the highest concentration being obtained with 5% inoculum. Additionally, *Kazachstania barnettii* D1 produced no more than 0.2 g/L of acetic acid at all inoculum levels. However, such amounts do not produce excessively pungent odours [61]. Overall, a higher level of inoculum does not correspond to a higher level of obtained organic compounds.

When it comes to ethanol production during fermentation, the highest concentrations were obtained in hydrolysed apple pomace, in particular for the samples fermented by *W. anomalus* D11. The lowest values were reached for *K. barnettii* D1. Regarding the fermentation of untreated apple pomace, the ethanol content was remarkably lower in all samples. The highest concentration of ethanol was observed for the samples fermented by *H. uvarum* with 1% and 10% inoculum. Lower ethanol concentrations in samples containing untreated apple pomace may result from the lower glucose content compared to the hydrolysed samples.

3.5. Biosynthesis of Volatile Organic Compounds

Volatile organic compounds are a huge group of by-products of yeast's primary and secondary metabolism [24,25]. They also naturally occur in plant biomass [62]. They are mainly composed of esters, organic acids, higher alcohols, aldehydes, and ketones. Their biosynthesis occurs mostly in the stationary phase of yeast growth, but some of them, such as acetaldehyde, are produced in the logarithmic growth phase as an intermediate compound after glycolysis [63]. VOCs can beneficially contribute to the sensory profiles of fermented foods and beverages due to their interesting aromas [26]. However, at concentrations that are too high, VOCs can spoil the quality of products [64]. Depending on the concentrations and compositions of VOCs, their impact can either be desired, improving the overall characteristics of the end product, or undesired, decreasing its quality [26]. As temperature heavily influences yeast's metabolism, it also has an impact on the biosynthesis of VOCs. This means that the fermentation temperature could affect the composition of key aroma compounds [65]. The SHS-GC-MS method was used for the purpose of quantifying VOCs [66]. The analysed samples were collected after 5 days of fermentation of the untreated and hydrolysed apple pomace at 15 °C. The concentrations of esters formed during fermentation at 15 °C are shown in Tables 5 and 6.

Table 5. SHS-GC-MS analysis of esters formed during fermentation of untreated apple pomace at 15 °C. Detailed report, including total chromatograms after qualitative analysis, is shown in Supplementary Materials ("Supplementary reports S1–S3").

Concentration [µg/L] AP	*K. barnettii* D1			*H. uvarum* D9			*W. anomalus* D11			*p*-Value
	1%	5%	10%	1%	5%	10%	1%	5%	10%	
Ethyl acetate	237,107.53 c	230,338.25 c	237,367.87 c	31,640.17 b	366,183.91 a	330,358.62 b	10,035.32 g	20,129.12 fg	8515.68 g	<0.0001
Isobutyl acetate	138.47 cd	114.43 cd	148.51 c	59.94 e	958.83 a	271.32 b	4.74 f	120.23 cd	3.70 f	<0.0001
3-Methylbutyl acetate	1411.28 c	837.60 e	1190.81 d	1158.22 d	2119.26 a	1133.25 d	119.27 h	1828.91 b	78.04 h	<0.0001
2-Methylbutyl acetate	76.14 cd	63.60 de	90.83 c	45.43 ef	334.31 a	126.64 b	6.08 hi	142.31 b	4.02 i	<0.0001
2-Phenethyl acetate	181.67 e	90.58 e	120.27 e	514.84 d	1927.60 a	863.74 c	42.56 e	90.76 e	24.02 e	<0.0001
Methyl acetate	926.72 e	797.21 f	847.14 ef	774.96 f	1066.15 d	561.58 g	106.21 h	1063.40 d	84.16 h	<0.0001
Propyl acetate	254.94 c	232.05 c	251.63 c	318.00 b	588.73 a	338.23 b	5.41 e	15.81 de	4.14 e	<0.0001
Butyl acetate	28.61 a	21.52 b	24.26 b	13.53 c	24.66 b	10.70 c	-	-	-	<0.0001
Ethyl formate	5.13 d	4.14 def	4.30 de	46.83 a	11.54 c	20.31 b	-	-	-	<0.0001

Table 5. *Cont.*

Concentration [μg/L] AP	*K. barnettii* D1			*H. uvarum* D9			*W. anomalus* D11			*p*-Value
	1%	5%	10%	1%	5%	10%	1%	5%	10%	
Ethyl propionate	376.45 d	329.63 d	359.18 d	639.58 c	845.25 a	709.76 b	77.64 gh	136.45 efg	61.08 h	<0.0001
Isoamyl propionate	3.68 d	0.83 g	1.03 g	4.70 c	1.45 fg	1.94 ef	4.61 c	16.25 a	2.35 e	<0.0001
Ethyl butyrate	2.25 ab	2.02 b	2.40 a	0.34 cd	0.12 d	0.15 d	0.49 c	0.23 cd	0.31 cd	<0.0001
Ethyl isobutyrate	1.84 de	1.48 ef	1.97 de	9.57 c	15.78 a	12.02 b	2.58 d	2.05 de	1.22 ef	<0.0001
Ethyl hexanoate	0.04 bc	0.04 bc	0.04 bc	0.23 a	0.04 bc	0.06 b	0.03 c	0.02 c	0.02 c	<0.0001
Ethyl octanoate	0.10 b	0.08 bc	0.07 bc	0.15 a	0.06 bc	0.07 bc	0.06 bc	0.05 c	0.04 c	<0.0001
Ethyl nonanoate	0.04 a	0.04 ab	0.03 abc	0.03 bcd	0.02 bcd	0.02 bcd	0.03 abcd	0.02 cd	0.02 d	<0.0001
Ethyl decanoate	0.05 a	0.04 ab	0.04 ab	0.04 ab	0.03 bc	0.02 cd	0.02 cd	0.02 cd	0.02 d	<0.0001
Ethyl dodecanoate	0.05 a	0.04 ab	0.04 abc	0.04 abc	0.03 cd	0.02 de	0.02 e	0.02 de	0.01 e	<0.0001

'-'—not detected; a–i—mean values in columns denoted by different letters differ statistically significantly (one-way ANOVA, $p < 0.0001$).

Table 6. SHS-GC-MS analysis of esters formed during fermentation of hydrolysed apple pomace at 15 °C. Detailed report, including ion chromatograms after qualitative analysis, is shown in Supplementary Materials ("Supplementary report S4–S6").

Concentration [μg/L] H-AP	*K. barnettii* D1			*H. uvarum* D9			*W. anomalus* D11			*p*-Value
	1%	5%	10%	1%	5%	10%	1%	5%	10%	
Ethyl acetate	309,289.05 a	114,625.79 f	186,110.41 d	323,599.38 a	232,286.38 c	163,362.05 e	2230.40 g	2430.36 g	3071.41 g	<0.0001
Isobutyl acetate	474.88 c	572.56 b	945.33 a	34.25 ef	63.37 e	108.01 d	5.62 f	7.41 f	10.98 f	<0.0001
3-Methylbutyl acetate	4618.64 b	3968.45 c	5661.64 a	94.77 e	202.52 de	378.28 d	7.93 e	9.35 e	13.10 e	<0.0001
2-Methylbutyl acetate	700.72 b	640.15 c	895.65 a	26.73 de	34.44 de	48.74 d	1.45 e	1.67 e	2.25 e	<0.0001
2-Phenethyl acetate	423.65 c	199.39 def	367.63 cd	297.80 cde	761.56 b	1595.53 a	64.97 f	42.79 f	48.24 f	<0.0001
Methyl acetate	70,312.88 b	75,479.62 b	95,368.50 a	38,637.27 cd	29,889.40 ef	26,668.54 f	252.88 g	308.75 g	392.99 g	<0.0001
Propyl acetate	1640.38 a	293.67 d	635.04 b	575.77 b	392.50 c	255.81 de	12.63 g	17.47 g	17.16 g	<0.0001
Butyl acetate	24.33 a	3.71 e	7.78 c	5.61 d	3.81 e	1.51 f	0.18 f	0.23 f	0.20 f	<0.0001
Ethyl formate	4.96 f	3.62 f	3.39 f	48.86 ab	44.17 bc	38.43 cd	33.63 d	35.71 d	56.29 a	<0.0001
Ethyl propionate	1717.82 a	369.70 g	648.51 f	1516.17 b	1413.36 bc	1110.23 d	19.75 h	18.05 h	18.53 h	<0.0001
Isoamyl propionate	12.19 a	4.44 c	7.19 b	0.53 efg	1.60 d	4.00 c	0.09 fg	0.07 g	0.07 g	<0.0001
Ethyl butyrate	5.14 d	3.82 e	3.88 e	1.76 g	2.64 f	2.64 f	11.39 a	9.55 b	8.56 c	<0.0001
Ethyl isobutyrate	6.05 c	8.12 b	5.16 cd	4.77 d	8.16 b	11.90 a	1.92 e	2.04 e	2.68 e	<0.0001
Ethyl hexanoate	0.24 fg	0.14 fg	0.11 g	0.95 e	0.35 f	0.18 fg	1.63 a	1.37 bc	1.27 cd	<0.0001
Ethyl octanoate	0.41 d	0.35 de	0.27 de	0.16 de	0.13 e	0.11 e	1.59 a	1.25 b	0.76 c	<0.0001
Ethyl nonanoate	0.41 a	0.46 a	0.27 a	0.13 a	0.11 a	0.12 a	0.42 a	0.50 a	0.25 a	0.0305
Ethyl decanoate	0.33 a	0.24 b	0.17 c	0.11 cd	0.09 d	0.10 cd	0.09 d	0.12 cd	0.09 d	<0.0001
Ethyl dodecanoate	0.26 a	0.20 b	0.17 bc	0.09 efg	0.08 fg	0.09 efg	0.06 g	0.06 g	0.06 g	<0.0001

a–h—mean values in columns denoted by different letters differ statistically significantly (one-way ANOVA, $p < 0.0001$).

From the data obtained after the fermentation of untreated apple pomace, it can be observed that esters were the biggest group of VOCs produced during the fermentation of untreated apple pomace. During fermentation, the biosynthesis of two main groups of esters occurs—acetate esters and ethyl esters. The first are made from acetic acid and ethyl alcohol, or alcohol produced when yeast metabolises amino acids. Ethyl esters are formed from ethyl alcohol and medium-chain fatty acids (MCFAs) [67,68]. In this study, esters formed during fermentation included ethyl acetate, ethyl formate, ethyl propionate, isobutyl acetate, ethyl butyrate, 3-methylbutyl acetate, 2-methylbutyl acetate, ethyl hexanoate, ethyl octanoate, 2-phenethyl acetate, ethyl nonanoate, ethyl decanoate, ethyl dodecanoate, methyl acetate, propyl acetate, ethyl isobutyrate, butyl acetate, and isoamyl propionate. Among them, ethyl acetate was formed in large quantities during fermentation, exceeding the threshold reported in the literature [64]. This compound is characterised by a fruity aroma. The highest concentration of ethyl acetate was obtained in the samples fermented by *H. uvarum* D9, and the lowest concentration was obtained in samples fermented by *W. anomalus* D11. This compound was the main product among biosynthesised acetate esters. The average odour threshold of ethyl acetate in the air is 18 ppm [69]; however, it depends on the matrix in which this ester occurs, the presence of other substances, as well as individual sensory sensitivity. For instance, the odour threshold of ethyl acetate in apple juice varies between 2 and 31 mg/kg [70]; in wine, it varies between 60 and 328 mg/L [71]; in beer, it is 16.8 mg/L [72]; and in plum distillates, it is 34 mg/L of ethyl alcohol 40% (*v*/*v*) [73].

In the group of ethyl esters, the highest concentration was obtained for ethyl propionate. The highest amount was determined in the samples fermented by *H. uvarum* D9, and the lowest in samples fermented by *W. anomalus* D11. Other esters were not formed in very large quantities; for example, ethyl hexanoate or ethyl octanoate were only present in trace amounts. Butyl acetate and ethyl formate were not detected at all in the samples fermented by *W. anomalus* D11. When analysing the total concentration of ester compounds, it can be seen that the highest concentrations were found in the samples fermented by *H. uvarum* D9, slightly lower concentrations (of approximately 100 mg/L) were found for the samples fermented by *K. barnettii* D1, and the lowest for samples containing *W. anomalus* D11 (below 25 mg/L). The level of inoculation did not significantly affect the biosynthesis of esters, and the concentration was higher only in the case where 5% of the inoculum *H. uvarum* D9 and *W. anomalus* D11 was used.

The concentrations of carbonyl compounds formed during fermentation at 15 °C are shown in Tables 7 and 8.

Table 7. SHS-GC-MS analysis of carbonyl compounds formed during fermentation of untreated apple pomace at 15 °C. Detailed report, including ion chromatograms after qualitative analysis, is shown in Supplementary Materials ("Supplementary reports S1–S3").

Concentration [µg/L] AP	*K. barnettii* D1			*H. uvarum* D9			*W. anomalus* D11			*p*-Value
	1%	5%	10%	1%	5%	10%	1%	5%	10%	
Acetaldehyde	5357.17 a	1632.13 ef	831.84 g	5471.13 a	5275.34 a	4048.57 c	1556.93 f	1980.04 de	1551.23 f	<0.0001
Acetone	73.10 de	77.99 cd	227.41 a	102.33 b	82.83 c	66.09 e	78.64 cd	52.69 f	65.67 e	<0.0001
Isobutyraldehyde	36.43 c	20.68 f	14.45 g	17.35 fg	65.35 a	47.74 b	31.03 d	26.61 de	29.94 d	<0.0001
3-methylbutanal	29.71 b	27.54 b	23.33 c	4.16 fg	9.57 e	3.17 g	28.62 b	38.85 a	19.53 d	<0.0001
2-methylbutanal	22.10 a	14.00 c	9.27 de	5.55 fg	17.90 b	8.34 e	10.78 d	13.89 c	8.34 e	<0.0001
2.3-Butanedione	19.60 h	24.21 g	25.48 g	22.58 gh	20.11 h	15.49 i	47.02 e	63.48 d	37.12 f	<0.0001

a–i—mean values in columns denoted by different letters differ statistically significantly (one-way ANOVA, $p < 0.0001$).

Table 8. SHS-GC-MS analysis of carbonyl compounds formed during fermentation of hydrolysed apple pomace at 15 °C. Detailed report, including ion chromatograms after qualitative analysis, is shown in Supplementary Materials ("Supplementary reports S4–S6").

Concentration [µg/L] H-AP	*K. barnettii* D1			*H. uvarum* D9			*W. anomalus* D11			*p*-Value
	1%	5%	10%	1%	5%	10%	1%	5%	10%	
Acetaldehyde	6000.60 e	3893.33 f	3816.90 f	15,357.92 c	19,560.95 b	10,944.65 d	14,405.02 c	15,849.47 c	26,577.55 a	<0.0001
Acetone	56.16 f	19.91 g	39.33 f	111.03 cd	123.96 c	109.00 cd	125.26 c	175.15 b	287.78 a	<0.0001
Isobutyraldehyde	8.24 c	2.21 g	3.60 fg	7.92 c	44.89 a	24.47 b	4.95 ef	3.66 efg	5.63 def	<0.0001
3-methylbutanal	2.20 d	2.38 d	2.07 d	4.52 c	9.98 a	8.12 b	1.21 d	1.07 d	1.53 d	<0.0001
2-methylbutanal	2.29 de	2.66 d	2.12 e	3.73 c	6.42 a	4.99 b	0.49 g	0.41 g	0.55 g	<0.0001
2.3-Butanedione	14.80 d	7.53 g	9.35 fg	42.13 a	26.66 b	19.68 c	7.90 g	10.33 efg	12.30 def	<0.0001

a–g—mean values in columns denoted by different letters differ statistically significantly (one-way ANOVA, $p < 0.0001$).

Carbonyl compounds include aldehydes and ketones. During fermentation, aldehydes are produced by the activity of microorganisms, and ketones by the metabolism of citrates and glucose. Most of them have a fruity or floral aroma [74,75]. In all samples, after the fermentation of untreated apple pomace, four aldehydes and two ketones were detected. These carbonyl compounds were isobutyraldehyde, 3-methylbutanal, 2-methylbutanal, 2,3-butanedione, acetone, and acetaldehyde. In all analysed samples, the biosynthesis of acetaldehyde was the most effective. This compound occurs naturally in fruits and is also produced during fermentation by yeast. It gives a pungent and fruity aroma [76]. In this study, the highest concentration was obtained in the samples fermented by *H. uvarum* D9. In the case of other yeasts, this level was similar, except for the test with the addition of 1% of *K. barnettii* D1 inoculum. In this sample, the concentration exceeded 5 mg/L. Similarly to the esters, the results do not suggest any strong connection between the amounts of these volatile compounds formed and the percentage of inoculum used for the fermentation [45]. A slight decrease in the content of carbonyl compounds was observed in samples fermented by *H. uvarum* D9 and *K. barnettii* D1. In the samples containing 1% inoculum, the total amount of carbonyl compounds was the highest, with the samples containing 10% inoculum having the lowest total amount.

The concentrations of higher alcohols formed during fermentation at 15 °C are shown in Tables 9 and 10.

Table 9. SHS-GC-MS analysis of higher alcohols formed during fermentation of untreated apple pomace at 15 °C. Detailed report, including ion chromatograms after qualitative analysis, is shown in Supplementary Materials ("Supplementary reports S1–S3").

Concentration [µg/L] AP	*K. barnettii* D1			*H. uvarum* D9			*W. anomalus* D11			*p*-Value
	1%	5%	10%	1%	5%	10%	1%	5%	10%	
Isobutyl alcohol	8068.30 b	8762.85 ab	9594.40 a	7728.63 b	4540.47 d	5718.72 c	1658.04 e	5278.23 cd	1466.72 e	<0.0001
3-Methylbutanol	16,952.44 c	21,213.11 b	25,047.58 a	8355.14 d	2088.60 ef	3928.08 e	9332.16 d	24,331.46 a	7889.52 d	<0.0001
2-Methylbutanol	3400.18 c	3809.53 bc	4515.96 a	3752.36 bc	1393.74 ef	2236.27 d	1599.58 e	4090.35 ab	1148.61 f	<0.0001
2-Phenylethanol	7963.36 abc	8676.85 ab	10,775.50 a	4750.72 cde	6606.54 bcd	7004.31 bcd	3852.06 de	3032.18 ef	2211.43 ef	<0.0001
1-Propanol	504.33 c	494.41 c	476.73 c	2899.02 a	272.25 d	684.17 b	228.10 d	221.51 d	188.05 de	<0.0001

a–f—mean values in columns denoted by different letters differ statistically significantly (one-way ANOVA, $p < 0.0001$).

Higher alcohols, also called fusel alcohols, are produced during fermentation, either directly from sugars or indirectly from amino acids in the Ehrlich reaction. Valine, leucine, isoleucine, methionine, and phenylalanine are amino acids and are the main precursors. Short-chain alcohols give a pungent, chemical, and intensive aroma, but as the aliphatic chain lengthens, the aroma changes into a fruity and fresh one [77]. Higher alcohols produced in this study during the fermentation of untreated apple pomace included isobutyl alcohol, 3-methylbutanol, 2-methylbutanol, 2-phenylethanol, and 1-propanol. The highest concentration was recorded in the case of 3-methylbutanol after fermentation by *Kazachstania barnettii* D1 with 5% and 10% inoculum, with 1% giving slightly lower results, and by *Wickerhamomyces anomalus* D11, especially at 5%. In all samples, 1-propanol was obtained in small concentrations. 2-phenylethanol is another interesting compound, the concentration of which ranged from 2 mg/L to 10 mg/L. The highest content of this compound was determined in samples fermented by *K. barnettii* D1, and the lowest in samples fermented by *W. anomalus* D11. The inoculum size was also important in the case of this compound—as the inoculum percentages of *K. barnettii* D1 and *H. uvarum* D9 increased, the concentration of 2-phenylethanol increased, while the opposite was true in the case of *W. anomalus* D11. The total amount of higher alcohols in the analysed samples did not exceed 50 mg/L. There is no significant relationship between the amount of inoculum and the total biosynthesis of these compounds.

Table 10. SHS-GC-MS analysis of higher alcohols formed during fermentation of hydrolysed apple pomace at 15 °C. Detailed report, including ion chromatograms after qualitative analysis, is shown in Supplementary Materials ("Supplementary reports S4–S6").

Concentration [μg/L] H-AP	*K. barnettii* D1			*H. uvarum* D9			*W. anomalus* D11			*p*-Value
	1%	5%	10%	1%	5%	10%	1%	5%	10%	
Isobutyl alcohol	17,141.45 a	9907.40 c	14,178.41 b	3176.63 f	4981.82 e	7024.62 d	4498.99 e	6533.78 d	9776.04 c	<0.0001
3-Methylbutanol	48,380.42 a	28,138.90 c	38,570.89 b	5034.51 fg	9762.78 e	18,812.46 d	4960.63 fg	6854.48 f	10,728.03 e	<0.0001
2-Methylbutanol	17,640.24 a	10,474.98 c	14,265.84 b	3183.62 e	3704.44 e	5601.32 d	1539.72 f	2005.29 f	3019.58 e	<0.0001
2-Phenylethanol	12,195.62 a	12,201.57 a	9234.97 a	5160.35 a	6219.84 a	9238.05 a	6829.96 a	6487.31 a	8175.67 a	0.0356
1-Propanol	1140.57 cde	706.45 f	571.25 f	1921.53 b	1090.56 de	560.58 f	1354.83 c	1759.63 b	2399.34 a	<0.0001

a–g—mean values in columns denoted by different letters differ statistically significantly (one-way ANOVA, $p < 0.0001$).

The summary concentrations of esters, carbonyl compounds, and higher alcohols quantified after the fermentation of untreated and hydrolysed apple pomace are shown in Table 11.

Table 11. Summary concentrations of VOCs formed during fermentation of untreated and hydrolysed apple pomace at 15 °C.

Concentration [mg/L]	*K. barnettii* D1			*H. uvarum* D9			*W. anomalus* D11		
	1%	5%	10%	1%	5%	10%	1%	5%	10%
					Esters				
Untreated AP	240.51	232.83	240.41	319.93	374.08	334.41	10.41	23.55	8.78
Hydrolysed AP	389.12	196.10	290.61	364.82	265.08	193.57	2.51	2.78	3.55
					Carbonyl compounds				
Untreated AP	5.54	1.80	1.13	5.62	5.47	4.19	1.75	2.18	1.71
Hydrolysed AP	6.08	3.93	3.87	15.53	19.77	11.11	14.54	16.04	26.89
					Alcohols				
Untreated AP	36.89	42.96	50.41	27.49	14.90	19.57	16.67	36.95	12.90
Hydrolysed AP	96.50	61.43	76.82	18.48	25.76	41.24	19.18	23.64	34.10

From the data obtained after the fermentation of untreated apple pomace, it can be observed that esters were formed generally in lower concentrations. The concentrations of ethyl acetate were also very high. Predictably, as in the case of untreated apple pomace, the concentration obtained from the sample after fermentation with *Wickerhamomyces anomalus* D11 was very low compared to the other examined strains. The samples collected from other strains gave higher amounts of ethyl acetate at 1% inoculation.

Considering other esters, the most notable changes in biosynthesis were observed for 3-methylbutyl acetate, 2-methylbutylacetate, 2-phenylethyl acetate, methyl acetate, and ethyl propionate. These compounds were produced generally in higher concentrations during the fermentation of hydrolysed apple pomace. In this fermentation medium, one additional compound, ethyl hexanoate, was detected. However, in all samples, it was quantified only in trace amounts. The summary concentrations of esters did not correlate with the inoculum level, as different amounts were detected at different inoculum levels with no increasing or decreasing tendencies. Only in the case of samples fermented by *H. uvarum* D9 was a decreasing tendency observed as the amount of inoculum increased.

Besides esters, the summary concentrations of carbonyl compounds quantified during the experiment were higher compared to the samples of untreated apple pomace. This is due to the higher concentration of acetaldehyde quantified in all analysed samples, particularly in samples with *H. uvarum* D9 and *W. anomalus* D11. The amounts of other compounds remained at a similar level or were even lower. A relationship between the total amount of carbonyl compounds and the inoculum size for given strains was also observed. For *K. barnettii* D1, with an increase in the amount of inoculum, a decreasing tendency in the concentration of compounds was visible. For *W. anomalus* D11, an increasing tendency in production was observed with an increase in the level of inoculation.

The composition of higher alcohols did not differ in both analysed media. An almost twofold increase in the concentrations of all alcohols was observed. The total concentration of these compounds was much higher in the analysed samples compared to the samples of untreated apple pomace. With increasing inoculation levels, increased productivity of these compounds fermented by *H. uvarum* D9 and *W. anomalus* D11 strains was demonstrated. Such fermented component should not contain high concentrations of fusel alcohols for further use in food.

Among the quantified VOCs, their precursors were, among others, valine, leucine, and glucose precursors. These findings are closely related to both the composition of the apple pomace and the enzymatic abilities of the analysed yeast strains. This indicates that the formation of specific volatile organic compounds is influenced by the available precursors in the substrate as well as by the metabolic pathways active in the yeast. For instance, leucine arylamidase and valine arylamidase hydrolyse peptide chains release amino acids that serve as precursors to various aroma compounds, including 2-methylpropanal, 3-methylpropanol, isobutyl acetate, 3-methylbutanal, 3-methylbutanol, and isoamyl acetate [78].

In this study, experiments were conducted at 15 °C. Nonetheless, significant amounts of VOCs were produced. Future studies could further inquire into the connection between the production of VOCs and the fermentation temperature, as well as into whether it is possible to modulate the composition of volatile compounds in fermented products depending on the temperature. This could also be connected to the specific product that manufacturers want to obtain since the desired combination of VOCs may vary from product to product, as each compound gives a different aroma impression. Overall, the amounts of VOCs produced during the experiment were high, which suggests that the selected yeast strains maintained good metabolic activity at the temperature of 15 °C. *H. uvarum* D9 presented the best productivity of aroma compounds during the fermentation of untreated apple pomace regardless of the inoculum level. However, the samples collected after the fermentation of the hydrolysed medium showed higher summary concentrations of higher alcohols, which are not desirable compounds in this experiment. This means that this yeast produces a better aroma bouquet in the medium without enzymatic treatment. A similar observation can be made for the sample fermented by *K. barnettii* D1. The other yeast, *W. anomalus* D11, produced carbonyl compounds and fusel alcohols at a similar level but produced significantly lower amounts of esters.

Further research is needed to determine how they would contribute to the sensory profiles of selected fermented products. However, there is evidence to suggest that fermentation performed at a low temperature by selected non-*Saccharomyces* yeast strains could positively contribute to the aroma of fermented products. Apple pomace was used as a fermentation medium. The composition of the medium could have influenced the fermentation process, especially in terms of the carbohydrates used as a carbon source for the yeasts [79]. However, the obtained results show the potential of low-temperature microbial fermentation of apple pomace in the production of valuable bioactive compounds. Furthermore, the process was designed in such a way that it eliminates the need for additional methods, such as extraction, to obtain the desired components. The utilisation of the ferment in this form has the potential to reduce the production cost of the final product and prevent the generation of unwanted by-products, such as solvents and waste microbial biomass. Therefore, the optimisation of research and subsequent production will minimise the scale of the issue of unused pomace, which currently causes an environmental issue [80]. This aligns with the assumptions of environmentally friendly processes based on the circular economy principles.

4. Conclusions

Yeast plays an important role in the fermentation process and biosynthesis of aroma compounds. Thanks to their metabolic abilities, it is possible to valorise waste plant materials and obtain valuable compounds by fermentation. This study proves the possibility of using unconventional yeasts to produce volatile compounds during the fermentation of apple pomace. The previous adaptation of yeast to a temperature of 15 °C enabled the use of these microorganisms for low-temperature biological processes. The conducted investigation of apple pomace valorisation by fermentation with non-*Saccharomyces* yeast at the temperature of 15 °C allowed us to draw the following conclusions:

- The use of commercial enzymatic preparations enabled satisfactory liquefaction of apple pomace, but enzymatic hydrolysis did not make the microbial growth process more efficient.
- The ability of the tested yeast to produce volatile organic compounds during the fermentation of both untreated and hydrolysed apple pomace was confirmed.
- The higher the inoculum size was, the greater biomass growth was observed. However, in the model experiment, a larger amount of inoculum did not make the biosynthesis of volatile organic compounds more efficient.
- *Hanseniaspora uvarum* D9 produced the highest amounts of volatile organic compounds during the fermentation of untreated and hydrolysed apple pomace. However, a significant increase in the higher alcohols content was observed in the hydrolysed medium.

Supplementary Materials: The following supporting information can be downloaded at https://www.mdpi.com/article/10.3390/app15052726/s1, Supplementary report S1_Apple_pomace_1_10; Supplementary report S2_Apple_pomace_9_10; Supplementary report S3_Apple_pomace_11_10; Supplementary report S4_Apple_pomace_hydrolysate_1_10; Supplementary report S5_Apple_pomace_hydrolysate_9_10; Supplementary report S6_Apple_pomace_hydrolysate_11_10.

Author Contributions: Conceptualization, W.L., I.M. and J.B.; methodology, W.L., K.P.-P. and S.M.; validation, I.M. and J.B.; formal analysis, K.P.-P.; investigation, W.L., J.D. and S.M.; resources, I.M.; data curation, W.L.; writing—original draft preparation, W.L. and J.D.; writing—review and editing,

W.L., I.M. and J.B.; supervision, J.B.; funding acquisition, J.B. All authors have read and agreed to the published version of the manuscript.

Funding: This work was supported by the Wojewodzki Fundusz Ochrony Srodowiska i Gospodarki Wodnej w Lodzi (Grant no. 157/BN/D/2020).

Data Availability Statement: DNA sequences (Genbank accessions OQ275007.1, OQ275152.1, and OQ275025.1) and the initial characterisation of cold-adapted isolated yeast are available online at https://bioresources.cnr.ncsu.edu/resources/plant-biomass-as-a-source-of-low-temperature-yeasts/ (accessed on 25 February 2025).

Acknowledgments: This research was completed while the first author was a doctoral candidate at the Interdisciplinary Doctoral School at the Lodz University of Technology, Poland. The authors thank John Speller for his collaboration regarding the English language revision of this manuscript.

Conflicts of Interest: The authors declare no conflicts of interest.

Abbreviations

The following abbreviations are used in this manuscript:

VOCs	Volatile organic compounds
AP	Apple pomace
H-AP	Hydrolysed apple pomace

References

1. Boloy, R.A.M.; da Cunha Reis, A.; Rios, E.M.; de Araújo Santos Martins, J.; Soares, L.O.; de Sá Machado, V.A.; de Moraes, D.R. Waste-to-Energy Technologies Towards Circular Economy: A Systematic Literature Review and Bibliometric Analysis. *Water Air Soil Pollut.* **2021**, *232*, 306. [CrossRef]
2. Ranjbari, M.; Shams Esfandabadi, Z.; Quatraro, F.; Vatanparast, H.; Lam, S.S.; Aghbashlo, M.; Tabatabaei, M. Biomass and Organic Waste Potentials towards Implementing Circular Bioeconomy Platforms: A Systematic Bibliometric Analysis. *Fuel* **2022**, *318*, 123585. [CrossRef]
3. Wang, K.; Tester, J.W. Sustainable Management of Unavoidable Biomass Wastes. *Green Energy Resour.* **2023**, *1*, 100005. [CrossRef]
4. AGRO Industry Raport o Rynku Jabłek A.D. 2022. Available online: https://issuu.com/prografika-raciborz/docs/agro_4_2022_prografika/s/17713612 (accessed on 7 November 2023).
5. Lyu, F.; Luiz, S.F.; Azeredo, D.R.P.; Cruz, A.G.; Ajlouni, S.; Ranadheera, C.S. Apple Pomace as a Functional and Healthy Ingredient in Food Products: A Review. *Processes* **2020**, *8*, 319. [CrossRef]
6. Martău, G.A.; Teleky, B.E.; Ranga, F.; Pop, I.D.; Vodnar, D.C. Apple Pomace as a Sustainable Substrate in Sourdough Fermentation. *Front. Microbiol.* **2021**, *12*, 742020. [CrossRef]
7. Gumul, D.; Ziobro, R.; Korus, J.; Kruczek, M. Apple Pomace as a Source of Bioactive Polyphenol Compounds in Gluten-Free Breads. *Antioxidants* **2021**, *10*, 807. [CrossRef]
8. Gołębiewska, E.; Kalinowska, M.; Yildiz, G. Sustainable Use of Apple Pomace (AP) in Different Industrial Sectors. *Materials* **2022**, *15*, 1788. [CrossRef]
9. Putra, N.R.; Rizkiyah, D.N.; Abdul Aziz, A.H.; Che Yunus, M.A.; Veza, I.; Harny, I.; Tirta, A. Waste to Wealth of Apple Pomace Valorization by Past and Current Extraction Processes: A Review. *Sustainability* **2023**, *15*, 830. [CrossRef]
10. Bhushan, S.; Kalia, K.; Sharma, M.; Singh, B.; Ahuja, P.S. Processing of Apple Pomace for Bioactive Molecules. *Crit. Rev. Biotechnol.* **2008**, *28*, 285–296. [CrossRef]
11. Varela, C. The Impact of Non-Saccharomyces Yeasts in the Production of Alcoholic Beverages. *Appl. Microbiol. Biotechnol.* **2016**, *100*, 9861–9874. [CrossRef]
12. Steensels, J.; Verstrepen, K.J. Taming Wild Yeast: Potential of Conventional and Nonconventional Yeasts in Industrial Fermentations. *Annu. Rev. Microbiol.* **2014**, *68*, 61–80. [CrossRef]
13. Spencer, J.; Ragout de Spencer, A.; Laluce, C. Non-Conventional Yeasts. *Appl. Microbiol. Biotechnol.* **2002**, *58*, 147–156. [CrossRef] [PubMed]
14. Walker, G.M. Pichia Anomala: Cell Physiology and Biotechnology Relative to Other Yeasts. *Antonie Van. Leeuwenhoek Int. J. Gen. General. Mol. Microbiol.* **2011**, *99*, 25–34. [CrossRef] [PubMed]
15. Parpinello, G.P.; Ricci, A.; Folegatti, B.; Patrignani, F.; Lanciotti, R.; Versari, A. Unraveling the Potential of Cryotolerant Saccharomyces Eubayanus in Chardonnay White Wine Production. *LWT* **2020**, *134*, 110183. [CrossRef]

16. Deed, R.C.; Pilkington, L.I. Lifestyle, Lineage, and Geographical Origin Influence Temperature-Dependent Phenotypic Variation across Yeast Strains during Wine Fermentation. *Microorganisms* **2020**, *8*, 1367. [CrossRef]
17. Molina, A.M.; Swiegers, J.H.; Varela, C.; Pretorius, I.S.; Agosin, E. Influence of Wine Fermentation Temperature on the Synthesis of Yeast-Derived Volatile Aroma Compounds. *Appl. Microbiol. Biotechnol.* **2007**, *77*, 675–687. [CrossRef]
18. Fleet, G.H. Yeast Interactions and Wine Flavour. *Int. J. Food Microbiol.* **2003**, *86*, 11–22. [CrossRef]
19. Mangiagalli, M.; Brocca, S.; Orlando, M.; Lotti, M. The "Cold Revolution". Present and Future Applications of Cold-Active Enzymes and Ice-Binding Proteins. *New Biotechnol.* **2020**, *55*, 5–11. [CrossRef]
20. Shi, X.; Liu, Y.; Ma, Q.; Wang, J.; Luo, J.; Suo, R.; Sun, J. Effects of Low Temperature on the Dynamics of Volatile Compounds and Their Correlation with the Microbial Succession during the Fermentation of Longyan Wine. *LWT* **2022**, *154*, 112661. [CrossRef]
21. Cavicchioli, R.; Charlton, T.; Ertan, H.; Omar, S.M.; Siddiqui, K.S.; Williams, T.J. Biotechnological Uses of Enzymes from Psychrophiles. *Microb. Biotechnol.* **2011**, *4*, 449–460. [CrossRef]
22. Margesin, R.; Schinner, F. Properties of Cold-Adapted Microorganisms and Their Potential Role in Biotechnology. *J. Biotechnol.* **1994**, *33*, 1–14. [CrossRef]
23. Liang, C.; Liu, L.X.; Liu, J.; Aihaiti, A.; Tang, X.J.; Liu, Y.G. New Insights on Low-Temperature Fermentation for Food. *Fermentation* **2023**, *9*, 477. [CrossRef]
24. Hung, R.; Lee, S.; Bennett, J.W. Fungal Volatile Organic Compounds and Their Role in Ecosystems. *Appl. Microbiol. Biotechnol.* **2015**, *99*, 3395–3405. [CrossRef]
25. Korpi, A.; Järnberg, J.; Pasanen, A.L. Microbial Volatile Organic Compounds. *Crit. Rev. Toxicol.* **2009**, *39*, 139–193. [CrossRef] [PubMed]
26. Carrau, F.; Boido, E.; Dellacassa, E. Yeast Diversity and Flavor Compounds. In *Fungal Metabolites*; Springer: Cham, Switzerland, 2017; pp. 569–597. [CrossRef]
27. Contarino, R.; Brighina, S.; Fallico, B.; Cirvilleri, G.; Parafati, L.; Restuccia, C. Volatile Organic Compounds (VOCs) Produced by Biocontrol Yeasts. *Food Microbiol.* **2019**, *82*, 70–74. [CrossRef]
28. Latimer, G.W., Jr. *Official Methods of Analysis of AOAC INTERNATIONAL*; AOAC INTERNATIONAL: Rockville, MD, USA, 2023. [CrossRef]
29. Pająk, P.; Fortuna, T. Ocena Właściwości Fizykochemicznych i Jakości Sensorycznej Wybranych Żeli z Owocami. *Żywność. Nauka. Technologia. Jakość* **2010**, *2*, 85–94.
30. Balcerek, M.; Dziekońska-Kubczak, U.; Pielech-Przybylska, K.; Oleszczak, A.; Koń, M.; Patelski, A.M. Waste Bread as Raw Material for Ethanol Production: Effect of Mash Preparation Methods on Fermentation Efficiency. *Appl. Sci.* **2024**, *14*, 9565. [CrossRef]
31. Pielech-Przybylska, K.; Balcerek, M.; Dziekonska-Kubczak, U.; Pacholczyk-Sienicka, B.; Ciepielowski, G.; Albrecht, Ł.; Patelski, P. The Role of Saccharomyces Cerevisiae Yeast and Lactic Acid Bacteria in the Formation of 2-Propanol from Acetone during Fermentation of Rye Mashes Obtained Using Thermal-Pressure Method of Starch Liberation. *Molecules* **2019**, *24*, 610. [CrossRef]
32. Antonic, B.; Jancikova, S.; Dordevic, D.; Tremlova, B. Apple Pomace as Food Fortification Ingredient: A Systematic Review and Meta-Analysis. *J. Food Sci.* **2020**, *85*, 2977–2985. [CrossRef] [PubMed]
33. Gustafsson, J.; Landberg, M.; Bátori, V.; Åkesson, D.; Taherzadeh, M.J.; Zamani, A. Development of Bio-Based Films and 3D Objects from Apple Pomace. *Polymers* **2019**, *11*, 289. [CrossRef]
34. Mateos-Aparicio, I.; De la Peña Armada, R.; Pérez-Cózar, M.L.; Rupérez, P.; Redondo-Cuenca, A.; Villanueva-Suárez, M.J. Apple By-Product Dietary Fibre Exhibits Potential Prebiotic and Hypolipidemic Effectsin High-Fat Fed Wistar Rats. *Bioact. Carbohydr. Diet. Fibre* **2020**, *23*, 100219. [CrossRef]
35. Li, F.; Foucat, L.; Bonnin, E. Effect of Solid Loading on the Behaviour of Pectin-Degrading Enzymes. *Biotechnol. Biofuels* **2021**, *14*, 107. [CrossRef] [PubMed]
36. Pakarinen, A.; Zhang, J.; Brock, T.; Maijala, P.; Viikari, L. Enzymatic Accessibility of Fiber Hemp Is Enhanced by Enzymatic or Chemical Removal of Pectin. *Bioresour. Technol.* **2012**, *107*, 275–281. [CrossRef] [PubMed]
37. Liu, L.; You, Y.; Deng, H.; Guo, Y.; Meng, Y. Promoting Hydrolysis of Apple Pomace by Pectinase and Cellulase to Produce Microbial Oils Using Engineered Yarrowia Lipolytica. *Biomass Bioenergy* **2019**, *126*, 62–69. [CrossRef]
38. Gama, R.; Van Dyk, J.S.; Pletschke, B.I. Optimisation of Enzymatic Hydrolysis of Apple Pomace for Production of Biofuel and Biorefinery Chemicals Using Commercial Enzymes. *3 Biotech* **2015**, *5*, 1075–1087. [CrossRef] [PubMed]
39. Luo, J.; Xu, Y. Comparison of Biological and Chemical Pretreatment on Coproduction of Pectin and Fermentable Sugars from Apple Pomace. *Appl. Biochem. Biotechnol.* **2020**, *190*, 129–137. [CrossRef] [PubMed]
40. Levantovsky, R.; Allen-Blevins, C.R.; Sela, D.A. Nutritional Requirements of Bifidobacteria. In *The Bifidobacteria and Related Organisms: Biology, Taxonomy, Applications*; Academic Press: Cambridge, MA, USA, 2018; pp. 115–129. [CrossRef]
41. Frîncu, M.; Dumitrache, C.; Begea, M.; Teodorescu, R.I.; Filofteia Diguță, C.; Daniel Baniță, C.; Tudor, V.; Cîrîc, A.I.; Mărculescu, S.-I.; Bărbulescu, I.D. Active Wine Yeast Biomass Obtained through Biotechnological Process. *J. Agroaliment. Process. Technol.* **2022**, *28*, 20–26.

42. Kaur, S.; Kaur, K.; Bhushan, B.; Kaur, M.; Hans, M. Inoculum Size and Age Studies on Single and Mixed Strain Fermentation of Grape Juice. *J. Pure Appl. Microbiol.* **2020**, *14*, 2137–2145. [CrossRef]
43. Priest, F.G.; Campbell, I. *Brewing Microbiology*, 2nd ed.; Priest, F.G., Campbell, I., Eds.; Springer US: New York, NY, USA, 2003; ISBN 978-1-4613-4858-0.
44. Price, E.J.; Linforth, R.S.T.; Dodd, C.E.R.; Phillips, C.A.; Hewson, L.; Hort, J.; Gkatzionis, K. Study of the Influence of Yeast Inoculum Concentration (*Yarrowia lipolytica* and *Kluyveromyces lactis*) on Blue Cheese Aroma Development Using Microbiological Models. *Food Chem.* **2014**, *145*, 464–472. [CrossRef]
45. Erten, H.; Tanguler, H.; Cabaroglu, T.; Canbas, A. The Influence of Inoculum Level on Fermentation and Flavour Compounds of White Wines Made from Cv. *Emir. J. Inst. Brew.* **2006**, *112*, 232–236. [CrossRef]
46. Bordet, F.; Romanet, R.; Bahut, F.; Ferreira, V.; Peña, C.; Julien-Ortiz, A.; Roullier-Gall, C.; Alexandre, H. Impact of Saccharomyces Cerevisiae Yeast Inoculation Mode on Wine Composition. *Food Chem.* **2024**, *441*, 138391. [CrossRef] [PubMed]
47. Narendranath, N.V.; Power, R. Effect of Yeast Inoculation Rate on the Metabolism of Contaminating Lactobacilli during Fermentation of Corn Mash. *J. Ind. Microbiol. Biotechnol.* **2004**, *31*, 581–584. [CrossRef]
48. Chen, H.; Mei, S.; Dai, C.; He, Y.; Yu, J. Improving Fermentation of Steamed Stalk to Feed Using Candida Utilis and Pachysolen Tannophilus. *Acta. Univ. Cibiniensis Ser. E: Food Technol.* **2018**, *22*, 51–58. [CrossRef]
49. Pieczywek, P.M.; Cybulska, J.; Zdunek, A. An Atomic Force Microscopy Study on the Effect of β-Galactosidase, α-l-Rhamnosidase and α-l-Arabinofuranosidase on the Structure of Pectin Extracted from Apple Fruit Using Sodium Carbonate. *Int. J. Mol. Sci.* **2020**, *21*, 4064. [CrossRef]
50. D'Amore, T.; Russell, I.; Stewart, G.G. Sugar Utilization by Yeast during Fermentation. *J. Ind. Microbiol.* **1989**, *4*, 315–323. [CrossRef]
51. Chanprasartsuk, O.O.; Prakitchaiwattana, C. Growth Kinetics and Fermentation Properties of Autochthonous Yeasts in Pineapple Juice Fermentation for Starter Culture Development. *Int. J. Food Microbiol.* **2022**, *371*, 109636. [CrossRef]
52. Endoh, R.; Horiyama, M.; Ohkuma, M. D-Fructose Assimilation and Fermentation by Yeasts Belonging to Saccharomycetes: Rediscovery of Universal Phenotypes and Elucidation of Fructophilic Behaviors in Ambrosiozyma Platypodis and Cyberlindnera Americana. *Microorganisms* **2021**, *9*, 758. [CrossRef] [PubMed]
53. Muzio, G.; Maggiora, M.; Paiuzzi, E.; Oraldi, M.; Canuto, R.A. Aldehyde Dehydrogenases and Cell Proliferation. *Free Radic. Biol. Med.* **2012**, *52*, 735–746. [CrossRef]
54. Wang, M.F.; Han, C.L.; Yin, S.J. Substrate Specificity of Human and Yeast Aldehyde Dehydrogenases. *Chem. Biol. Interact.* **2009**, *178*, 36–39. [CrossRef] [PubMed]
55. Escribano, R.; González-Arenzana, L.; Garijo, P.; Berlanas, C.; López-Alfaro, I.; López, R.; Gutiérrez, A.R.; Santamaría, P. Screening of Enzymatic Activities within Different Enological Non-Saccharomyces Yeasts. *J. Food Sci. Technol.* **2017**, *54*, 1555–1564. [CrossRef]
56. Dygas, D.; Nowak, S.; Olszewska, J.; Szymańska, M.; Mroczyńska-Florczak, M.; Berłowska, J.; Dziugan, P.; Kręgiel, D. Ability of Yeast Metabolic Activity to Reduce Sugars and Stabilize Betalains in Red Beet Juice. *Fermentation* **2021**, *7*, 105. [CrossRef]
57. Bezus, B.; Cavello, I.; Contreras-Esquivel, J.C.; Cavalitto, S. Pectinases Produced by Extremophilic Yeasts: From Cold Environments to the Food Industry. In *Value-Addition in Food Products and Processing Through Enzyme Technology*; Elsevier: Amsterdam, The Netherlands, 2022; pp. 437–452. [CrossRef]
58. Liszkowska, W.; Motyl, I.; Pielech-Przybylska, K.; Szulc, J.; Sypka, M.; Dziugan, P.; Berłowska, J. Plant Biomass as a Source of Low-Temperature Yeasts. *Bioresources* **2023**, *18*, 599–612. [CrossRef]
59. Lin, M.M.H.; Boss, P.K.; Walker, M.E.; Sumby, K.M.; Jiranek, V. Influence of *Kazachstania* Spp. on the Chemical and Sensory Profile of Red Wines. *Int. J. Food Microbiol.* **2022**, *362*, 109496. [CrossRef] [PubMed]
60. Merli, G.; Becci, A.; Amato, A.; Beolchini, F. Acetic Acid Bioproduction: The Technological Innovation Change. *Sci. Total Environ.* **2021**, *798*, 149292. [CrossRef]
61. Liang, R.; Lin, S.; Chen, D.; Sun, N. Differentiation of Penaeus Vannamei from Different Thermal Processing Methods in Physico-Chemical, Flavor and Sensory Characteristics. *Food Chem.* **2022**, *378*, 132092. [CrossRef]
62. Madrera, R.R.; Bedriñana, R.P.; Valles, B.S. Production and Characterization of Aroma Compounds from Apple Pomace by Solid-State Fermentation with Selected Yeasts. *LWT Food Sci. Technol.* **2015**, *64*, 1342–1353. [CrossRef]
63. Rębas, J. Drożdże i Ich Metabolity w Procesie Fermentacji Alkoholowej. *Przemysł Ferment. I Owocowo-Warzywny* **2021**, *1*, 11–13. [CrossRef]
64. Sumby, K.M.; Grbin, P.R.; Jiranek, V. Microbial Modulation of Aromatic Esters in Wine: Current Knowledge and Future Prospects. *Food Chem.* **2010**, *121*, 1–16. [CrossRef]
65. Peng, B.; Li, F.; Cui, L.; Guo, Y. Effects of Fermentation Temperature on Key Aroma Compounds and Sensory Properties of Apple Wine. *J. Food Sci.* **2015**, *80*, S2937–S2943. [CrossRef] [PubMed]
66. Warburton, A.; Silcock, P.; Eyres, G.T. Impact of Sourdough Culture on the Volatile Compounds in Wholemeal Sourdough Bread. *Food Res. Int.* **2022**, *161*, 111885. [CrossRef]

67. Saerens, S.M.G.; Delvaux, F.R.; Verstrepen, K.J.; Thevelein, J.M. Production and Biological Function of Volatile Esters in Saccharomyces Cerevisiae. *Microb. Biotechnol.* **2010**, *3*, 165–177. [CrossRef] [PubMed]
68. Fu, Z.; Sun, B.; Li, X.; Fan, G.; Teng, C.; Alaa, A.; Jia, Y. Isolation and Characterization of a High Ethyl Acetate-Producing Yeast from Laobaigan Daqu and Its Fermentation Conditions for Producing High-Quality Baijiu. *Biotechnol. Biotechnol. Equip.* **2018**, *32*, 1218–1227. [CrossRef]
69. National Center for Biotechnology Information PubChem Compound Summary for CID 8857, Ethyl Acetate. 2025. Available online: https://pubchem.ncbi.nlm.nih.gov/compound/Ethyl-Acetate (accessed on 10 February 2025).
70. Niu, Y.; Wang, R.; Xiao, Z.; Zhu, J.; Sun, X.; Wang, P. Characterization of Ester Odorants of Apple Juice by Gas Chromatography-Olfactometry, Quantitative Measurements, Odour Threshold, Aroma Intensity and Electronic Nose. *Food Res. Int.* **2019**, *120*, 92–101. [CrossRef]
71. Miranda, A.; Pereira, V.; Pontes, M.; Albuquerque, F.; Marques, J.C. Acetic Acid and Ethyl Acetate in Madeira Wines: Evolution with Ageing and Assessment of the Odour Rejection Threshold. *CiÊNcia E TÉCnica VitivinÍCola* **2017**, *32*, 1–11. [CrossRef]
72. Engan, S. Organoleptic Threshold Values of Some Alcohols and Esters in Beer. *J. Inst. Brew.* **1972**, *78*, 33–36. [CrossRef]
73. Pielech-Przybylska, K.; Balcerek, M.; Nowak, A.; Patelski, P.; Dziekońska-Kubczak, U. Influence of Yeast on the Yield of Fermentation and Volatile Profile of 'Węgierka Zwykła' Plum Distillates. *J. Inst. Brew.* **2016**, *122*, 612–623. [CrossRef]
74. Pesis, E. The Role of the Anaerobic Metabolites, Acetaldehyde and Ethanol, in Fruit Ripening, Enhancement of Fruit Quality and Fruit Deterioration. *Postharvest Biol. Technol.* **2005**, *37*, 1–19. [CrossRef]
75. Tylewicz, U.; Inchingolo, R.; Rodriguez-Estrada, M.T. Food Aroma Compounds. In *Nutraceutical and Functional Food Components: Effects of Innovative Processing Techniques*; Elsevier: Amsterdam, The Netherlands, 2017; pp. 297–334. [CrossRef]
76. Zea, L.; Serratosa, M.P.; Mérida, J.; Moyano, L. Acetaldehyde as Key Compound for the Authenticity of Sherry Wines: A Study Covering 5 Decades. *Compr. Rev. Food Sci. Food Saf.* **2015**, *14*, 681–693. [CrossRef]
77. Stanzer, D.; Hanousek Čiča, K.; Blesić, M.; Smajić Murtić, M.; Mrvčić, J.; Spaho, N. Alcoholic Fermentation as a Source of Congeners in Fruit Spirits. *Foods* **2023**, *12*, 1951. [CrossRef]
78. Maoz, I.; Lewinsohn, E.; Gonda, I. Amino Acids Metabolism as a Source for Aroma Volatiles Biosynthesis. *Curr. Opin. Plant Biol.* **2022**, *67*, 102221. [CrossRef]
79. Flores, C.-L.; Rodríguez, C.; Petit, T.; Gancedo, C. Carbohydrate and Energy-Yielding Metabolism in Non-Conventional Yeasts. *FEMS Microbiol. Rev.* **2000**, *24*, 507–529. [CrossRef] [PubMed]
80. Gospodarka o Obiegu Zamkniętym: Definicja, Znaczenie i Korzyści (Wideo) | Tematy | Parlament Europejski. Available online: https://www.europarl.europa.eu/topics/pl/article/20151201STO05603/gospodarka-o-obiegu-zamknietym-definicja-znaczenie-i-korzysci-wideo (accessed on 29 February 2024).

applied sciences

MDPI

Article

Residues from the Oil Pressing Process as a Substrate for the Production of Alternative Biochar Materials

Bogdan Saletnik *, Radosław Czarnota, Mateusz Maczuga, Aneta Saletnik, Marcin Bajcar, Grzegorz Zaguła and Czesław Puchalski

Department of Bioenergetics, Food Analysis and Microbiology, Institute of Food Technology and Nutrition, College of Natural Science, Rzeszow University, Ćwiklińskiej 2D, 35-601 Rzeszow, Poland; asaletnik@ur.edu.pl (A.S.); gzagula@ur.edu.pl (G.Z.); cpuchalski@ur.edu.pl (C.P.)

* Correspondence: bsaletnik@ur.edu.pl

Abstract: The purpose of this study was to evaluate the feasibility of using residues from cooking oil production to produce alternative biochar fuels along with optimizing the pyrolysis process. The work consisted of carrying out the pyrolysis process at varying temperatures and holding times at the final temperature, and then evaluating the energy potential of the materials studied. Taking into account aspects of environmental emissions, the content of selected oxides in the flue gases generated during the combustion of cakes and the biochar obtained from them was evaluated. Plant biomass derived from a variety of oilseeds, i.e., fennel flower *(Nigella sativa* L.), rapeseed *(Brassica napus* L. var. Napus), flax *(Linum usitatissimum* L.), evening primrose *(Oenothera biennis* L.), milk thistle *(Silybum marianum* L. *Gaertn.*) and hemp *(Cannabis sativa* L.), was used to produce biochar. The experimental data have shown that the obtained biochar can have a calorific value of nearly 27 MJ kg^{-1}. The use of pyrolysis allowed for a maximum increase in the calorific value of nearly 41% compared to non-thermally processed cakes and a several-fold decrease in carbon monoxide, nitrogen oxides and sulfur dioxide emissions. According to these results, it can be concluded that the pyrolysis process can be an attractive method for using residues from the production of various cooking oils to produce alternative biofuels, developing the potential of the circular economy.

Keywords: biochar; pyrolysis; oilseed cakes; calorific value; gas emissions; alternative biofuels; waste management

1. Introduction

In the face of increasing climate change, environmental degradation and depletion of fossil fuels, humanity faces a major challenge. However, understanding these problems allows you to approach them as an opportunity to make beneficial changes. Countries are focusing on intensifying research to reduce greenhouse gas emissions. A key area of this research is the development of zero-emission technologies. In the context of growing interest in alternative energy sources and the need to reduce greenhouse gas emissions, unconventional fuels with high energy efficiency are being sought. Between 2004 and 2016, energy consumption in the EU fell by 10%. How much countries invest in renewable energy in many cases depends on environmental and geographic factors. To harness hydropower, we need many rivers and significant land gradients. Austrians, Slovenians or Scandinavian countries are eager to use such solutions. If one wants to invest in wind energy, open spaces are necessary and favorable wind conditions, such as in Ireland or the Netherlands. Solar energy requires many sunny days a year, so countries such as Spain, Cyprus and Malta have favorable conditions for this type of energy [1].

Through the diversity of its sources, and its widespread availability around the world, biomass represents a promising, sustainable and renewable energy source. It is one of the three most important fuel sources for electric and thermal power generation, as up to 50% of the world's population relies on biomass [2].

An industry sector capable of supplying large quantities of raw material for biochar production is proving to be the oil market, whose continued growth is caused by increasing consumerism among people and growing demand for food. Converting cakes through the pyrolysis process, based on the controlled thermal decomposition of biomass, can provide an alternative form of energy generation and positively influence the development of more sustainable and efficient energy systems. With the above in mind, the optimization of the pyrolysis process of waste biomass, i.e., cakes produced in the process of pressing oil from the seeds of six different types of oil crops, was chosen as the goal of this work. This work consisted of carrying out the pyrolysis process at varying temperatures and holding times at the final temperature, and then evaluating the energy potential of the materials studied. Taking into account aspects of environmental emissions, the content of selected oxides in the flue gases generated during the combustion of cakes and the biochar obtained from them was evaluated.

2. Literature Review

Biomass is a renewable source of energy, but just as importantly, it is available virtually anywhere in the world and in large quantities. The development of technology toward biomass as a renewable source of heat makes it possible to reduce emissions, which has a direct impact on the natural environment. Biomass energy can be in the form of biogas, liquid and solid biofuels, enabling its wider use in place of fossil fuels in the energy and transportation sectors. How biomass is converted can generate different types of energy. Among other things, the diversity of biomass transformation methods is related to the variety of morphology and physical characteristics of biomass. It can be in wet or dry form, thick or fluffy, with high or low ash content, small or large in shape, being homogeneous or heterogeneous. This complexity means that the use of biomass fuels in most cases requires some form of feedstock processing. Energy production mainly uses biomass from low-quality wood residues, straw and other agricultural production wastes, animal manure, sewage sludge, organic wastes such as beet pulp, corn stalks, grasses, vegetable oils and animal fats [3–6].

The oil market both in Poland and around the world is developing very rapidly. This is mainly due to the fact that there is an increasing demand for food, for the production of which oil is needed; another factor is the increasing population. Cold-pressed oils are also gaining more popularity due to consumer interest driven by their increasing concern for health. The main raw material in Poland used for oil production is rapeseed, which accounted for between 9 and 97% of the total crop area as of 2021. Soybeans, sunflower or flax are less important in the oil industry. Rapeseed oil sales in 2021 were 0.3 million tons. Other popular oils in Poland are sunflower oil at 28,900 tons and soybean oil at 55 tons [7]. The oil pressing process produces pomace (cakes), a byproduct of the process. Cakes are treated as waste and used mainly as a feed additive for livestock. The pomace can be a high-value feed product, such as rapeseed cakes, which has a higher protein content than whole seeds [8]. Flax pomace, on the other hand, can be used in the production of cookies as a substitute for oat, which is high in fat [9]. Cakes from various materials can also be used for food production, such as protein isolates. The amount of protein, carbohydrates or fiber in cakes depends on the raw material from which they were obtained and the method of pressing the oil. Rapeseed pomace, which is mainly used for animal feed, can also be used as a natural fertilizer due to its rich mineral composition. Rapeseed cakes can also be used as an ingredient in food production, for example, to make crisps [10].

Despite the various possibilities for the use of cakes, there is still a need to manage a large portion of these residues. A possible solution can be provided by the energy sector. There are several types of biomass used for energy purposes. One of these is combustion or co-combustion. The co-combustion process can be carried out indirectly, directly or in parallel. Another method is gasification, which is a thermochemical process that aims to convert organic matter or carbonized materials into a combustible gas, which mainly contains methane, carbon dioxide and hydrogen. The gasification medium is air [11,12].

Another method is hydrothermal liquefaction, which is the process of converting wet or dried lignocellulosic raw materials into liquid products that can be used as fuels after valorization [13–16]. Torrefaction, on the other hand, involves roasting biomass in the temperature range of 250–350 °C at atmospheric pressure and without oxygen, resulting in moisture-free solid products that, unlike raw biomass, have hydrophobic properties. Moreover, they are more homogeneous and more durable, which is important when using waste from the agricultural sector. Biochar is characterized by properties similar to brittle substances, which means it can be mixed with coal in amounts as high as 30% [17].

Pyrolysis is a thermochemical process of biomass transformation. It occurs under anaerobic conditions, or in a small amount of oxygen sufficient for combustion. This process involves the thermal decomposition of a substance rich in organic carbon. The pyrolysis process is classified by many factors, one of which is the temperature at which the process is carried out. In this case, a distinction is made between low-temperature pyrolysis, which is carried out in the temperature range from 450 °C to 700 °C, and high-temperature pyrolysis, or so-called coking, in the temperature range from 900 °C to 1200 °C. Another division of the pyrolysis process concerns its speed. In this case, a distinction is made between slow and fast pyrolysis. To distinguish between them, two basic parameters are used, namely temperature and time. Most often, the process is carried out at a temperature of 400 °C to 1000 °C, and the time the raw material spends in the reactor ranges from a few minutes for slow pyrolysis to up to 0.5 s for very fast pyrolysis. Running the process at a low temperature, as well as slowly heating the raw material, are more effective in terms of solid product formation, while to obtain more liquid raw material, higher temperatures are used. Using very fast pyrolysis at 900 °C, we can obtain gaseous products. The conditions under which the pyrolysis process is carried out are chosen depending on the type of raw material; in this way, one can obtain the desired product. Pyrolysis with a slow increase in temperature was used for thousands of years, and the main product was charcoal. During the slow pyrolysis process, the biomass is heated to 500 °C, and the time it stays in the reactor is between 5 and 30 min. The residence of the particles in the chamber for a long time causes the additional transformation of the originally formed products. The components of the gas phase react with each other to form a solid phase in the form of carbon and a liquid phase. The material processed by pyrolysis can be kept at a constant temperature or gradually heated, and the emitted gases can be regularly removed during the process. Fast pyrolysis is a process carried out at high temperatures. The biomass is heated at a high rate and there is no access to air. The products of this process are gases, aerosols and a charred residue. If the gas phase and aerosols condense, a brown liquid is obtained, which has half of the heating value of traditionally obtained diesel fuel. The advantage of fast pyrolysis is that it generates no waste. Bio-oil and charred products can be used as fuel, and gases can be recycled back into the process. In this process, an important aspect is the rate of heat transfer, which is facilitated by the finely ground material. A feature of fast pyrolysis is also the residence time of the vapors in the reactor, which is very short (less than 2 s). Rapid cooling followed by the condensation of vapors and aerosols helps to produce bio-oil. During the process of rapid pyrolysis, i.e., at temperatures between 500 °C and 700 °C and a short residence time of primary products in the conversion zone, the content of individual products is as follows: liquid fraction—75%, biochar—12% and gas—13%. Comparing this to slow pyrolysis, carried out at temperatures below 550 °C and a long residence time of primary products in the conversion zone, the content of each product is distributed differently: liquid fraction—30%, biochar—35% and gas—35%. Knowing what material is to be subjected to the pyrolysis process and what products are to be obtained, the type of pyrolysis to be carried out should be selected accordingly. Thermal conversion of biomass opens up prospects for creating a more sustainable and environmentally friendly energy system. The move toward zero-carbon energy sources and the efficient use of biomass as a renewable resource is not only a technological challenge, but also a strategic step toward a more sustainable future [18–26].

The steadily growing global economy and increasing energy intensity of global industry is contributing to excess CO_2 emissions into the atmosphere. Carbon balance issues are an important problem for modern industry and the subject of ongoing research work. Measures are therefore needed to balance carbon in the atmosphere using carbon capture and storage, for example, in the soil. One solution to this problem may be the use of biochar produced from different types of biomass [27,28].

Biochar is produced from organic matter, by thermal treatment without oxygen. As reported in the scientific literature, the wide range of applications for biocarbon continues to grow and, as it stands, mainly concerns industry, agriculture, cosmetics production, pharmaceuticals and environmental aspects. It can be introduced into soils as a natural additive, added to feedstuffs and silage and also used for water purification [29–31]. Biochar can also be used for greenhouse gas reduction, carbon sequestration, soil contaminant removal, wastewater treatment and as a supporting material in composting as well as methane fermentation [32–37]. It should also be emphasized that biochar products can be versatile materials used in the field of modern energy storage and conversion. It seems desirable to develop new, efficient, complementary and fully safe methods for the production of carbonaceous materials from renewable resources, which have high efficiency and limited environmental impact [38]. The energetic use of plant biomass residues as a substrate for the production of biochar can also be a viable alternative for the waste management sector. The structure and properties of the carbonate strictly depend on the raw material used in the pyrolysis process. Biochar can be produced from materials of different origins, and these can include energy crops and forestry waste as well as agricultural biomass [39–42]. Waste from agri-food processing, e.g., fermenting oats, chicken manure and cattle manure, can also be used to produce pyrolysates [43,44].

3. Materials and Methods

3.1. Research Object

Waste plant biomass in the form of fennel flower (*Nigella sativa* L.), rapeseed (*Brassica napus* L. var. *Napus*), flax (*Linum usitatissimum* L.), evening primrose (*Oenothera biennis* L.), milk thistle (*Silybum marianum* L. *Gaertn.*) and hemp (*Cannabis sativa* L.) cakes was used to conduct a study on the possibility of using residues from the oil pressing process to produce biochar (Figure 1). The material intended for testing was dried to a moisture level of less than 10% and then crushed to a fraction of less than 10 mm.

Materials for this study were obtained from an oil mill located in Podkarpackie Province, Poland. The post-process biomass created during oil extrusion (cakes) at this plant is intended as an additive for animal feed, but its full use for this purpose is difficult to realize due to the developing oil market in Poland.

(a) (b) (c)

Figure 1. *Cont.*

(d) (e) (f)

Figure 1. Cakes of fennel flower (**a**), rapeseed (**b**), flax (**c**), evening primrose (**d**), milk thistle (**e**) and hemp (**f**).

3.2. *Pyrolysis Process*

A retort furnace FCF 2 R designed for heat treatment in the atmosphere of inert gas and equipped with a post-process gas cooler with water well (CZYLOK, Jastrzębie-Zdrój, Poland) was used to carry out the pyrolysis process.

Pyrolysis tests of the test samples were performed in nitrogen atmosphere of 99.99% purity with a gas flow of 10 L/min at temperatures of 400, 450 and 500 °C, respectively, and in times of 5, 10 and 15 min. The obtained pyrolysates were then sifted through a sieve with a diameter of holes equal to 1 mm. The samples were rinsed several times with distilled water and then dried for 12 h (at 80 °C) in order to remove potential contaminants.

3.3. *Combustion Process*

The process of combustion of post-production residues from the oil pressing process and the biochar produced from them was carried out using a TGA 701 from LECO (LECO Corporation, Saint Joseph, MI, USA). This test consisted of burning 100 g of the sample under controlled conditions at 800 °C. The process took 20 min until the selected material was completely burned.

3.4. *Analysis of Samples*

The materials were analyzed with the aim of determining the basic physical and chemical parameters, e.g., the total content of carbon, ash, nitrogen, hydrogen, volatile substances and calorific value. The thermogravimetric method, with use of TGA 701 apparatus from LECO (LECO Corporation, Saint Joseph, MI, USA), was used to analyze the contents of ash and volatile substances in the samples. The TrueSpec CHN analyzer from LECO (LECO Corporation, Saint Joseph, MI, USA) was used to test the contents of the total carbon, hydrogen and nitrogen. The AC500 calorimeter from LECO (LECO Corporation, Saint Joseph, MI, USA) was used to determine the calorific value of the materials analyzed. Analysis of the content of carbon monoxide (CO), nitrogen oxides (NOx) and sulfur oxide (SO_2) was carried out using an ULTRAMAT 23 gas analyzer (SIEMENS AG, Munich, Germany). An algorithm that takes into account the density of each gas was used to calculate the content of the individual gases. Table 1 presents an example of the calculations for CO.

Table 1. Algorithm for calculating gas quantities (example for CO).

Calculated Parameters	Method of Calculation
The amount of carbon monoxide (CO)	$SCO = 0.5 \times (\lvert X_{n-1} - X_n \rvert) \times (t_{n-1} - t_n)$
Amount of gas mixture taken	$S_{gases} = t_c \times 100$
Percentage of carbon dioxide (CO) in the analyzed gas mixture	$\%CO = SCO/S_{gases}$
Amount of gas mixture taken during the test	$AL. = t_c \times 0.025$ [L/s]
Amount of carbon monoxide (CO) taken during the test	$ACO = AL. \times \%CO$
Carbon monoxide (CO) density at room temperature (20 °C) [mg L^{-1}]	1250
Amount of carbon monoxide (CO) taken during measurement [mg]	$ACO \times 1250$

where
SCO—the calculated amount of CO;
S_{gases}—the amount of gas mixture taken;
$X_{n-1} - X_n$—CO concentration results recorded by the analyzer during the intervals;
$t_{n-1} - t_n$—measured time intervals;
$t_c \times 100$—100-percentage sum of absorbed gases;
AL.—the amount of gas mixture taken during the test;
ACO—the amount of CO taken during the test;
0.5—the complement of the curve integrating the measurement result;
0.025—the amount of air suctioned per unit time [L/s].

Table 2 summarizes the density parameters of the analyzed gases, which were used in the calculations.

Table 2. The densities of the analyzed gases at room temperature (20 °C).

Gases Analyzed	Gas Density [mg L^{-1}]
Carbon monoxide (CO)	1250
Nitrogen oxides (NO_x)	1340
Sulfur dioxide (SO_2)	2830

Samples of biomass and biochars were subjected to laboratory analyses using current analytical standards (Table 3).

Table 3. Parameters analyzed with research methods.

Parameter	Research Method	
Content of carbon, nitrogen and hydrogen	PN-EN ISO 16948:2015-07 [45]	Solid biofuels—determination of total carbon, hydrogen and nitrogen content
Ash content	PN-EN ISO 18122:2023-05 [46]	Solid biofuels—determination of ash content
Volatile matter	PN-EN ISO 18123:2023 [47]	Solid biofuels—determination of ash content
Calorific value	PN-EN ISO 18125:2017-07 [48]	Solid biofuels—determination of ash content

3.5. Names of Tests

Biomass samples were described with the use of symbols depending on the type of material, temperature and duration of the pyrolysis process and used for further identification (Table 4).

Table 4. Names of tests.

Number	Symbol	Type of Material	Pyrolysis Temperature [°C]	Duration of the Pyrolysis [min.]
1.	NC	fennel flower cakes	-	-
2.	NC400/5		400	5
3.	NC400/10			10
4.	NC400/15			15
5.	NC450/5		450	5
6.	NC450/10			10
7.	NC450/15			15
8.	NC500/5		500	5
9.	NC500/10			10
10.	NC500/15			15

Table 4. *Cont.*

Number	Symbol	Type of Material	Pyrolysis Temperature [°C]	Duration of the Pyrolysis [min.]
11.	RC	rapeseed cakes	-	-
12.	RC400/5		400	5
13.	RC400/10			10
14.	RC400/15			15
15.	RC450/5		450	5
16.	RC450/10			10
17.	RC450/15			15
18.	RC500/5		500	5
19.	RC500/10			10
20.	RC500/15			15
21.	FC	flax cakes	-	-
22.	FC400/5		400	5
23.	FC400/10			10
24.	FC400/15			15
25.	FC450/5		450	5
26.	FC450/10			10
27.	FC450/15			15
28.	FC500/5		500	5
29.	FC500/10			10
30.	FC500/15			15
31.	EPC	evening primrose cakes	-	-
32.	EPC400/5		400	5
33.	EPC400/10			10
34.	EPC400/15			15
35.	EPC450/5		450	5
36.	EPC450/10			10
37.	EPC450/15			15
38.	EPC500/5		500	5
39.	EPC500/10			10
40.	EPC500/15			15
41.	MTC	milk thistle cakes	-	-
42.	MTC400/5		400	5
43.	MTC400/10			10
44.	MTC400/15			15
45.	MTC450/5		450	5
46.	MTC450/10			10
47.	MTC450/15			15
48.	MTC500/5		500	5
49.	MTC500/10			10
50.	MTC500/15			15
51.	HC	hemp cakes	-	-
52.	HC400/5		400	5
53.	HC400/10			10
54.	HC400/15			15
55.	HC450/5		450	5
56.	HC450/10			10
57.	HC450/15			15
58.	HC500/5		500	5
59.	HC500/10			10
60.	HC500/15			15

For instance, RC-non-heat-treated rapeseed cakes, HC500/10–hemp cakes after pyrolysis at 500 °C and a time of 10 min (Figure 2).

Figure 2. Selected biochar (500 °C/10 min) from fennel flower (**a**), rapeseed (**b**), flax (**c**), evening primrose (**d**), milk thistle (**e**) and hemp (**f**) cakes.

3.6. Statistical Analysis

The effects of experimental factors reflected by the relevant parameters and the relationships between them were examined by Analysis of Variance (ANOVA) with the use of Duncan's test. In order to compute the statistical analyses, STATISTICA version 12.0 (StatSoft Inc., Tulsa, OK, USA) was applied. A significance threshold of ≤0.05 was set for all analyses. The data were analyzed separately for each type of materials.

4. Results and Discussion

4.1. Elemental Analysis

There are many ways and processes to improve the energy properties of materials. One of them is pyrolysis, the main purpose of which is to increase the calorific value. This parameter is one of the most important when it comes to determining fuel efficiency.

In this study, the pyrolysis process was carried out at temperatures of 400, 450 and 500 °C and holding times of 5, 10 and 15 min at these temperatures. The test material consisted of six types of residues from the oil pressing process in the form of cakes. The biochar obtained was statistically significantly different from the control sample (non-thermally processed biomass) in terms of the content of basic biogenic elements. In contrast to the typical lignocellulosic plant biomass used in the production of biochar, the cakes produced after oil pressing were characterized by a relatively high content of total nitrogen. Increasing the time and temperature of the pyrolysis process further affected the concentration of this element. In the case of fennel flower cakes, the value changed statistically significantly against the initial value (4.77%) from the moment of the pyrolysis process carried out at 400 °C and holding time of 15 min. The highest content of total nitrogen, i.e., 7.68%, was recorded for the pyrolysis process carried out at 450 °C and a temperature holding time of 15 min. In the case of rapeseed cakes, the nitrogen content of the control sample was at the level of 4.34%, and the highest value, i.e., 6.50%, was measured for biochar made from this material by pyrolysis at 500 °C and a temperature holding time of 15 min. The same parameters of the pyrolysis process yielded the highest nitrogen content, i.e., 6.92%, in the biochar prepared on the basis of cakes formed from the production of flax seed oil. The content of total nitrogen in the residue from the production of evening primrose, milk thistle and hemp seed oil was 3.49, 3.12 and 4.7%, respectively. Each time, the pyrolysis process increased the concentration of this element in the materials obtained. In the case of pyrolysates from evening primrose, the content of total nitrogen was in the range of 3.60–4.32%, milk thistle of 3.25–3.77% and hemp of 4.88–5.74%. Pyrolysates prepared at 500 °C and in 15 min had the highest values, recording statistically significant changes relative to the control sample. In the case of research by Ferens (2017), the author notes that both for biochar from coffee brewing waste as well as from sunflower shells, the highest nitrogen value was measured for the pyrolysis process conducted at 450 °C. These values were 4.83% for coffee brewing waste and 3.57% for sunflower shells [49]. In contrast, Hmid et al. (2014) report the nitrogen content of biochar from olive pomace being waste after oil production to be less than 1%. These values were obtained after using a pyrolysis process at 430–530 °C [50]. The study by Martínez-Gómez et al. again indicates a significantly lower concentration of nitrogen in biochar than in our own study. The authors report that in the biochar obtained from grape pomace (pyrolysis 350 and 700 °C), the nitrogen content was in the range of 1.8–2.24% [51]. On the other hand, results coinciding with those obtained in this study are reported by Mazurek et al. (2024). The authors determined nitrogen content in the range of 4.5 to 5.6% for biochar from rapeseed cakes prepared at 700 °C and chemically modified [52]. In a study by Sładeczek and Głodek-Bucyk (2017) on biochar from rye straw, the authors obtained a total nitrogen content result of 1.25%, which was more than four times lower than the maximum total nitrogen content of the biochar analyzed in our own work. In a study by the same authors on biochar obtained from chicken manure, the content of total nitrogen was equal to 3.6%, which was within the range of the content in our own study [53].

The percentage of hydrogen decreased in the obtained biochar as the temperature and time of the anaerobic thermal treatment increased. The hydrogen content changed statistically significantly with respect to the values for the control samples, i.e., fennel flower cakes—7.43%, rapeseed cakes—7.25%, flax cakes—7.16%, evening primrose cakes—5.71%, thistle cakes—6.9% and flax cakes—6.34%. The lowest hydrogen content was recorded each time for pyrolysis carried out at 500 °C and a holding time of 15 min, respectively, for biochar from fennel flower, rapeseed, flax, evening primrose, milk thistle and hemp at 3.93, 3.81, 3.94, 3.66, 3.87 and 3.67%. There was an inverse relationship related to the change in temperature and pyrolysis process time for this parameter than for the nitrogen content. Similar correlations were noted by Ferens (2017) when it came to biochar from coffee brewing waste and sunflower shells. For biochar obtained at the highest temperature, the values were 1.28% for coffee brewing waste and 1.05% for biochar from sunflower shells [49]. In the results reported by Li et al. (2023), the percentage of hydrogen increased with the increase in the temperature of performing the pyrolysis process. In the case of biochar from wheat straw, the increase was more than five times, with a final value of 2.73%. As for corn straw in this case, the difference was even greater, more than 20 times, and the final value was 3.68% [54]. In a study by Crombie et al. (2013) on rice husk biochar, the authors obtained a hydrogen content of 1.97% in the biochar, which was lower than in the analyses conducted in our own study [55]. Mazurek et al. (2024) determined the hydrogen content of rapeseed cake biochar in the range of 1.5 to 2.6% [52].

All pyrolysates obtained were statistically significantly different from the control sample with respect to the total carbon content. The average content of total carbon in non-heat-treated cakes of fennel flower, rapeseed, flax, evening primrose, milk thistle and hemp was at the level of 48.79%. The highest content of total carbon in the pyrolysate from fennel flower, i.e., 64.06%, was recorded at a pyrolysis temperature of 500 °C and a holding time of 15 min. Pyrolysates prepared from rapeseed cakes had the highest total carbon content at 66.43%. Thermal treatment of flax cakes made it possible to obtain a material characterized by a total carbon content of a maximum of 65.71%, again obtained under temperature parameters of 500 °C and a time of 15 min. Pyrolysates prepared from evening primrose cakes had carbon concentrations in the range of 58.45–64.47%, from milk thistle of 62.5–71.67% and from hemp of 59.29–66.83%. Similar relationships are reported by Li et al. (2023) on the pyrolysis process in the 300–650 °C temperature range for materials such as wheat and corn straw. The authors report that the percentage of carbon increases with the increase in the temperature at which the pyrolysis process is carried out [54]. Convergent results were also reported by Ferens (2017) when performing tests on sunflower shells and coffee brewing waste. The biochar was obtained by pyrolysis in the temperature range of 250–850 °C with a heat-up rate of 10 °C/min. The author determined the highest total carbon content for biochar obtained at the highest temperature. The increase in content for coffee brewing waste was 34%, while for sunflower shells, it was 21% [49]. In contrast, in a study by Yao Y. et al. (2012), the total carbon content of bamboo biochar was 76.9%, which is higher than in our own study [56]. Ferreira et al. (2021), in a study on the characterization of biochars from grape pomace, determined the carbon content of the unprocessed material to be 51.1%. The biochar obtained from this material was characterized in turn by a concentration of this element in the range of 67–81% (pyrolysis at 300–800 °C) [57]. Angın et al. (2014), when producing biochar from safflower seed cakes at 500 °C, achieved a total carbon content of 62.45%. Subjecting this material to chemical activation with the use of KOH and a temperature of 800 °C increased the content of this element to 72.09% [58]. Sun et al. (2021), in a study of the pyrolysis of flax seed residues as solid waste generated during flex seed oil production, reported an increase in the total carbon content from 44.9% for unprocessed material to less than 55% for the biochars obtained [59]. Similar correlations as in our own study are reported by Mukhambet et al. (2022), noting an increase in the carbon content with the increasing flax straw pyrolysis temperatures at 400, 500 and 600 °C [60].

The percentage of ash increased as the temperature and time of the pyrolysis process increased. After processing the biomass into biochar, the ash content was statistically significantly different from the control sample (fennel flower—6.15%, rapeseed—5.32%, flax—6.35%, evening primrose—12%, milk thistle—99% and hemp—6.55%). After thermal processing, the ash content of the pyrolysates was statistically significantly different from the control sample and was in the following ranges: fennel flower 12.86–23.22%, rapeseed 11.21–22.09%, flax 11.06–21.91%, evening primrose 16.79–19.53%, milk thistle 15.75–17.22% and hemp 20.39–22.41%. The highest percentage of ash among all analyzed biochars, i.e., 23.22%, was recorded for fennel flower pyrolysates prepared at 500 °C and a temperature holding time of 15 min. Tag et al. (2016), in a study on biochar produced from corn kernels, report a percentage content of ash at 18.7%, which meant a result within the range of ash content variations in the analyzed biochars from the cakes in our own study [61]. On the other hand, Mulimani and Navindgi (2017) report that biochar made from mahua de-oiled seed cakes has an ash content of 16.4% [62]. An inverse relationship was observed for the volatile matter, which decreased as the temperature of the pyrolysis process increased. Changes from an average content for the control samples from 73.57 to 16.36% were recorded for biochars obtained at the maximum pyrolysis parameters tested (500°C, 15 min). The lowest volatile matter content (13.93%) of all analyzed materials was in the biochar prepared from flax cakes (Table 5). The statistically significant value has already changed for the biochar with the smallest transformation (400 °C/5 min). These values were practically three times lower than the initial value for raw biomass. Similar conclusions were reached by Sładeczek and Głodek-Bucyk (2017), where the proportion of volatile matter also decreased during the conversion of biomass to biochar [53]. In a study by Kazimiersi et al. (2022) on biochar from pistachio nut shells, it was reported that the volatile matter content was equal to 30.62%, giving a similar but slightly higher volatile matter content compared to the analyses in our own work [63]. In a research paper by Wu et al. (2012) on rice straw, a content of volatile matter ranging from 40.2 to 13.1% was reported. The authors noted a significant decrease in the content of volatile matter as the pyrolysis temperature increased from 300 to 700 °C [64].

Table 5. Average content of total nitrogen, total carbon, hydrogen, ash and volatile substances for cakes of fennel flower *nigella (Nigella sativa* L.), rapeseed *(Brassica napus* L. var. *Napus*), flax *(Linum usitatissimum* L.), evening primrose *(Oenothera biennis* L.), milk thistle *(Silybum marianum* L. *Gaertn.*), hemp *(Cannabis sativa* L.) and prepared biochars.

Research Material	General Nitrogen	Total Carbon	Hydrogen	Ash	Volatile Substances
	%				
NC	4.77 ^{a} ± 0.04	49.55 ^{a} ± 0.11	7.43 ^{d} ± 0.02	6.15 ^{a} ± 0.02	73.41 ^{f} ± 0.11
NC400/5	4.84 ^{a} ± 0.02	57.66 ^{b} ± 0.10	4.94 ^{c} ± 0.01	12.86 ^{b} ± 0.06	27.13 ^{e} ± 0.05
NC400/10	4.90 ^{a} ± 0.03	61.61 ^{bc} ± 0.07	4.91 ^{c} ± 0.01	13.96 ^{b} ± 0.11	25.19 ^{e} ± 0.07
NC400/15	5.57 ^{b} ± 0.07	61.76 ^{bc} ± 0.08	4.92 ^{c} ± 0.02	15.62 ^{c} ± 0.09	24.48 ^{de} ± 0.08
NC450/5	6.64 ^{c} ± 0.07	61.49 ^{bc} ± 0.13	4.53 ^{b} ± 0.01	16.81 ^{c} ± 0.10	23.25 ^{cd} ± 0.07
NC450/10	7.40 ^{d} ± 0.05	62.01 ^{bc} ± 0.07	4.53 ^{b} ± 0.01	18.92 ^{d} ± 0.11	22.41 ^{cd} ± 0.06
NC450/15	7.68 ^{d} ± 0.05	62.53 ^{bc} ± 0.09	4.53 ^{b} ± 0.01	20.09 ^{de} ± 0.10	21.63 ^{c} ± 0.10
NC500/5	6.84 ^{cd} ± 0.04	62.95 ^{bc} ± 0.13	4.10 ^{a} ± 0.01	21.24 ^{de} ± 0.16	19.44 ^{b} ± 0.10
NC500/10	6.96 ^{cd} ± 0.01	63.57 ^{c} ± 0.08	4.09 ^{a} ± 0.01	22.77 ^{e} ± 0.08	17.66 ^{ab} ± 0.04
NC500/15	7.30 ^{d} ± 0.03	64.06 ^{c} ± 0.09	3.93 ^{a} ± 0.00	23.22 ^{e} ± 0.09	16.50 ^{a} ± 0.06
RC	4.34 ^{a} ± 0.06	51.97 ^{a} ± 0.06	7.25 ^{c} ± 0.00	5.32 ^{a} ± 0.02	79.33 ^{f} ± 0.07
RC400/5	4.65 ^{ab} ± 0.05	60.33 ^{b} ± 0.16	4.74 ^{b} ± 0.00	11.21 ^{b} ± 0.06	28.16 ^{e} ± 0.09
RC400/10	4.89 ^{b} ± 0.04	62.32 ^{b} ± 0.06	4.66 ^{b} ± 0.00	12.48 ^{b} ± 0.05	27.14 ^{de} ± 0.11
RC400/15	5.40 ^{c} ± 0.04	62.80 ^{b} ± 0.03	4.66 ^{b} ± 0.01	13.89 ^{c} ± 0.05	24.84 ^{d} ± 0.02
RC450/5	5.60 ^{c} ± 0.05	63.65 ^{b} ± 0.10	4.34 ^{b} ± 0.01	14.72 ^{c} ± 0.05	21.46 ^{c} ± 0.09
RC450/10	6.01 ^{d} ± 0.02	64.32 ^{b} ± 0.06	4.26 ^{ab} ± 0.04	17.03 ^{d} ± 0.14	20.28 ^{c} ± 0.10
RC450/15	6.22 ^{d} ± 0.05	65.06 ^{b} ± 0.05	4.25 ^{ab} ± 0.01	18.57 ^{d} ± 0.07	18.62 ^{bc} ± 0.07
RC500/5	5.77 ^{cd} ± 0.04	65.89 ^{b} ± 0.08	4.06 ^{ab} ± 0.15	19.86 ^{de} ± 0.06	17.53 ^{ab} ± 0.05
RC500/10	5.95 ^{cd} ± 0.02	66.45 ^{b} ± 0.05	3.89 ^{a} ± 0.05	21.13 ^{e} ± 0.03	17.07 ^{ab} ± 0.05
RC500/15	6.50 ^{d} ± 0.02	66.43 ^{b} ± 0.14	3.81 ^{a} ± 0.00	22.09 ^{e} ± 0.02	16.33 ^{a} ± 0.08
FC	4.92 ^{a} ± 0.03	49.29 ^{a} ± 0.05	7.16 ^{c} ± 0.01	6.35 ^{a} ± 0.02	75.69 ^{f} ± 0.08
FC400/5	5.37 ^{ab} ± 0.06	59.47 ^{b} ± 0.05	4.89 ^{b} ± 0.01	11.06 ^{b} ± 0.05	27.36 ^{e} ± 0.02
FC400/10	5.75 ^{b} ± 0.06	62.43 ^{b} ± 0.05	4.84 ^{b} ± 0.01	12.25 ^{b} ± 0.08	25.87 ^{de} ± 0.08
FC400/15	6.10 ^{b} ± 0.03	62.75 ^{b} ± 0.05	4.84 ^{b} ± 0.01	13.80 ^{c} ± 0.07	24.37 ^{d} ± 0.03
FC450/5	5.91 ^{b} ± 0.05	63.04 ^{b} ± 0.11	4.49 ^{ab} ± 0.01	14.80 ^{c} ± 0.04	21.07 ^{c} ± 0.02
FC450/10	5.96 ^{b} ± 0.02	63.63 ^{b} ± 0.04	4.44 ^{ab} ± 0.02	17.01 ^{d} ± 0.12	20.05 ^{bc} ± 0.06
FC450/15	6.49 ^{cd} ± 0.04	64.27 ^{b} ± 0.03	4.44 ^{ab} ± 0.01	18.36 ^{d} ± 0.08	18.84 ^{b} ± 0.08
FC500/5	6.43 ^{cd} ± 0.04	64.89 ^{b} ± 0.08	4.11 ^{a} ± 0.09	19.59 ^{de} ± 0.05	15.50 ^{a} ± 0.04
FC500/10	6.63 ^{cd} ± 0.03	65.48 ^{b} ± 0.07	4.01 ^{a} ± 0.05	20.98 ^{e} ± 0.04	14.38 ^{a} ± 0.04
FC500/15	6.92 ^{d} ± 0.01	65.71 ^{b} ± 0.04	3.94 ^{a} ± 0.03	21.91 ^{e} ± 0.03	13.93 ^{a} ± 0.05
EPC	3.49 ^{a} ± 0.01	49.23 ^{a} ± 0.05	5.71 ^{d} ± 0.02	7.12 ^{a} ± 0.02	70.32 ^{e} ± 0.04
EPC400/5	3.60 ^{ab} ± 0.02	58.45 ^{b} ± 0.37	4.81 ^{c} ± 0.02	16.79 ^{b} ± 0.02	30.15 ^{d} ± 0.07
EPC400/10	3.65 ^{ab} ± 0.03	60.97 ^{b} ± 0.04	4.64 ^{c} ± 0.02	16.85 ^{b} ± 0.03	29.22 ^{d} ± 0.08
EPC400/15	3.76 ^{ab} ± 0.03	61.06 ^{b} ± 0.07	4.75 ^{c} ± 0.01	17.52 ^{b} ± 0.07	28.45 ^{d} ± 0.10
EPC450/5	3.90 ^{bc} ± 0.05	58.69 ^{b} ± 0.03	4.31 ^{bc} ± 0.02	18.6 ^{bc} ± 0.05	27.92 ^{cd} ± 0.19
EPC450/10	4.49 ^{c} ± 0.05	61.64 ^{b} ± 0.05	4.3 ^{bc} ± 0.01	19.37 ^{bc} ± 0.06	25.74 ^{c} ± 0.07
EPC450/15	4.77 ^{c} ± 0.05	62.22 ^{b} ± 0.10	4.25 ^{bc} ± 0.01	19.63 ^{bc} ± 0.06	24.17 ^{b} ± 0.12
EPC500/5	3.95 ^{bc} ± 0.03	58.63 ^{b} ± 0.04	3.98 ^{ab} ± 0.01	18.79 ^{bc} ± 0.04	23.99 ^{b} ± 0.10
EPC500/10	4.05 ^{bc} ± 0.01	64.28 ^{b} ± 0.05	3.94 ^{ab} ± 0.01	18.85 ^{bc} ± 0.03	22.47 ^{a} ± 0.02
EPC500/15	4.32 ^{c} ± 0.02	64.47 ^{b} ± 0.07	3.66 ^{a} ± 0.04	19.53 ^{c} ± 0.03	21.45 ^{a} ± 0.02

Table 5. *Cont.*

Research Material	General Nitrogen	Total Carbon	Hydrogen	Ash	Volatile Substances
			%		
MTC	3.12 ᵃ ± 0.06	47.80 ᵃ ± 0.07	6.90 ᵈ ± 0.01	5.99 ᵃ ± 0.02	69.54 ᵍ ± 0.12
MTC400/5	3.25 ab ± 0.02	62.50 b ± 0.04	5.41 c ± 0.02	15.75 b ± 0.06	25.80 f ± 0.12
MTC400/10	3.36 ab ± 0.01	65.40 bc ± 0.08	5.24 c ± 0.01	16.14 b ± 0.07	25.80 f ± 0.09
MTC400/15	3.39 ab ± 0.01	66.83 bc ± 0.06	5.08 c ± 0.03	16.49 bc ± 0.04	24.20 e ± 0.13
MTC450/5	3.42 ab ± 0.01	67.03 bc ± 0.07	4.53 bc ± 0.01	16.29 bc ± 0.05	22.95 de ± 0.09
MTC450/10	3.43 ab ± 0.01	67.44 bc ± 0.08	4.51 bc ± 0.03	16.34 bc ± 0.02	21.70 d ± 0.11
MTC450/15	3.47 ab ± 0.02	69.56 bc ± 0.06	4.34 bc ± 0.01	16.87 bc ± 0.04	19.33 c ± 0.06
MTC500/5	3.54 ab ± 0.02	68.44 bc ± 0.06	4.16 ab ± 0.01	16.81 bc ± 0.04	17.47 b ± 0.06
MTC500/10	3.65 b ± 0.02	70.82 c ± 0.03	4.13 ab ± 0.01	16.92 bc ± 0.05	16.71 ab ± 0.02
MTC500/15	3.77 b ± 0.06	71.67 c ± 0.07	3.87 a ± 0.01	17.22 c ± 0.02	15.67 a ± 0.12
HC	4.70 a ± 0.05	47.89 a ± 0.03	6.34 d ± 0.02	6.55 a ± 0.05	73.12 f ± 0.04
HC400/5	4.88 a ± 0.03	59.29 b ± 0.12	4.86 c ± 0.04	20.39 b ± 0.02	23.92 e ± 0.07
HC400/10	5.09 a ± 0.01	62.06 bc ± 0.05	4.74 c ± 0.02	20.50 b ± 0.04	23.46 de ± 0.08
HC400/15	5.19 a ± 0.02	62.74 bc ± 0.07	4.68 c ± 0.01	21.12 b ± 0.03	22.26 de ± 0.10
HC450/5	5.23 ab ± 0.03	61.69 bc ± 0.05	4.19 b ± 0.004	21.54 b ± 0.04	21.35 d ± 0.19
HC450/10	5.32 ab ± 0.02	63.39 bc ± 0.06	4.16 b ± 0.02	21.93 b ± 0.02	19.65 c ± 0.07
HC450/15	5.42 ab ± 0.01	64.75 bc ± 0.03	4.07 ab ± 0.01	22.36 b ± 0.03	17.71 b ± 0.12
HC500/5	5.46 ab ± 0.02	62.40 bc ± 0.04	3.85 ab ± 0.03	21.89 b ± 0.05	16.63 b ± 0.10
HC500/10	5.63 ab ± 0.02	66.42 bc ± 0.02	3.81 ab ± 0.01	22.00 b ± 0.04	15.51 ab ± 0.02
HC500/15	5.74 b ± 0.02	66.83 c ± 0.04	3.67 a ± 0.01	22.41 b ± 0.02	14.32 a ± 0.02

Differences between average values marked with the same letters are not statistically significant at the level of $p \leq 0.05$ according to the Duncan test. The data were analyzed separately for each type of materials.

According to the Duncan test results, statistically significant differences were noted in the content of carbon, hydrogen, ash and volatile substances between the control sample and biochars for all the analyzed materials. In the case of the total nitrogen content, smaller discrepancies in the results were noted and only some samples differed statistically significantly. It was found that the time and temperature of the pyrolysis process affected statistically significant changes in the values of the parameters tested.

4.2. Calorific Value

One of the main parameters defining the energy usefulness of a substance is its calorific value, which is why the post-production residues from the oil pressing process in the form of cakes from fennel flower *nigella (Nigella sativa* L.), rapeseed *(Brassica napus* L. var. *Napus*), flax *(Linum usitatissimum* L.), evening primrose *(Oenothera biennis* L.), milk thistle *(Silybum marianum* L. *Gaertn.)* and hemp *(Cannabis sativa* L.), as well as biochar prepared from them, were subjected to calorific value analysis. There was an upward trend, i.e., as the temperature and time of the pyrolysis process increased, the calorific value also increased, and the maximum increase in this parameter was recorded each time for a temperature of 500 °C and a process time of 15 min. The lowest calorific value at 18.88 MJ kg^{-1}, among the six types of cakes analyzed, was characterized by milk thistle seed residues, while the highest, i.e., 22.03 MJ kg^{-1}, was that of rapeseed residues. On the other hand, the calorific value of the prepared pyrolysates was in the following ranges: fennel flower 21.77–24.33 MJ kg^{-1}; rapeseed 22.57–25.22 MJ kg^{-1}; flax 21.46–24.67 MJ kg^{-1}; evening primrose 21.06–24.22 MJ kg^{-1}; milk thistle 23.93–26.59 MJ kg^{-1} and hemp 22.2–25.13 MJ kg^{-1}. The highest increase in the calorific value after applying anaerobic thermal modification was obtained for the biochar from the milk thistle seed residues produced by pyrolysis at 500 °C and a temperature holding time of 15 min (Figure 3). In a study conducted by Hmid et al. (2014), the authors prepared biochars from the waste of olive processing at temperatures of 430, 480 and 530 °C. The highest calorific value for biochar made from olive cakes was obtained for pyrolysis carried out at 430 °C, and the value obtained was 31 MJ kg $^{-1}$ [50]. On the other hand, in Mulimani and Navindgi's (2017) study of biochar from butter tree seed cakes, a calorific value of 26.43 M/kg was obtained, which was almost identical to the results of biochar from milk thistle cakes obtained in our own study [62]. According to Martínez-Gómez et al. (2023), the calorific value of the grape pomace tested was at 20.4 MJ kg^{-1}. The pyrolysis process of this material increased the calorific value to 27.5 MJ kg^{-1} after pyrolysis at 700 °C [51]. Ozcimen and Karaosmanoglu (2004), in their study of the liquid fraction obtained from the pyrolysis of rapeseed cake waste, showed, on the other hand, that the bio-oil obtained in this fraction is a fuel with a high combustion value of 36 MJ kg^{-1} [is]. In contrast, the calorific value of biochar from rapeseed cake obtained at 500 °C was 25.3 MJ kg^{-1} [65]. In a study by Sun et al. (2021) on the pyrolysis of flax seed residues, the authors identify the calorific value of this material at 25 MJ kg^{-1}. They also state that this parameter decreased after the pyrolysis process. The calorific value of the obtained biochars decreased to 19.5 MJ kg^{-1} after pyrolysis at 650 °C [59]. The literature also reports the production of biochar from mahua de-oiled cakes. The calorific value of such biomass was at the level of 19.97 MJ kg^{-1} and increased to the value of 26.43 MJ kg^{-1} after pyrolysis under temperature conditions of 400–600 °C [62]. According to the results of Durak and Genel (2020), the energy values of liquid products from cakes of fennel flower, after the hydrothermal liquefaction method, ranged from 21.87–31.78 MJ kg^{-1}, and the energy values of solid products (biochar) ranged from 16.22 to 18.28 MJ kg^{-1} [66]. The literature reports that hemp biochar obtained by pyrolysis at 400–600 °C can be treated as a material with high potential for applications in the solid biofuels sector. The above thesis is correlated with high energy density, associated with high calorific values, i.e., reaching nearly 31 MJ kg^{-1} [67–73].

According to the Duncan test results, statistically significant differences in the calorific value were noted between the control sample and biochars for evening primrose cakes, milk thistle cakes and hemp cakes. In the case of the remaining three types of cakes, statistically significant changes were noted between the control sample and biochars obtained at the highest temperature and the longest time. In most cases, the change in the temperature and time of pyrolysis did not affect a statistically significant change in the calorific value between pyrolysates.

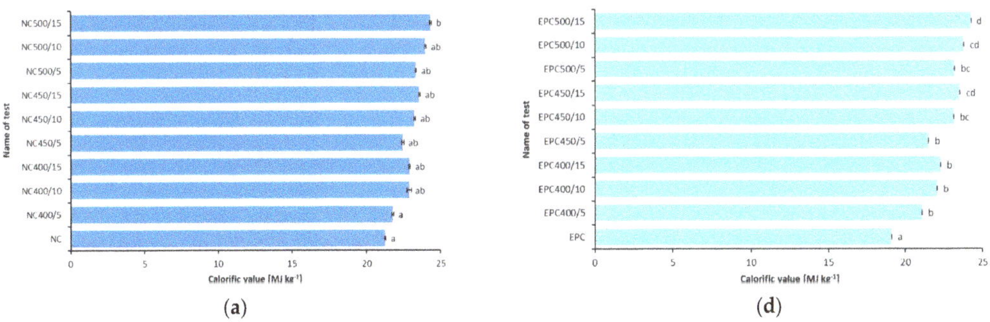

Figure 3. *Cont.*

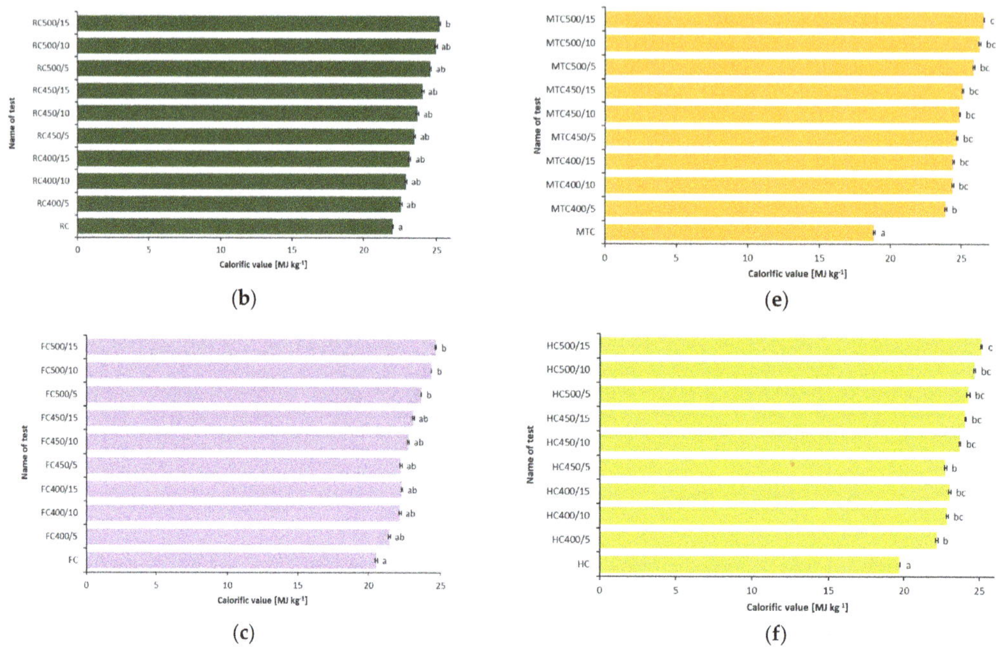

Figure 3. Calorific value of the tested cakes and biochars prepared from (**a**) fennel flower *(Nigella sativa* L.), (**b**) rapeseed *(Brassica napus* L. var. *Napus*), (**c**) flax *(Linum usitatissimum* L.), (**d**) evening primrose *(Oenothera biennis* L.), (**e**) milk thistle *(Silybum marianum* L. *Gaertn.*), (**f**) hemp *(Cannabis sativa* L.).

4.3. Emissions of CO, NOx and SO_2

Analyses of carbon monoxide (CO) emissions have indicated high emissions of this gas during the combustion of oil pressing residues. The highest value at 5712.3 g kg^{-1} among the analyzed cakes was recorded for flax residues, while the lowest value was recorded for milk thistle residues (2523 g kg^{-1}). Biochar materials, on the other hand, were characterized by significantly lower carbon monoxide emissions. Each time, the lowest values were recorded for pyrolysates obtained at the maximum parameters of the pyrolysis process. Analyzing all the biochars obtained, the emission levels during their combustion were approximated in the range of 35.33–2821.67 g kg^{-1}. The lowest value characterized the pyrolysates obtained from fennel flower cakes. The content of nitrogen oxides in the cakes subjected to combustion was in the range of 78.96–116.33 g kg^{-1}. The highest emission value was recorded during the combustion of evening primrose seed residues, while the lowest was recorded for rapeseed cake. In the case of the pyrolysates obtained, inconclusive results regarding nitrogen oxide emissions during their combustion were obtained. In the case of cakes from fennel flower, rapeseed, evening primrose and milk thistle, significantly lower values were recorded than in the control sample. In turn, the pyrolysates obtained from flax and hemp had higher NOx emissions than the control samples. The lowest NOx emissions were determined in the case of the combustion of pyrolysates of fennel flower, with an average value of 17.14 g kg^{-1} and a minimum value of 2.67 g kg^{-1}. Subjecting the obtained cakes to the combustion process made it possible to determine the emission of sulfur dioxide in the range of decreasing values, i.e., 291.33, 142.33, 123.32, 112.67, 87.66 and 94.31 g kg^{-1} for rapeseed, hemp, fennel flower, evening primrose, flax and milk thistle, respectively. Pyrolysis of cakes at 400 °C resulted in a several-fold decrease in SO_2 in the pyrolysates. Pyrolysates obtained at temperatures of 450 and 500 °C showed no sulfur dioxide emissivity in most of the variants analyzed. Only pyrolysates obtained from rapeseed were found to emit SO_2 during their combustion. The values ranged from 103.67 (pyrolysates obtained at 400 °C and in 5 min) to 13.65 g kg^{-1} (pyrolysates obtained at 500 °C and in 15 min) (Figure 4). According to Wei et al. (2012), CO emissions during the combustion of firewood and crop residues were 47.8 and 52 g kg^{-1}, respectively. In contrast, the combustion of pine and corn pellets generated 4.4 and 17.9 g kg^{-1} CO, respectively [74]. Jelonek et al. (2021), in their study on the emission of carbon monoxide, total hydrocarbons and particulate matter during wood combustion in an operating stove, report CO emissions in the range of 795.2 to 1983.1 ppm [75]. In contrast, in a study by Vicente et al. (2021) on the emission of carbon monoxide, total hydrocarbons and particulate matter during wood combustion in a stove operating under distinct conditions, CO emissions did not exceed 73.9 g kg^{-1} for the materials analyzed, i.e., pine *(Pinus pinaster)* and beech *(Fagus sylvatica)* [76]. According to Martínez-Gómez et al. (2023), biochar from grape pomace (pyrolysis in 350 °C) heated to 900 °C generates 2155 μmol g^{-1} CO. In contrast, an increase in the pyrolysis temperature (biochar 700 °C) resulted in a decrease in emissions to a level of 1154 μmol g^{-1} [51]. According to Sher et al. (2020), NOx emissions depend on a number of factors including the temperature, combustion time, volatile matter and burner configuration. The authors report that NOx emissions were over 700 ppm when coal was burned. Beech wood generated significantly less NOx in comparison to coal. On the other hand, the smallest values in the range of 251–269 ppm were obtained for torrefied beech wood. The authors noted a similar trend for the emissions of SO_2, which amounted, respectively, to 593, 13.47 and 1.07–1.98 ppm for coal, beech wood and beech biochar. The authors correlate this relationship with the lower content of siraka in biomass that has undergone torrefaction [77]. Lasek et al. (2017) found similar results in their study; they indicate the emissions of SO_2 during the combustion of coal and torrefied willow at 1184 and 2 ppm, respectively [78]. Numerous authors indicate, in turn, that burning mixtures of "raw" fuels such as biomass or coal with torrefied biomass can significantly reduce CO_2, SOx and NOx emissions into the atmosphere [79–82].

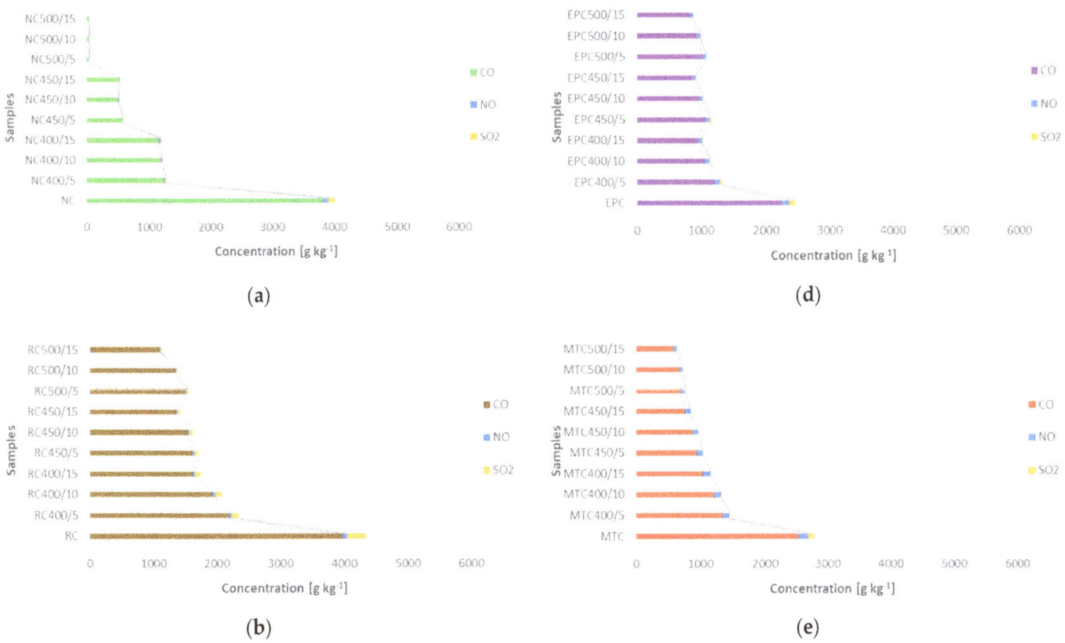

Figure 4. *Cont.*

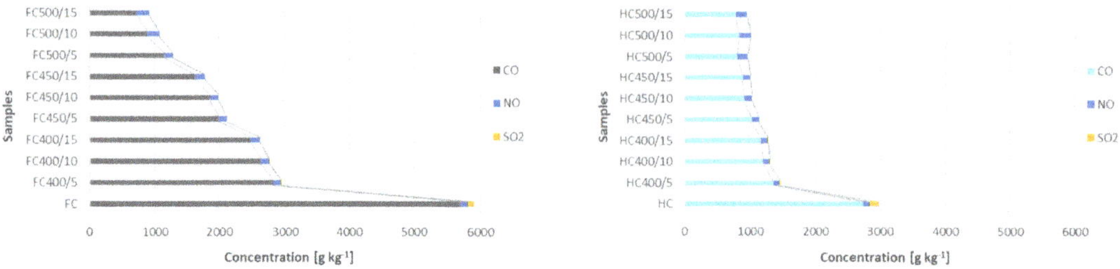

(c) (f)

Figure 4. Comparison of emission parameters of tested cakes and biochars: (**a**) fennel flower (*Nigella sativa* L.), (**b**) rapeseed (*Brassica napus* L. var. Napus), (**c**) flax (*Linum usitatissimum* L.), (**d**) evening primrose (*Oenothera biennis* L.), (**e**) milk thistle (*Silybum marianum* L. *Gaertn.*), (**f**) hemp (*Cannabis sativa* L.).

The obtained results allowed for the comparison of the energy potential of a wide group of production residues from the edible oil production sector. Additionally, the parameters of the pyrolysis process of waste biomass generated from six different types of oilseeds were optimized. The results achieved complement the existing knowledge on the impact of the pyrolysis process on gas emissions during the combustion of a wide group of biochars generated from oil cakes.

5. Conclusions

In recent years, increasing emphasis has been placed on research related to the use of biochar as a substitute for fossil fuels. Biochar is treated as a carbon-neutral fuel. When it is burned, 15× less sulfur and 10× less chlorine are released into the atmosphere compared to coal, while maintaining similar heating values. Biochar is undoubtedly a material with high potential for application in the fight against climate change. However, it should be noted that the use of biochar on a global scale involves huge costs in terms of technology, the search for new substrates and the adaptation of production facilities to modern solutions. The obtained results indicate directions for the management of biomass residues from the production of edible oils together with the determination of processing parameters for biocarbon fuels. Appropriate use of the knowledge obtained could allow for the development of the food production and energy production sectors along with technological development and employment growth.

Promoting closed-loop production and the use of a variety of post-production residues seems to be extremely important in times of energy crisis and large-scale energy transition in many countries. Given the above, it seems important to deepen knowledge and conduct further research in the field of the management of waste biomass being a residue of oil production from a variety of oilseeds. Due to possible differences in the composition of the residue of oilseeds depending on its origin, the conducted research requires further replication. Such material heterogeneity could have a significant, uncontrolled impact on the energy parameters of the obtained biocarbon fuels. Additionally, in the next stages, an analysis of the durability of the produced biochar fuels and safety tests during storage and processing of biomass should be carried out. Future research directions should include assessment of dust explosion hazard and assessment of biomass material stability during storage. The possibility of producing biochars from oilseed cakes in conditions close to real ones, i.e., on an industrial scale, should also be assessed. Taking into account the development of energy policy, it would be necessary to develop a detailed scheme of procedure for the management of post-processing residues from the edible oil production sector for use as energy carriers.

The conducted tests allowed us to conclude that the use of the pyrolysis process increased the calorific value of the analyzed materials, and the highest calorific value was characterized by biochar produced in the pyrolysis process conducted at a temperature of 500 °C and a holding time of 15 min. This translated into an increase against the "raw" material of 14.4; 14.5; 19.9; 26.7; 40.8 and 27.3%, respectively, for the cakes of fennel flower, rapeseed, flax, evening primrose, milk thistle and hemp. Among all the prepared pyrolysates, the highest calorific value was the biochar based on milk thistle seed residue, i.e., 26.59 MJ kg^{-1}. It is worth noting that the use of the pyrolysis process of cakes resulted in a several-fold decrease in CO and NOx emissions during their combustion. In the case of sulfur dioxide, there was a total reduction in emissions for biochar produced at higher temperatures.

Based on the analysis performed, it can be concluded that the production of biochar from cakes made from the pressing of various oilseeds offers the prospect of producing alternative biofuels, which can indirectly have a positive impact on the natural environment and the fight against progressive climate change. The use of residues from oil production to make biochar can also be crucial in sustainable development and production and fit in with closed-loop ideas. However, it should be noted that the production of biochar from oilseed residues depends on many factors. Maintaining the quality of the feedstock, reducing costs and energy inputs, and producing high-quality biofuel involves, among other things, properly storing oilseed waste, developing appropriate storage techniques, ensuring optimal logistics processes or optimizing the pyrolysis process on a large scale. The establishment of appropriate production practices, standards and procedures based on modern technologies appears to be essential to realize the energy potential of alternative biochar fuels based on seed residues from cooking oil production.

Author Contributions: Conceptualization, B.S., R.C. and M.M.; methodology, B.S. and A.S.; validation, B.S. and A.S.; formal analysis, B.S., G.Z. and M.B.; investigation, B.S. and A.S.; data curation, B.S., A.S. and M.B.; writing—original draft preparation, B.S., R.C. and M.M.; writing—review and editing, B.S.; visualization, B.S.; supervision, B.S.; project administration, B.S. and C.P.; All authors have read and agreed to the published version of the manuscript.

Funding: This research received no external funding.

Data Availability Statement: Data are contained within the article.

Conflicts of Interest: The authors declare no conflicts of interest.

References

1. Rokicki, T.; Michalski, K.; Ratajczak, M.; Szczepaniuk, H.; Golonko, M. Use of renewable energy sources in European Union countries. Szkoła Główna Gospodarstwa Wiejskiego. *Rocz. Ochr. Sr.* **2018**, *20*, 1318–1334.
2. Chen, S.T.; Kuo, H.I.; Chen, C.C. Modelling the Relationship between the Oil Price and Global Food Prices. *Appl. Energy* **2010**, *87*, 2517–2525. [CrossRef]
3. Perea-Moreno, M.A.; Samerón-Manzano, E.; Perea-Moreno, A.J. Biomass as Renewable Energy: Worldwide Research Trends. *Sustainability* **2019**, *11*, 863. [CrossRef]
4. Kilkis, S.; Krajacic, G.; Duic, N.; Rosen, M.A.; Al-Nimr, M.A. Advancements in sustainable development of energy, water and environment systems. *Energy Convers. Manag.* **2018**, *176*, 164–183. [CrossRef]
5. Brewer, C.E.; Chuang, V.J.; Masiello, C.A.; Gonnermann, H.; Gao, X.; Dugan, B. New approaches to measuring biochar density and porosity. *Biomass Bioenergy* **2014**, *66*, 176–185. [CrossRef]
6. Al-Wabel, M.I.; Al-Omran, A.; El-Naggar, A.H.; Nadeem, M.; Usman, A.R.A. Pyrolysis temperature induced changes in characteristics and chemical composition of biochar produced from cono-carpus wastes. *Bioresour. Technol.* **2013**, *131*, 374–379. [CrossRef] [PubMed]
7. Bochniak, M. Quality of Two-Stage Pressed Rapeseed Oil after Low-Temperature Bleaching Process. Ph.D. Thesis, Uniwersytet Ekonomiczny We Wrocławiu, Wrocław, Poland, 2020. [CrossRef]
8. Grela, R.E.; Zaworska-Zakrzewska, A.; Milewski, S. Nutritional value and feed suitability of by-products from organic farms in animal nutrition. *Wiadomości Zootech.* **2023**, *R. LXI*, 38–48.
9. Babicz, M.; Kropiwiec-Domańska, K.; Szymanowska, U. *Selected Issues in Respect to the Production of Raw Materials, Food and Cosmetics*; Wydawnictwo Uniwersytetu Przyrodniczego W Lublinie: Lublin, Poland, 2023; ISBN 978-83-7259-342-9. [CrossRef]
10. Kita, A.; Górnicka, E.; Wojdyło, A.; Drożdż, W. Effect of the addition of rapeseed pomace on the properties of fried potato chips. *Rośliny Oleiste Oilseed Crops* **2012**, *33*, 286–288.

11. Karatas, H.; Olgun, H.; Akgun, F. Experimental results of gasification of waste tire with air in a bubbling fluidized bed gasifier. *Fuel Process. Technol.* **2012**, *102*, 166–174. [CrossRef]
12. Wilk, M.; Magdziarz, A.; Gajek, M.; Zajemska, M.; Jayaraman, K.; Gokalp, I. Combustion and kinetics parameters estimation of torrefied pine, acacia and Miscanthus giganteus using experimental and modeling techniques. *Bioresour. Technol.* **2017**, *243*, 304–314. [CrossRef]
13. Erdei, B.; Frankó, B.; Galbe, M.; Zacchi, G. Separate hydrolysis and co fermentation for improved xylose utilization in integrated ethanol production from wheat meal and wheat straw. *Biotechnol. Biofuels* **2012**, *5*, 12. [CrossRef]
14. Beims, R.F.; Hu, Y.; Shui, H.; Xu, C. Hydrothermal liquefaction of biomass to fuels and value-added chemicals: Product applications and challenges to develop largescale operations. *Biomass Bioenergy* **2020**, *135*, 105510. [CrossRef]
15. Yang, F.; Li, G.; Shi, H.; Wang, Y. Effects of phosphogypsum and superphosphate on compost maturity and gaseous emis-sions during kitchen waste composting. *Waste Manag.* **2015**, *36*, 70–76. [CrossRef] [PubMed]
16. Biller, P.; Madsen, R.B.; Klemmer, M.; Becker, J.; Iversen, B.B.; Glasius, M. Effect of hydrothermal liquefaction aqueous phase recycling on bio-crude yields and composition. *Bioresour. Technol.* **2016**, *220*, 190–199. [CrossRef] [PubMed]
17. Kobyłecki, R.; Bis, Z. Autothermal thermolysis as an efficient technology to produce clean and high energy fuels. *Archiwum Spalania* **2006**, *6*, 114–119.
18. Saletnik, B.; Zaguła, G.; Bajcar, M.; Tarapatskyy, M.; Bobula, G.; Puchalski, C. Biochar as a Multifunctional Component of the Environment—A Review. *Appl. Sci.* **2019**, *9*, 1139. [CrossRef]
19. Saletnik, B.; Zaguła, G.; Saletnik, A.; Bajcar, M.; Puchalski, C. Biochar and Ash Fertilization Alter the Chemical Properties of Basket Willow (*Salix viminalis* L.) and Giant Miscanthus (*Miscanthus* x *giganteus*). *Agronomy* **2020**, *10*, 660. [CrossRef]
20. Retajczyk, M.; Wróblewska, A. Biomass pyrolysis as an energy source. *Wiadomości Chemiczne* **2018**, *72*, 127–146.
21. Liu, Q.; Chmely, S.C.; Abdoulmoumine, N. Biomass Treatment Strategies for Thermochemical Conversion. *Energy Fuels* **2017**, *31*, 3525–3536. [CrossRef]
22. Jahirul, M.I.; Rasul, M.G.; Chowdhury, A.A.; Ashwath, N. Biofuels production through biomass pyrolysis—A technological review. *Energies* **2012**, *5*, 4952–5001. [CrossRef]
23. Chhiti, Y.; Kemiha, M. Thermal Conversion of Biomass, Pyrolysis and Gasification: A Review. *Int. J. Eng Sci.* **2013**, *2*, 75–85.
24. Varma, A.; Shankar, R.; Mondal, P. A Review on Pyrolysis of Biomass and the Impacts of Operating Conditions on Product Yield, Quality, and Upgradation. *Recent Adv. Biofuels Bioenergy Util.* **2018**, 227–259.
25. Rhodes, A.H.; Carlin, N.A.; Semple, O.R. Impact of black carbon in the extraction and mineralization of phenanthrene in soil. *Environ. Sci. Technol.* **2008**, *42*, 740–745. [CrossRef] [PubMed]
26. Lewandowski, W.M.; Radziemska, E.; Ryms, M.; Ostrowski, P. Modern methods of thermochemical biomass conversion into gas, liquid and solid fuels. *Proc. ECOpole* **2010**, *4*, 2.
27. Bis, Z. Biowęgiel–powrót do przeszłości, szansa dla przyszłości. *Czysta Energia* **2012**, *6*.
28. Matovic, D. Biochar as a viable carbon sequestration option: Global and Canadian perspective. *Energy* **2011**, *36*, 2011–2016. [CrossRef]
29. Bednik, M.; Medyńska–Juraszek, A.; Dudek, M.; Kloc, S.; Kręt, A.; Łabaz, B.; Waroszewski, J. Wheat Straw Biochar and NPK Fertilization Efficiency in Sandy Soil Reclamation. *Agronomy* **2020**, *10*, 496. [CrossRef]
30. Malińska, K. Prawne i jakościowe aspekty dotyczące wymagań dla biowęgla. *Inżynieria I Ochr. Sr.* **2015**, *18*, 359–371.
31. Pereira, R.C.; Muetzel, S.; Arbestain, M.C.; Bishop, P.; Hina, K.; Hedley, M. Assessment of the influence of biochar on rumen and silage fermentation: A laboratory-scale experiment. *Anim. Feed Sci. Technol.* **2014**, *196*, 22–31. [CrossRef]
32. Belluati, M.; Tabasso, S.; Gaudino, E.C.; Cravotto, G.; Manzoli, M. Biomass-derived carbon-based catalysts for lignocellulosic biomass and waste valorisation: A circular approach. *Green Chem.* **2024**, *26*, 8642–8668. [CrossRef]
33. Garcia, B.; Alves, O.; Rijo, B.; Lourinho, G.; Nobre, C. Biochar: Production, Applications, and Market Prospects in Portugal. *Environments* **2022**, *9*, 95. [CrossRef]
34. Tang, J.; Zhy, W.; Kookana, R.; Katayama, A. Characteristics of biochar and its application in remediation of contaminated soil. *J. Biosci. Bioeng.* **2013**, *116*, 653–659. [CrossRef] [PubMed]
35. Mohan, D.; Sarswat, A.; Ok, S.Y.; Pittman, C.U., Jr. Organic and inorganic contaminants removal from water with biochar, a renewable, low cost and sustainable adsorbent—A critical review. *Bioresour. Technol.* **2014**, *160*, 191–202. [CrossRef] [PubMed]
36. Steiner, C.; Melear, N.; Harris, K.; Das, K.C. Biochar as bulking agent for poultry litter composting. *Carbon Manag.* **2011**, *2*, 227–230. [CrossRef]
37. Malińska, K.; Zabochnicka–Świątek, M.; Dach, J. Effects of biochar amendment on ammonia emission during composting of sewage sludge. *Ecol. Eng.* **2014**, *71*, 474–478. [CrossRef]
38. Wu–Jun, L.; Hong, J.; Han–Qing, Y. Emerging applications of biochar-based materials for energy storage and conversion. *Energy Environ. Sci.* **2019**, *12*, 1751–1779.
39. Kwapinski, W.; Byrne, C.M.P.; Kryachko, E.; Wolfram, P.; Adley, C.; Leahy, J.J.; Novotny, E.H.; Hayes, M.H.B. Biochar from Biomass and Waste. *Waste Biomass Valorization* **2010**, *1*, 177–189. [CrossRef]
40. Sànchez, M.E.; Lindao, E.; Margaleff, D.; Martínez, O.; Morán, A. Pyrolysis of agricultural residues from rape and sunflower: Production and characterization of biofuels and biochar soil management. *J. Anal. Appl. Pyrolysis* **2019**, *85*, 142–144. [CrossRef]
41. Xu, X.; Cao, X.; Zhao, L. Comparison of rice husk-and dairy manure-derived biochars for simultaneously removing heavy metals from aqueous solutions: Role of mineral components in biochars. *Chemosphere* **2013**, *92*, 955–961. [CrossRef] [PubMed]

42. Yao, Y.; Gao, B.; Chen, H.; Jiang, L.; Inyang, M.; Zimmerman, A.R.; Cao, X.; Yang, L.; Xue, Y.; Li, H. Adsorption of sulfamethoxazole on biochar and its impact on reclaimed water irrigation. *J. Hazard. Mater* **2012**, *209*, 408–413. [CrossRef]
43. Ibarrola, R.; Shackely, S.; Hammond, J. Pyrolysis biochar systems for recovering biodegradable materials: A life cycle carbon assessment. *Waste Manag.* **2012**, *32*, 859–868. [CrossRef]
44. Song, W.; Guo, M. Quality variations of poultry litter biochar generated at different pyrolysis temperatures. *J. Anal. Appl. Pyrolysis* **2012**, *94*, 138–145. [CrossRef]
45. *PN-EN ISO 16948:2015-07*; Solid Biofuels—Determination of Total Carbon, Hydrogen and Nitrogen Content. ISO: Geneva, Switzerland, 2015.
46. *PN-EN ISO 18122:2023-05*; Solid Biofuels—Determination of Ash Content. ISO: Geneva, Switzerland, 2023.
47. *PN-EN ISO 18123:2023*; Solid Biofuels—Determination of Volatile Matter. ISO: Geneva, Switzerland, 2023.
48. *PN-EN ISO 18125:2017-07*; Solid Biofuels—Determination of Calorific Value. ISO: Geneva, Switzerland, 2017.
49. Ferens, W. Analysis of the change in the composition of solid products of biomass pyrolysis. *Zesz. Energetyczne* **2017**, *IV*, 63–71.
50. Hmid, A.; Mondelli, D.; Fiore, S.; Fanizzi, F.P.; Al Chami, Z.; Dumontet, S. Production and characterization of biochar from three-phase olive mill waste through slow pyrolysis. *Biomass Bioenergy* **2014**, *71*, 330–339. [CrossRef]
51. Martínez-Gómez, A.; Fe Andrés, M.; Barón-Sola, Á.; Díaz-Manzano, F.E.; Yousef, I.; Mena, I.F.; Díaz, E.; Gómez-Torres, Ó.; González-Coloma, A.; Hernández, L.E.; et al. Biochar from grape pomace, a waste of vitivinicultural origin, is effective for root-knot nematode control. *Biochar* **2023**, *3*, 30. [CrossRef]
52. Mazurek, K.; Drużyński, S.; Kiełkowska, U.; Wróbel-Kaszanek, A.; Igliński, B.; Cichosz, M. The Application of Pyrolysis Biochar Obtained from Waste Rapeseed Cake to Remove Copper from Industrial Wastewater: An Overview. *Energies* **2024**, *17*, 498. [CrossRef]
53. Sładeczek, F.; Głodek-Bucyk, E. Study on the use of low-temperature pyrolysis to convert waste biomass into biochar in a test plant. *Pr. Inst. Ceram. I Mater. Bud.* **2017**, *10*, 50–61.
54. Li, Y.; Grupta, R.; Zhang, Q.; You, S. Review of biochar production via crop residue pyrolysis: Development and perspectives. *Bioresour. Technol.* **2023**, *369*, 128423. [CrossRef]
55. Crombie, K.; Mašek, O.; Sohi, S.P.; Brownsort, P.; Cross, A. The effect of pyrolysis conditions on biochar stability as determined by three methods. *Gcb Bioenergy* **2013**, *5*, 122–131. [CrossRef]
56. Yao, Y.; Gao, B.; Zhang, M.; Inyang, M.; Zimmerman, A.R. Effect of biochar amendment on sorption and leaching of nitrate, ammonium, and phosphate in a sandy soil. *Chemosphere* **2012**, *89*, 1467–1471. [CrossRef] [PubMed]
57. Ferreira, A.F.; Ribau, J.P.; Costa, M. A decision support method for biochars characterization from carbonization of grape pomace. *Biomass Bioenergy* **2021**, *145*, 105946. [CrossRef]
58. Angın, D.; Köse, T.E.; Selengil, U. Production and characterization of activated carbon prepared from safflower seed cake biochar and its ability to absorb reactive dyestuff. *Appl. Surf. Sci.* **2013**, *280*, 705–710. [CrossRef]
59. Sun, Y.; Li, C.; Li, Q.; Zhang, S.; Hu, X. Pyrolysis of flaxseed residue: Exploration of characteristics of the biochar and bio-oil products. *J. Energy Inst.* **2021**, *97*, 1–12. [CrossRef]
60. Mukhambet, Y.; Shah, D.; Tatkeyeva, G.; Sarbassov, Y. Slow pyrolysis of flax straw biomass produced in Kazakhstan: Characterization of enhanced tar and high-quality biochar. *Fuel* **2022**, *324*, 124676. [CrossRef]
61. Tag, A.T.; Duman, G.; Ucar, S.; Yanik, J. Effects of feedstock type and pyrolysis temperature on potential applications of biochar. *J. Anal. Appl. Pyrolisis* **2016**, *120*, 200–206. [CrossRef]
62. Mulimani, H.V.; Navindgi, M.C. Production of solid fuel biochar from de-oiled seed cake by pyrolysis. *J. Mech. Civ. Eng.* **2017**, *14*, 57–61. [CrossRef]
63. Kazimierski, P.; Januszewicz, K.; Godlewski, W.; Fijuk, A.; Suchocki, T.; Chaja, P.; Barczak, B.; Kardaś, D. The Course and the Effects of Agricultural Biomass Pyrolysis in the Production of High-Calorific Biochar. *Materials* **2022**, *15*, 1038. [CrossRef]
64. Wu, W.; Yang, M.; Feng, Q.; McGrouther, K.; Wang, H.; Lu, H.; Chen, Y. Chemical characterization of rice straw-derived biochar for soil amendment. *Biomass Bioenergy* **2012**, *47*, 268–276. [CrossRef]
65. Ozcimen, D.; Karaosmanoglu, F. Production and characterization of bio-oil and biochar from rapeseed cake. *Renew. Energy* **2004**, *29*, 779–787. [CrossRef]
66. Durak, H.; Genel, S. Characterization of bio-oil and bio-char obtained from black cumin seed by hydrothermal liquefaction: Investigation of potential as an energy source. *Energy Sources Part A Recovery Util. Environ. Eff.* **2020**, *45*, 3205–3215. [CrossRef]
67. Prade, T.; Svensson, S.E.; Mattsson, J.E. Energy balances for biogas and solid biofuel production from industrial hemp. *Biomass Bioenergy* **2012**, *40*, 36–52. [CrossRef]
68. Prade, T.; Finell, M.; Svensson, S.E.; Mattsson, J.E. Effect of harvest date on combustion related fuel properties ofindustrial hemp (*Cannabis sativa* L.). *Fuel* **2012**, *102*, 592–604. [CrossRef]
69. Parvez, A.M.; Lewis, J.D.; Afzal, M.T. Potential of industrial hemp (*Cannabis sativa* L.) for bio-energy production in Canada: Status, challenges and outlook. *Renew. Sustain. Energy Rev.* **2021**, *141*, 110784. [CrossRef]
70. Zouari, M.; Devallance, D.B.; Marrot, L. Effect of Biochar Addition on Mechanical Properties, Thermal Stability, and Water Resistance of Hemp-Polylactic Acid (PLA) Composites. *Materials* **2022**, *15*, 2271. [CrossRef] [PubMed]
71. Marrot, L.; Candelier, K.; Valette, J.; Lanvin, C.; Horvat, B.; Legan, L.; David, B.; DeVallance, D.B. Valorization of hemp stalkwaste through thermochemical conversion for energy and electrical applications. *Waste Biomass Valorization* **2022**, *13*, 2267–2285. [CrossRef]

The exploitation of food wastes was for a long time as feed additives for animals [5]. Recently, other applications arose. For instance, biorefineries started exploiting wastes to produce natural renewable gas [6], and textile sector relied more on natural dyes, such as pomegranate seeds, for tissue staining [7].

Castanea sativa, known in Europe as sweet chestnut, is economically valuable for its nutrient-rich fruit, the chestnuts. Their abundance in carbohydrates, fibers, starch, fatty acids, minerals (potassium, phosphorous, calcium, and magnesium), and vitamins (B9, C, and E) [8–12] valorizes them in a Mediterranean diet [8,13] and for treating illnesses like diarrhea and cough [12].

The nuts are encased in a hard, spiky shell, and a typical glossy brown nut with a pointed tip is found within the husk. When cooked, chestnuts have a starchy and unique sweet flavor, and lower oil content compared to other nuts [14]. Interest in chestnut production is currently rising, especially in Spain, France, Brazil, the USA, and Switzerland, holding the lead in chestnut exportation rate (77.6%) [15]. Despite the relevance of *C. sativa* as a fruit crop, its bark in particular is used to extract tannins (vescalgin, castalin, and vescalin) and it accounts for 60% of the total amount of tannic substances [12].

Several studies have shown that chestnuts are rich in tannins in the bark and leaves; furthermore, chestnut byproducts (like the burr) represent a source of biomolecules with antioxidant, anticarcinogenic, and cardioprotective activity [11].

Burrs must be removed from the brush woods to ensure optimal growth of the tree (usually by burning them), causing damage in terms of time and financial resources. This practice is not only dangerous, but it also contributes to environmental pollution by releasing carbon dioxide.

From this perspective, the interest in a more sustainable chestnut exploitation process rose. Previous research already exploited chestnut shells to withhold pollutants from the environment, to produce energy through carbon combustion, and to generate active carbons as adsorbent molecules [16].

Another current application of agri-food wastes, mainly crop residues, sees them a source of various bioactive molecules. For instance, coffee husks are a fount of phenolic compounds, notoriously displaying antioxidant activity [1]. In the chestnut burrs of *C. sativa* particularly abundant is ellagic acid with antioxidant properties, as well as an antimicrobial effect against *Helicobacter pylori* [17]. *Enterococcus faecalis* is a Gram-positive bacterium able to cause hospital-acquired infections (HAI), urinary tract infections (UTIs), bacteremia, valve endocarditis, and wound infections [18,19]. Currently, endocarditis, bacteremia, and UTI treatments are represented by combination therapy, consisting of the coadministration of cell wall-active agents, such as β-lactams, and aminoglycosides [19]. However, *E. faecalis* presents with a decreased susceptibility to penicillin and ampicillin [20], and intrinsic low-level resistance to aminoglycosides, leading to worrying and growing antimicrobial resistance levels [21].

In this setting, an innovative in vitro and in silico pipeline was developed to investigate the antimicrobial activity of a previously characterized chestnut burr biomass extract [22].

In vitro MIC assays showed significant antimicrobial activity against Gram-positive *E. faecalis* but not against other tested strains (*Staphylococcus aureus*, *Escherichia coli*, and *Pseudomonas aeruginosa*).

In silico analysis identified potential bioactive compounds and *E. faecalis* targets, revealing pharmacodynamic features through molecular dynamic simulations.

Overall, this approach focuses on potentially innovative applications within the circular bioeconomy, utilizing natural waste as a source of antimicrobial bioactive compounds.

2. Materials and Methods

2.1. Preparation of C. sativa Burrs (CSB) Extract

C. sativa burrs extract was prepared as previously described [22]. The spiny burrs of Castanea *sativa* (Mill.), certified under the PGI (Protected Geographical Indication) Castagna del Monte Amiata (Reg. CEE n. 2081/92), were obtained from Tuscany, a notable chestnut-producing region in Italy. The burrs were meticulously cleaned and air-dried at room temperature until they reached a stable weight. Subsequently, they were ground into a fine powder and stored in sealed, dark plastic bags at −80 °C to be preserved until extraction. The extraction process involved immersing 330 g of the powdered burrs in 1 L of water and treating the mixture with 20 kHz ultrasonic waves for three hours at room temperature. The resulting aqueous extract, referred to as CSB, was then freeze-dried using a lyophilizer (Lyovapor L-200, Bhuchi, India) and kept at −32 °C for further experiments.

2.2. *In Vitro Antibacterial Susceptibility Testing*

Enterococcus faecalis ATCC 29212, *Staphylococcus aureus* ATCC 25392, *Escherichia coli* ATCC 25922, and *Pseudomonas aeruginosa* ATCC 27853 were obtained from the American Type Culture Collection (Manassas, VA, USA) and conserved at −80 °C in Brain–Heart Infusion (BHI) medium supplemented with 30% glycerol. All bacteria were routinely grown aerobically on Mueller–Hinton agar plates incubated at 37 °C for 18–24 h. The extracts were resuspended in DMSO at a final concentration of 50 mg/mL and subsequently diluted in the culture medium. Minimum inhibitory concentrations (MICs) of the extracts were determined in triplicate in Mueller–Hinton medium using the microdilution broth method and a bacterial inoculum of 5×10^4 CFU/well, as recommended by the Clinical Laboratory Standards Institute (CLSI). Briefly, serial twofold dilutions of the extract were prepared extemporaneously in a 96-well plate (final volume in well, 90 µL), prior to the addition (10 µL) of the bacterial suspension containing 5×10^6 CFU/mL. Final DMSO concentration in the well did not exceed 1%. Growth controls were carried out in parallel using up to 1% DMSO in the well (no growth inhibition could be observed with any of the tested organism). MICs were recorded after 18 h of incubation at 35 ± 2 °C. Vancomycin and colistin (tested at concentrations ranging 32-0.015 µg/mL) were used as antibiotic controls for *E. faecalis* and *S. aureus*, or *E. coli* and *P. aeruginosa*, respectively.

2.3. *In Silico Methods*

2.3.1. Sequence Resources and Multiple Sequence Alignment

The *E. faecalis* ATCC 29212 strain protein primary structures (197 sequences) were downloaded from the NCBI genome database (NCBI RefSeq assembly: GCF_000742975.1) in FASTA format, while the *P. aeruginosa* (110.903 sequences), *S. aureus* (108.451 sequences), and *E. coli* (543.617 sequences) protein primary structures were retrieved and extracted from the entire "Non-redundant protein sequences (nr)" database, which contains nearly all protein sequences available at NCBI [23]. The Blastp tool performed a multiple sequence alignment (MSA) between the *E. faecalis* sequences and the *P. aerugimosa*, *S. aureus*, and *E. coli* sequences (collected in a unique FASTA file containing 762.971 sequences) using BLOSUM 62 matrix, a word size of 6, E-value of 0.01, adding the flag -sorthits 3 and -sorthsps 0 to sort the MSA output for identity and E-value, respectively; all other parameters were set by default [23]. MSA results provided an output of 150.305.287 sequence alignments; thus, a python script written in-house extracted all *E. faecalis* sequences (38 sequences), which did not show any identity/similarity with the sequences of the other bacteria strains (with a similarity/identity value less than 20%). The 38 sequences of *E. faecalis* were analyzed in the UniProt database [24] and all sequences annotated as "obsolete" were discarded. From the UniProt searches, only one sequence was obtained: "endocarditis and biofilm-associated pilus tip protein EbpA" (NCBI Reference Sequence: WP_148304834.1).

2.3.2. Molecular Modeling and Structural Optimization

The von Willebrand factor type A (vWA) structural domain 3D structure of the *Eca* endocarditis and biofilm-associated pilus tip protein EbpA was generated by AlpfaFold [25] using the NCBI Reference Sequence WP_148304834.1 as a target. AlphaFold predicted 5 different 3D structures using different weights and ranked them from best to worst by their mean pLDDT; the uncertainty metric ranges between 0 and 100, where 0 means high certainty and 100 means high uncertainty (lower is better). We selected the first model as the best 3D structure generated by AlphaFold, showing an uncertainty value of 2.81% with a mean pLDDT of 85.75, suggesting high confidence of the whole 3D structure. The Mg^{2+} ion was added to the structure, through structural superimposition, using as template the "Crystal structure of pilus adhesin, SpaC from Lactobacillus rhamnosus GG—open conformation" of *Lacticaseibacillus rhamnosus* (PDB code: 6M3Y) [26].

2.3.3. In Silico Docking and Molecular Dynamics Simulations

The 3D structure was optimized using PyMOD3.0 [27] with MODELLER 10.5 and validated through PROCHECK [28] to prevent errors during the molecular dynamic (MD) simulations. The parameters of the force field were assigned by CHARMM-GUI platforms [29] and GROMACS 2019.3 [30] carried out the MD run as suggested in a previous works [31,32], where the structure was enclosed, in a triclinic box filled with TIP3P water molecules and counter ions to neutralize the net charge of the system. The system was minimized 5.000 steps using steepest descent algorithm to a minimum energy with forces less than 100 kJ/mol/nm [33]. A cMD simulation was integrated each 2 fs, while a V-rescale thermostat and Nose–Hoover barostat led the temperature and pressure at 300 K and 1 atm, respectively. MD run was of 150 ns.

The "gmx cluster" function extracted the frame most representative of the MD run and it was used as target for the docking simulation of compounds extracted by the chestnut burrs. The 3D structures of gallic acid, quinic acid, protocatechuic acid, brevifolin acid, and ellagic acid were retrieved from the PubChem database [34] (with compound CID: 370, 6508, 72, 9838995, and 5281855, respectively) and downloaded in sdf format. A box with dimensions of $20 \times 20 \times 20$ Å was created around the sensing region of the target using DockingPie2.0 [35] implemented in PyMOL 3.0 using Autodock/Vina as a docking tool [36,37]. Docking simulation was set with exhaustiveness of 32, selecting only the binding poses with RMSD lower of 2 Å with a binding free energy value of at least −7 Kcal/mol. All other parameters were used as default. P.L.I.P. Tool [38,39] provided the interaction network. The alignment of sequences were performed with ClustalW [40]. The best complexes were selected and subjected to cMD run, with the same protocols previously described.

All MD analyses were explored with GROMACS 2019.3 packages. GRACE generated the MD graphs and PyMOL 3.0 was used as a molecular graphic interface and produced the biological system pictures. All simulations and analyses were computed using a workstation with CPU AMD Ryzen Threadripper PRO 5965WX 24 Core (4.5 GHz, 140 MB CACHE), DDR4 Kingston 3200 MHz 128 GB RAM, Nvidia GeForce RTX4090 24 GB graphic card, and SSD NVME M.2 of 2 TB + 2x HDD S-ATA3 of 14 TB as storage partition with OS Linux Mint 21.1.

3. Results

3.1. Extract Susceptibility in Bacterial Strains

Results from the MIC assay are reported in Table 1. The MIC value measured with *E. faecalis* ATCC 29212 was 64 µg/mL, thereby proving an antimicrobial activity, although modest, of the chestnut extract against this strain. No such activity was confirmed for the other strains. Indeed, in *S. aureus* ATCC 25392, the MIC value was 512 µg/mL. In *E. coli* strain $CCUG^T$ and *P. aeruginosa* ATCC 27853, the observed MIC values were >512 µg/mL. The MIC values of such control antibiotics are reported in Table 1.

Table 1. Determination of MIC values of the chestnut burrs extract (CBE) on *E. faecalis*, *S. aureus*, *E. coli*, and *P. aeruginosa*. The MIC values of the reference antibiotics (vancomycin or colistin, showing the expected activity on either Gram-positive or Gram-negative organisms, respectively) are also indicated.

Organism	MIC (μg/mL)		
	CBE	Vancomycin	Colistin
E. faecalis	64	2	>32
S. aureus	512	1	>32
E. coli	>512	>32	0.25
P. aeruginosa	>512	>32	0.5

3.2. Multiple Sequence Alignments

In silico results showed a susceptibility of the *E. faecalis* strain against the *C. sativa* extract (gallic acid, quinic acid, protocatechuic acid, brevifolin acid, and ellagic acid), differently, the other strains (*P. aeruginosa*, *S. aureus*, and *E. coli*) did not exhibit any vulnerability after the treatment with the extract.

Based on the evidence that our extract was active only on the *E. faecalis* strain, we detected the potential *E. faecalis* targets, involved in the interaction with our compounds, by performing an MSA of the entire *E. faecalis* protein pool against the entire "Non-redundant protein sequences (nr)" database of the *P. aeruginosa*, *S. aureus*, and *E. coli* strains to identify the *E. faecalis* non-homologue sequences aligned with the sequences of the other strains.

A Python script written in-house analyzed 150.305.287 alignment of sequences and provided 38 *E. faecalis* protein sequences that showed appropriate coverage and a similarity/identity value of less than 20% (non-homologous sequences). The sequences were explored on the UniProt database and all sequences that were annotated as "obsolete" were discarded; our analysis identified the endocarditis and biofilm-associated pilus tip protein EbpA (EbpA) as the only *E. faecalis* non-homologue sequence.

3.3. Molecular Modeling and Docking Simulations

AlphaFold generated and optimized the target 3D structure, while molecular modeling solved potential structural gaps and steric clashes. Gallic acid, quinic acid, protocatechuic acid, and brevifolin acid ellagic acid were tested against the target through in silico virtual screening. Two different strategies were applied to select the best complexes: (i) selecting only the compounds that showed a binding free energy (docking score) lower than −7 kcal/mol and (ii) the ability of the compound to trigger strong interactions with target *consensus* binding residues.

The docking results showed that the best complexes were EbpA/brevifolin and EbpA/ellagic acid providing a binding free energy of −7.8 kcal/mol and −7.3 kcal/mol, respectively. Interaction network analyses showed that the brevifolin and ellagic acid docked in a similar binding pose and formed a polar and hydrophobic interaction network within the target binding pocket (Figure 1A,B). Furthermore, docking results revealed the ability only of the ellagic acid to trigger H-bonds with target key residues, Asp-326 and Asp-429, both involved in the coordination of Mg^{2+} metal cofactor. Our results suggested the ability of the compounds to spontaneously bind on *E. faecalis* EbpA by forming a significant interaction network. The best two complexes were further explored in silico to define in detail the pharmacodynamic features of compounds against the target.

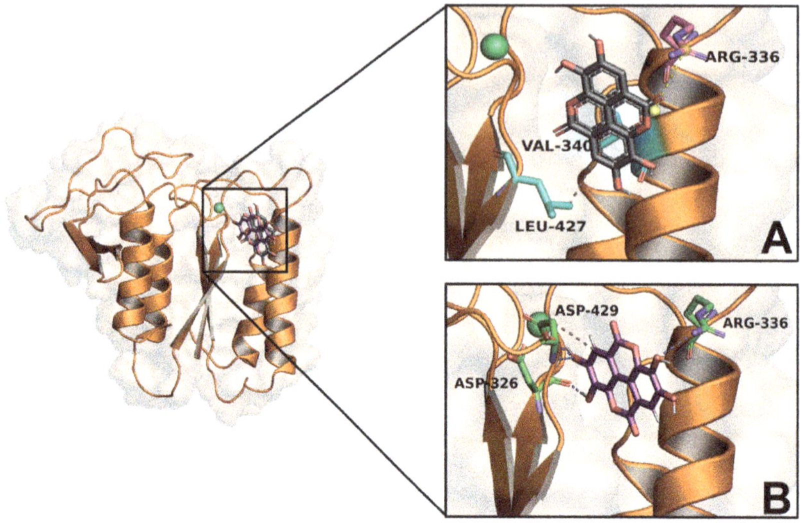

Figure 1. Overview of the docked pose of ellagic acid and brevifolin acid in complex with EbpA. EbpA3D structure is depicted in transparency orange cartoon/surface, while the magnesium ion was reported as a green sphere. Interaction network of (**A**) brevifolin (in gray) acid and (**B**) ellagic acid (in purple) in complex with EbpA after the docking simulation. The binding residues forming hydrophobic interactions (grey dashed lines), hydrogen bonds (blue dashed lines), and salt bridges (yellow dashed lines) are shown as cyan, green, and pink sticks, respectively.

3.4. cMD: EbpA/Ellagic Acid Complex (RMSD and Total Interaction Energy)

The target/compound interaction was dissected by analyzing structural and energy features performing cMDs for each complex and free state EbpA. To define the stability of the binding pose of compounds within the target binding pocket, the distance between the center of mass of the ligand and the center of mass of the binding pocket was evaluated. Ellagic acid showed a stable distance trend, suggesting high binding pose stability; surprisingly, brevifolin left the target binding pocket in the first frames of the MD run (Supplementary Materials Figure S1), likely due to the different interactions of the starting docking pose; thus, the EbpA/brevifolin complex was not considered for further MD analysis.

The target backbone structural integrity was assessed by evaluating the RMSD values for the EbpA/ellagic acid complex and free state EbpA. RMSD analyses suggested good structural stability, showing a stable RMSD trend of 0.15 Å and 0.2 Å for the complex and the free state, respectively. However, the EbpA/ellagic acid complex RMSD trend showed a higher RMSD value (moreover in the last frames of the MD run) than free state EbpA, exhibiting a significant increase of RMSD value until 0.25 Å. The binding pose of the compound within the target binding pocket did not significantly show fluctuations during the MD run, providing a very stable RMSD range between 0.03 Å and 0.05 Å, confirming the high stability and reliability of the selected starting docking pose (Figure 2). To strengthen the impact of the docking and cMD simulation, the target/compound interaction energy was evaluated. From energy analyses, we noticed that our compound was able to spontaneously bind on the target with a total interaction energy of -38.1 ± 3.8 kcal/mol (Figure S2).

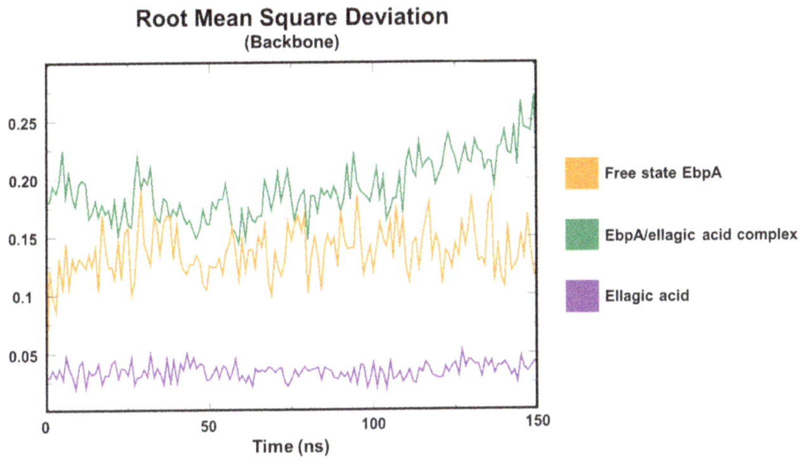

Figure 2. RMSD trends. RMSD profiles were evaluated for free state and EbpA/ellagic acid backbone along the entire MD run.

3.5. cMD: EbpA/Ellagic Acid Complex Interaction Network and Occupancy Analysis

To further explore the potential inhibitory activity of the ellagic acid in complex with the target, the dynamics interaction network was deeply explored during the MD run. The Prolif tool was used to detect both all target/compound interaction types and the occupancy of each interaction identified. Interestingly, interaction network analyses provided several polar and non-polar interactions (Figure 3A) and very high occupancy values. In detail, ellagic acid triggered van der Waals contacts with Arg-336 (50% of occupancy) and Leu-427 (81.25% of occupancy); hydrophobic interactions with Met-331 (87.5% of occupancy), Asp-339 (56.25% of occupancy), and Val-340 (100% of occupancy); and hydrogen bonds with Asp-326 (43.75% of occupancy) and Asp-429 (87.5% of occupancy) (Figure 3B).

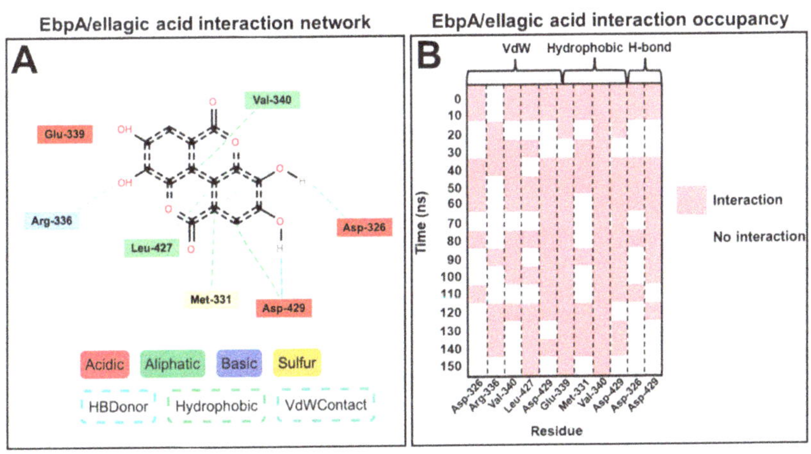

Figure 3. Interaction network and MD interaction network occupancy. (**A**) 2D interaction network among target binding residues and ellagic acid (in black). The involved target residues are labelled as acidic (in red), aliphatic (in green), basic (in blue), and sulfur (in yellow). The hydrogen bonds, hydrophobic interactions, and Wan der Waals contacts are represented as light blue, green, and cyan dashed lines, respectively. (**B**) EbpA/ellagic acid MD interaction network occupancy. The binding residues and the molecular dynamics run time are reported on the x- and y-axis, respectively. The target

residues are grouped on the interaction type (VdW, hydrophobic and H-bonds). The occupancy indicates the presence of each interaction over the simulation time; the pink color indicates the interaction presence; grey color means no interaction.

The role of the binding residues involved in the interaction with the ellagic acid was further explored by applying an evolutionary approach to define the conservation of residues along the protein evolution. The BLASTp tool performed MSA between the EbpA primary structure and the entire "non-redundant protein sequences (nr)" database excluding the *E. faecalis* organism and the Skylign tool analyzed the MSA results. Surprisingly, from the MSA analyses, we noted that Asp-326, Met-331, and Asp-429 showed a full conservation exhibiting a Shannon's entropy of 1 (Figure S3).

4. Discussion

Nowadays, climate change has been recognized as a major global challenge. Some strategies, such as the advancement of bioeconomy and sustainable forest management have proven to be crucial in tackling this global problem. The primary economic sector, driven by biological and "Protected Geographical Indication" products, has been acquiring a growing economic/social/environmental importance as shown by the latest trends in national agricultural metrics [2].

A plethora of natural sources, such as plants and microorganisms, stand as alternative sources of bioproducts with medicinal potential. Indeed, from these organisms, innovative compounds with biological activity can be isolated. For centuries, natural compounds have been harnessed in Chinese medicine as modulators of various signaling pathways like cell proliferation, angiogenesis, survival, and apoptosis, exhibiting different biological activities like antioxidant and anti-inflammatory properties.

Chestnut (*Castanea sativa*) cultivation has been deep sitting in Italy, particularly within its hilly and mountainous regions. As of 2017, Italy emerged as the leading producer of chestnuts in the European Union, contributing 52,356 tons, representing 38% of the total chestnut production in the EU (FAOSTAT, Food and Agriculture Organization of the United States). Nevertheless, this trend has been threatened in recent decades due to diseases such as chestnut blight and chestnut gall wasp, alongside the progressive depopulation of rocky areas.

In response to these issues, there has been a drive towards novel approaches to utilize various parts of the chestnut plant, such as the burrs, accounting for 10–15% of the total weight of chestnuts and currently employed in fuel and fermentable sugar production [3]. Burrs also exhibit multiple biological activities (anti-inflammatory, antioxidant, and antimicrobial) [11], hence the interest in investigating their potential as green sources of bioactive compounds.

Recently, the rising levels of resistant bacteria result in a threat for public health. This condition raised growing interest in antimicrobial resistance (AMR) and several approaches to investigate the antimicrobial activity of bio-compounds extracted from natural wastes have successfully been applied. Olive mill waste and artichoke bracts were both found to be successfully active against foodborne pathogens [41,42].

The current literature body already suggests antimicrobial activity from chestnut burrs, against *S. aureus*, *E. faecalis*, *Listeria monocytogenes* [43], and *E. coli* [44], confirmed by in vitro antimicrobial susceptibility assays. Here, however, an integrated in vitro and in silico workflow was adopted. After evaluating the antimicrobial activity of our chestnut burr extract [22,45] on different bacterial strains, a potential bacterial target of extract compounds (gallic acid, quinic acid, protocatechuic acid, brevifolin acid, and ellagic acid) was sought.

Firstly, an in vitro MIC assay was performed on both Gram-positive (*E. faecalis* and *S. aureus*) and Gram-negative (*E. coli* and *P. aeruginosa*) bacterial strains, yielding a measurable antimicrobial activity on *E. faecalis* (MIC = 64 μg/mL) only. Hence, the antimicrobial activity of the extract could be attributable to the interaction of the bioactive compound(s) with a specific target solely present in *E. faecalis* and not in all the other strains.

Multiple sequence alignment of *E. faecalis* proteins identified "endocarditis and biofilm-associated pilus tip protein EbpA" after removing homologous and obsolete sequences. The 3D structure of EbpA was created with AlphaFold. Subsequent computational screening revealed ellagic acid and brevifolin acid as potential antimicrobial compounds, with ellagic acid being the most abundant in the extract, but molecular dynamics simulations showed a stable binding only between ellagic acid binding and EbpA, while brevifolin rapidly dissociated. Network interaction analyses confirmed strong binding between ellagic acid and EbpA.

Asp-326 and Asp-429, involved in the coordination with the Mg^{2+} cofactor, were found as conserved residues by evolutionary analyses, further confirming docking and MD simulation. Furthermore, Met-331 binding residue was found located closely to the metal cofactor site, which is fully conserved. Despite the current absence of data on its functional/structural role in EbpA, our computational study implemented with the evolution approach would however propose it as a key residue for EbpA.

5. Conclusions

Our research was framed in a circular bioeconomy optics, where natural waste became a source of bioactive compounds targeting molecules involved in disease onset (bacterial infections, in this case). Through a combined in vitro and in silico approach, it was possible to discover novel bioactive material and their targets, simultaneously boosting the process's efficiency and cost-effectiveness.

In vitro and in silico data suggested an antimicrobial activity of *C. sativa* burrs extract against *E. faecalis*. Indeed, in vitro MIC assays showed bacterial susceptibility to the extract, with a MIC value of 64 ug/mL. Then, the unique bacterial target was pointed out, and the ligand–target interaction was investigated with docking and MD simulations. Those combined findings on the possible interaction between ellagic acid in the extract and the target protein EbpA of *E. faecalis* prepared the ground for further experimental validation with in vitro binding assays.

Overall, this initial step of target discovery proposes a green and sustainable scientific pipeline to investigate the biological activity of compounds extracted from agro-forest wastes. This would noticeably reduce time, costs, and chemical waste, thus addressing the detrimental impact of pollutants on the planet. Ultimately, if supported by further research, wastes would be employed as a cost-effective antimicrobial material for various applications (agro-food or chemical industry, healthcare, etc.).

Supplementary Materials: The following supporting information can be downloaded at: https://www.mdpi.com/article/10.3390/app14156552/s1, Figure S1: Minimum distance profiles; Figure S2: Interaction energy trends of ellagic acid bound within target binding pocket; Figure S3: LOGO representation of MSA of the EbpA primary structure against the "bacteria" database in BLASTp.

Author Contributions: Conceptualization, A.T.; methodology, A.T., R.B. and J.-D.D.; software, A.T.; validation, A.T., A.S. and J.-D.D.; formal analysis, A.T. and J.-D.D.; investigation, A.T., R.B., F.S. and J.-D.D.; resources, A.S.; data curation, A.T., J.-D.D. and A.S.; writing—original draft preparation, A.T., R.B., J.-D.D. and A.S.; writing—review and editing, A.T., R.B., M.G., L.F., J.-D.D. and A.S.; visualization, A.T., R.B., M.G., L.F., T.O., J.-D.D. and A.S.; supervision, A.S.; project administration, A.S.; funding acquisition, A.S. All authors have read and agreed to the published version of the manuscript.

Funding: This research received no external funding.

Institutional Review Board Statement: Not applicable.

Informed Consent Statement: Not applicable.

Data Availability Statement: The original contributions presented in the study are included in the article/supplementary material, further inquiries can be directed to the corresponding author/s.

Acknowledgments: The authors thank PRIN: PROGETTI DI RICERCA DI RILEVANTE INTERESSE NAZIONALE—Bando 2022 Prot. 2022LW54KC; PRIN: PROGETTI DI RICERCA DI RILEVANTE INTERESSE NAZIONALE—Bando 2022 PNRR Prot. P2022RYR5W; CHEBAPACK Programma di Sviluppo Rurale 2014–2020 Regione Toscana; PSR 2014–2020 Regione Toscana OPENRICCIO; F-Cur funds to M.G.; UE—FSE REACT-EU, PON Ricerca e Innovazione 2014–2020; Progetto ERICA ARTES 4.0, E87G23000100001 Agenzia Coesione Territoriale.

Conflicts of Interest: The authors declare no conflicts of interest.

References

1. Bongaarts, J. Human population growth and the demographic transition. *Philos. Trans. R. Soc. Lond. B Biol. Sci.* **2009**, *364*, 2985–2990. [CrossRef]
2. Di Vaio, A.; Hasan, S.; Palladino, R.; Hassan, R. The transition towards circular economy and waste within accounting and accountability models: A systematic literature review and conceptual framework. *Environ. Dev. Sustain.* **2023**, *25*, 734. [CrossRef]
3. Venkatesh, G.; Se, V.G. Circular Bio-economy-Paradigm for the Future: Systematic Review of Scientific Journal Publications from 2015 to 2021. *Circ. Econ. Sustain.* **2022**, *2*, 231–279. [CrossRef]
4. Ghangrekar, M.M.; Das, S.; Das, S. Microbial Electrochemical Technologies for CO_2 Sequestration. In *Biomass, Biofuels, Biochemicals. Circular Bioeconomy—Current Status and Future Outlook*; Elsevier: Amsterdam, The Netherlands, 2021; pp. 413–443. [CrossRef]
5. Zhu, Y.; Luan, Y.; Zhao, Y.; Liu, J.; Duan, Z.; Ruan, R. Current Technologies and Uses for Fruit and Vegetable Wastes in a Sustainable System: A Review. *Foods* **2023**, *12*, 1949. [CrossRef] [PubMed]
6. Pham, T.P.T.; Kaushik, R.; Parshetti, G.K.; Mahmood, R.; Balasubramanian, R. Food waste-to-energy conversion technologies: Current status and future directions. *Waste Manag.* **2015**, *38*, 399–408. [CrossRef]
7. Popescu, V.; Blaga, A.C.; Pruneanu, M.; Cristian, I.N.; Pîslaru, M.; Popescu, A.; Rotaru, V.; Crețescu, I.; Cașcaval, D. Green Chemistry in the Extraction of Natural Dyes from Colored Food Waste, for Dyeing Protein Textile Materials. *Polymers* **2021**, *13*, 3867. [CrossRef]
8. Silva, V.; Falco, V.; Dias, M.I.; Barros, L.; Silva, A.; Capita, R.; Alonso-Calleja, C.; Amaral, J.S.; Igrejas, G.; Ferreira, I.C.F.R.; et al. Evaluation of the Phenolic Profile of Castanea sativa Mill. By-Products and Their Antioxidant and Antimicrobial Activity against Multiresistant Bacteria. *Antioxidants* **2020**, *9*, 87. [CrossRef]
9. Corregidor, V.; Antonio, A.L.; Alves, L.C.; Cabo Verde, S. Castanea sativa shells and fruits: Compositional analysis by proton induced X-ray emission. *Nucl. Instrum. Methods Phys. Res. Sect. B Beam Interact. Mater. Atoms* **2020**, *477*, 98–103. [CrossRef]
10. Neves, J.M.; Matos, C.; Moutinho, C.; Queiroz, G.; Gomes, L.R. Ethnopharmacological notes about ancient uses of medicinal plants in Trás-os-Montes (northern of Portugal). *J. Ethnopharmacol.* **2009**, *124*, 270–283. [CrossRef] [PubMed]
11. Comandini, P.; Lerma-García, M.J.; Simó-Alfonso, E.F.; Toschi, T.G. Tannin analysis of chestnut bark samples (*Castanea sativa* Mill.) by HPLC-DAD–MS. *Food Chem.* **2014**, *157*, 290–295. [CrossRef] [PubMed]
12. de Vasconcelos, M.C.B.M.; Bennett, R.N.; Rosa, E.A.S.; Ferreira-Cardoso, J.V. Composition of European chestnut (Castanea sativa Mill.) and association with health effects: Fresh and processed products. *J. Sci. Food Agric.* **2010**, *90*, 1578–1589. [CrossRef] [PubMed]
13. Ozkan, G.; Günal-Köroğlu, D.; Capanoglu, E. Valorization of fruit and vegetable processing by-products/wastes. *Adv. Food Nutr. Res.* **2023**, *107*, 1–39. [CrossRef] [PubMed]
14. Santos, M.J.; Pinto, T.; Vilela, A. Sweet Chestnut (*Castanea sativa* Mill.) Nutritional and Phenolic Composition Interactions with Chestnut Flavor Physiology. *Foods* **2022**, *11*, 4052. [CrossRef] [PubMed]
15. Fernandes, P.; Colavolpe, M.B.; Serrazina, S.; Costa, R.L. European and American chestnuts: An overview of the main threats and control efforts. *Front. Plant Sci.* **2022**, *13*, 951844. [CrossRef] [PubMed]
16. Hu, W.; Zhang, X.; Chen, M.; Rahman, S.T.; Li, X.; Wang, G. Enhancing Cr (VI) Adsorption of Chestnut Shell Biochar through H3PO4 Activation and Nickel Doping. *Molecules* **2024**, *29*, 2220. [CrossRef] [PubMed]
17. Sangiovanni, E.; Piazza, S.; Vrhovsek, U.; Fumagalli, M.; Khalilpour, S.; Masuero, D.; Di Lorenzo, C.; Colombo, L.; Mattivi, F.; De Fabiani, E.; et al. A bio-guided approach for the development of a chestnut-based proanthocyanidin-enriched nutraceutical with potential anti-gastritis properties. *Pharmacol. Res.* **2018**, *134*, 145–155. [CrossRef] [PubMed]
18. Weiner, L.M.; Webb, A.K.; Limbago, B.; Dudeck, M.A.; Patel, J.; Kallen, A.J.; Edwards, J.R.; Sievert, D.M. Antimicrobial-Resistant Pathogens Associated with Healthcare-Associated Infections: Summary of Data Reported to the National Healthcare Safety Network at the Centers for Disease Control and Prevention, 2011–2014. *Infect. Control Hosp. Epidemiol.* **2016**, *37*, 1288–1301. [CrossRef] [PubMed]
19. Baddour, L.M.; Wilson, W.R.; Bayer, A.S.; Fowler, V.G.; Bolger, A.F.; Levison, M.E.; Ferrieri, P.; Gerber, M.A.; Tani, L.Y.; Gewitz, M.H.; et al. Infective endocarditis: Diagnosis, antimicrobial therapy, and management of complications: A statement for healthcare professionals from the Committee on Rheumatic Fever, Endocarditis, and Kawasaki Disease, Council on Cardiovascular Disease in the Young, and the Councils on Clinical Cardiology, Stroke, and Cardiovascular Surgery and Anesthesia, American Heart Association: Endorsed by the Infectious Diseases Society of America. *Circulation* **2005**, *111*, e394–e434. [CrossRef] [PubMed]
20. Murray, B.E. The life and times of the Enterococcus. *Clin. Microbiol. Rev.* **1990**, *3*, 46. [CrossRef]
21. Hollenbeck, B.L.; Rice, L.B. Intrinsic and acquired resistance mechanisms in enterococcus. *Virulence* **2012**, *3*, 421. [CrossRef]

22. Trezza, A.; Geminiani, M.; Cutrera, G.; Dreassi, E.; Frusciante, L.; Lamponi, S.; Spiga, O.; Santucci, A. A Drug Discovery Approach to a Reveal Novel Antioxidant Natural Source: The Case of Chestnut Burr Biomass. *Int. J. Mol. Sci.* **2024**, *25*, 2517. [CrossRef] [PubMed]
23. Johnson, M.; Zaretskaya, I.; Raytselis, Y.; Merezhuk, Y.; McGinnis, S.; Madden, T.L. NCBI BLAST: A better web interface. *Nucleic Acids Res.* **2008**, *36*, W5–W9. [CrossRef] [PubMed]
24. UniProt Consortium. UniProt: The Universal Protein Knowledgebase in 2023. *Nucleic Acids Res.* **2023**, *51*, D523–D531. [CrossRef] [PubMed]
25. Jumper, J.; Evans, R.; Pritzel, A.; Green, T.; Figurnov, M.; Ronneberger, O.; Tunyasuvunakool, K.; Bates, R.; Žídek, A.; Potapenko, A.; et al. Highly accurate protein structure prediction with AlphaFold. *Nature* **2021**, *596*, 583–589. [CrossRef] [PubMed]
26. Kant, A.; Palva, A.; von Ossowski, I.; Krishnan, V. Crystal structure of lactobacillar SpaC reveals an atypical five-domain pilus tip adhesin: Exposing its substrate-binding and assembly in SpaCBA pili. *J. Struct. Biol.* **2020**, *211*, 107571. [CrossRef] [PubMed]
27. Janson, G.; Paiardini, A. PyMod 3: A complete suite for structural bioinformatics in PyMOL. *Bioinformatics* **2021**, *37*, 1471–1472. [CrossRef]
28. Laskowski, R.A.; Rullmannn, J.A.; MacArthur, M.W.; Kaptein, R.; Thornton, J.M. AQUA and PROCHECK-NMR: Programs for checking the quality of protein structures solved by NMR*. *J. Biomol. NMR* **1996**, *8*, 477–486. [CrossRef] [PubMed]
29. Jo, S.; Kim, T.; Iyer, V.G.; Im, W. CHARMM-GUI: A web-based graphical user interface for CHARMM. *J. Comput. Chem.* **2008**, *29*, 1859–1865. [CrossRef] [PubMed]
30. Abraham, M.J.; Murtola, T.; Schulz, R.; Páll, S.; Smith, J.C.; Hess, B.; Lindah, E. GROMACS: High performance molecular simulations through multi-level parallelism from laptops to supercomputers. *SoftwareX* **2015**, *1–2*, 19–25. [CrossRef]
31. Trezza, A.; Birgauan, A.; Geminiani, M.; Visibelli, A.; Santucci, A. Molecular and Evolution In Silico Studies Unlock the h4-HPPD C-Terminal Tail Gating Mechanism. *Biomedicines* **2024**, *12*, 1196. [CrossRef]
32. Trezza, A.; Spiga, O.; Mugnai, P.; Saponara, S.; Sgaragli, G.; Fusi, F. Functional, electrophysiology, and molecular dynamics analysis of quercetin-induced contraction of rat vascular musculature. *Eur. J. Pharmacol.* **2022**, *918*, 174778. [CrossRef]
33. Carullo, G.; Ahmed, A.; Trezza, A.; Spiga, O.; Brizzi, A.; Saponara, S.; Fusi, F.; Aiello, F. A multitarget semi-synthetic derivative of the flavonoid morin with improved in vitro vasorelaxant activity: Role of CaV1.2 and KCa1.1 channels. *Biochem. Pharmacol.* **2021**, *185*, 114429. [CrossRef]
34. Sayers, E.W.; Bolton, E.E.; Brister, J.R.; Canese, K.; Chan, J.; Comeau, D.C.; Connor, R.; Funk, K.; Kelly, C.; Kim, S.; et al. Database resources of the national center for biotechnology information. *Nucleic Acids Res.* **2022**, *50*, D20–D26. [CrossRef]
35. Rosignoli, S.; Paiardini, A. DockingPie: A consensus docking plugin for PyMOL. *Bioinformatics* **2022**, *38*, 4233–4234. [CrossRef]
36. Trott, O.; Olson, A.J. AutoDock Vina: Improving the speed and accuracy of docking with a new scoring function, efficient optimization and multithreading. *J. Comput. Chem.* **2010**, *31*, 455–461. [CrossRef]
37. Fusi, F.; Durante, M.; Spiga, O.; Trezza, A.; Frosini, M.; Floriddia, E.; Teodori, E.; Dei, S.; Saponara, S. In vitro and in silico analysis of the vascular effects of asymmetrical N,N-bis(alkanol)amine aryl esters, novel multidrug resistance-reverting agents. *Naunyn-Schmiedeberg's Arch. Pharmacol.* **2016**, *389*, 1033–1043. [CrossRef]
38. Salentin, S.; Schreiber, S.; Haupt, V.J.; Adasme, M.F.; Schroeder, M. PLIP: Fully automated protein–ligand interaction profiler. *Nucleic Acids Res.* **2015**, *43*, W443. [CrossRef]
39. Fusi, F.; Trezza, A.; Spiga, O.; Sgaragli, G.; Bova, S. Cav1.2 channel current block by the PKA inhibitor H-89 in rat tail artery myocytes via a PKA-independent mechanism: Electrophysiological, functional, and molecular docking studies. *Biochem. Pharmacol.* **2017**, *140*, 53–63. [CrossRef]
40. Thompson, J.D.; Higgins, D.G.; Gibson, T.J. CLUSTAL W: Improving the sensitivity of progressive multiple sequence alignment through sequence weighting, position-specific gap penalties and weight matrix choice. *Nucleic Acids Res.* **1994**, *22*, 4673–4680. [CrossRef]
41. Salam, M.A.; Al-Amin, M.Y.; Salam, M.T.; Pawar, J.S.; Akhter, N.; Rabaan, A.A.; Alqumber, M.A.A. Antimicrobial Resistance: A Growing Serious Threat for Global Public Health. *Healthcare* **2023**, *11*, 1946. [CrossRef]
42. Bueno-Gavilá, E.; Abellán, A.; Girón-Rodríguez, F.; Cayuela, J.M.; Salazar, E.; Gómez, R.; Tejada, L. Bioactivity of hydrolysates obtained from bovine casein using artichoke (*Cynara scolymus* L.) proteases. *J. Dairy Sci.* **2019**, *102*, 10711–10723. [CrossRef] [PubMed]
43. Ciríaco, M.; Patarata, L.; Moura-Alves, M.; Nunes, F.; Saraiva, C. Antimicrobial Properties of Chestnut Shell Extract as an Ecofriendly Approach for Food Preservation. *Biol. Life Sci. Forum* **2023**, *26*, 123. [CrossRef]
44. Štumpf, S.; Hostnik, G.; Langerholc, T.; Pintarič, M.; Kolenc, Z.; Bren, U. The Influence of Chestnut Extract and Its Components on Antibacterial Activity against Staphylococcus aureus. *Plants* **2023**, *12*, 2043. [CrossRef] [PubMed]
45. Frusciante, L.; Geminiani, M.; Olmastroni, T.; Mastroeni, P.; Trezza, A.; Salvini, L.; Lamponi, S.; Spiga, O.; Santucci, A. Repurposing Castanea sativa Spiny Burr By-Products Extract as a Potentially Effective Anti-Inflammatory Agent for Novel Future Biotechnological Applications. *Life* **2024**, *14*, 763. [CrossRef]

applied sciences

MDPI

Article

The Effect of Natural Plant and Animal Fibres on PLA Composites Degradation Process

Ewa Szczepanik [1], Piotr Szatkowski [1], Edyta Molik [2] and Kinga Pielichowska [1,*]

1 Department of Biomaterials and Composites, Faculty of Materials Science and Ceramics, AGH University of Krakow, Al. Mickiewicza 30, 30-059 Krakow, Poland; eszczepanik@agh.edu.pl (E.S.); pszatko@agh.edu.pl (P.S.)

2 Department of Animal Nutrition and Biotechnology, and Fisheries, Faculty of Animal Science, University of Agriculture in Krakow, Al. Mickiewicza 24/28, 31-059 Krakow, Poland; rzmolik@cyf-kr.edu.pl

* Correspondence: kingapie@agh.edu.pl

Abstract: One of the methods to reduce long-term excessive plastic waste is the development and use of composite materials based on biodegradable polymers and natural fibres. Composites with natural fibres can exhibit very good mechanical properties, and the presence of natural fibres can significantly accelerate the degradation of the material. This study aimed to manufacture and analyse the biodegradation process of composites based on biodegradable polylactide (PLA) filled with flax and sheep wool fibres. The effect of flax and wool fibres and their content on the degradation rate compared to that of pure PLA was investigated. The degradation progress and properties of the composites were studied using an optical microscope, SEM, measurement of surface roughness, and contact angle. Additionally, flexural strength tests, a dynamic mechanical analysis (DMA), and a thermogravimetric analysis (TGA) were conducted. The effect of natural fibres on the phase transition and degree of crystallinity was analysed using differential scanning calorimetry (DSC). The results showed that PLA degrades only under UV light, but not in the composter simulating the natural environment. However, the incorporation of both types of fibres accelerated degradation of PLA/fibres composites in soil. Flax fibre composites exhibited better mechanical properties than pure PLA. For composites with wool fibres, although they showed a significant acceleration of the degradation process in the soil, their large content in the composite caused a reduction of mechanical properties. This research showed the positive effect of the addition of natural fibres on the biodegradation of PLA.

Keywords: wool; flax; PLA; composites; biodegradation

1. Introduction

In recent decades, there has been a very rapid development of polymeric materials, but more attention has been paid to the problem of storing and processing plastic waste. Conventional polymers commonly used lose some of their properties during reprocessing, and this significantly limits the possibilities of their reuse. In addition, the degradation time of the most commonly used polymers, such as polyethylene (PE), polypropylene (PP), and polystyrene (PS), is very long because their chain structure consists of only carbon linkages. Unfortunately, these polymers are mainly used to produce packaging, which has a very short lifetime and quickly becomes a waste. The packing industry generates the greatest demand for plastics, which is around 45% of all polymers produced [1]. Another serious problem is the huge amount of waste dumped in landfills that sometimes, due to poor management, can enter the seas and oceans, where it causes significant pollution of the ecosystem. Eight million tonnes of plastic is estimated to enter the oceans every year [2]. One possible solution to this difficult issue is to return to natural materials and, at the same time, reduce the use of conventional polymers.

In biocomposites, at least one of the components is of natural origin or is biodegradable. The role of the matrix is to transfer tension and protect against mechanical damage, while the most commonly used fillers (natural fibres) serve to improve mechanical properties, reduce density, and accelerate biodegradation [3]. The most popular biocomposites consist of polymer matrix and natural fibre filler.

We can divide natural fibres, according to their origin, into animal, vegetable, and mineral fibres [4]. An important advantage of natural fibres is that they reduce the density of the composite made with them. On the other hand, attention should be paid to the low thermal stability of these fibres and the difficulty of obtaining a homogeneous dispersion in the polymer [5].

One of the most studied plant fibres is flax. Its advantages are its low price and low density, as well as its very promising mechanical properties and the fact that it does not release harmful residues when burnt [6]. The properties of flax, such as its low density (1.5 g/cm^3), tensile strength between 345 and 1500 MPa, and Young's modulus between 30 and 80 GPa, are good enough for it to compete with or be used as a replacement for glass fibres [7]. Studies [8] show that the combination of natural fibres and biodegradable polymers can produce promising results. Plant fibres can significantly increase tensile strength. It is important to find the right amount of fibre addition, as too high a value may contribute to a gradual deterioration of properties, mainly due to the difficulty in maintaining a good degree of bonding between the composite components.

In recent years, animal fibres have not attracted as much attention compared to plant fibres. The development of materials based on sheep's wool fibres and their use in industry is one of the leading methods to the revival of declining pastoralism in mountainous regions (e.g., in the Tatra Mountains in Poland and the southern part of the country) [9], as well as in places where the inefficiency of using large amounts of wool causes significant economic losses [10]. Wool fibres do not have as high strength properties as plant fibres, but they are often considered for their potential use as a flame retardant [9]. Wool soaks up water, which can greatly accelerate the degradation of a composite made with it, but sometimes the degradation rate can be too high [11]. Wool fibres also have a high capacity for sound and thermal insulation. All of these characteristics have led to the use of wool in the textile industry for hundreds of years, but its use in other industries is constantly being developed [12–14]. Additionally, low-quality animal fibre waste amounts to approximately 3 million tons per year. The problem of disposal of this amount of fibrous waste is increasingly being discussed from an environmental-protection point of view [15].

Of the many biodegradable polymers, polylactide (PLA) has secured its status as the most popular on the market. This is supported by its easy availability and attractive value for money. PLA can be moulded as other thermoplastic polymers using melt processing methods, such as extrusion, blow moulding, or injection moulding. Moreover, the thermal stability of PLA, although slightly lower than that of conventional thermoplastics, is higher than that of other biodegradable polymers when subjected to melt processing and forming [16].

The combination of PLA with natural fibre is a very promising research direction. The use of both components makes it possible to reduce the disadvantages of the individual components and offers the possibility of creating a material alternative to the current solutions on the market.

Hence, there has been a great amount of research conducted on natural fibre composites. A vast majority of them are related to plant fibres because of their good mechanical properties. Pawłowska et al. [17] used the addition of modified and non-modified flax fibres in injection-moulded composites to improve their mechanical properties. The use of flax fibres resulted in an improved Young's modulus compared to that of pure PLA. Sit et al. [18] investigated the effect of weather conditions on composites with flax fibres compared to PVC. The initial samples showed properties better than PVC, and, despite ageing, their properties were still comparable. During testing in high humidity, they noticed swelling of the fibres, which caused cracking of the matrix. Pantaloni et al. [19] compared flax composites with biodegradable and PP matrixes. They confirmed in their study the possibility of designing materials with a mechanical performance suitable for industry and are additionally able to offer faster biodegradation. In another study, Pantaloni et al. [20] investigated the effect of humidity on flax composites, where only at high humidity values (98%RH) did the mechanical properties decrease significantly. Nassiopoulos and Njuguna [21] tested flax/PLA composites against flax/epoxy for their thermo-mechanical properties. The use of epoxy resin has advantages in terms of thermal stability, but the high Young's modulus obtained for PLA suggests that its composites have a good chance of being used in automotive applications.

Much fewer studies address the use of wool fibres in composites. Sharma et al. [22] studied the mechanical and thermal properties of wool/epoxy composites. They obtained composites with good interfacial adhesion and improved tensile strength. Abdellah et al. [23] conducted a study in which they created hybrid composites with date palm fibre/sheep wool in a polyester matrix. The compound containing 20 wt.% of the date palm fibre and wool with a mass ratio 50/50 showed the best properties in mechanical tensile and impact tests. Its properties were good enough for use in materials of daily use, while keeping the material very lightweight, which is one of the greatest advantages of using sheep wool. Haque et al. [24] made sheep wool/PCL (polycaprolactone) composites suitable for 3D printing. The composite with the addition of 10% wool showed increased biodegradability and higher yield strength.

The main objective of this research was to determine and compare the properties of biocomposites based on PLA and wool or flax fibres before and after the degradation process and to compare the effects of different natural fibres on the degradation process of the polymer matrix.

2. Materials and Methods

2.1. Materials

The PLA, in the form of granulates, that was used during this study was NatureWorks' Ingeo 3260 HP. According to the manufacturer data sheet, the density of the material is 1.24 g/cm^3, the tensile strength is 65 MPa, the elongation at break is 3.4%, and the melting point is 200 °C. The flax fibres were obtained from Safilin's Flax Low Twist Roving Tex 2000. According to the manufacturer, the fibre density is 1.44 g/cm^3, the tensile strength is 212 MPa, the elongation at break is 1.9% and the Young's modulus is 12 GPa. The wool fibres used were obtained from the breeding of mountain sheep carried out at the Experimental Station in Bielany by the University of Agriculture in Krakow. Polish mountain sheep are a typical species found in Southern Poland. As a solvent for PLA, dichloromethane (DCM) was used as received.

2.2. Manufacturing Methods

The first step in the preparation of the composite material was to clean the wool fibres by washing them (pure soap and water 60 °C) and then cutting the fibres into short pieces of several millimetres. Composites with 10 wt.% and 20 wt.% flax or wool fibre were obtained via mechanical mixing of the polymer granules with the fibre and adding a solvent (DCM p. a. 99.7%, PureLand®) to dissolve the polymer. Then, the material was poured onto a flat glass. After evaporation of the solvent, the material was cut (Figure 1a). Next, the granulate was used to obtain the samples using injection moulding process. A Zamak injection moulding machine was used. During the tests, the following parameters were set: injection time for pure PLA—10 s; injection time for natural fibre composites—12 s; mould temperature for PLA—50 °C; and mould temperature for composites—80 °C. Common parameters to all materials were a cylinder (chamber) temperature of 210 °C and injection force of 12 kN. Figure 1b shows specimens with 20 wt.% of wool added after the injection process.

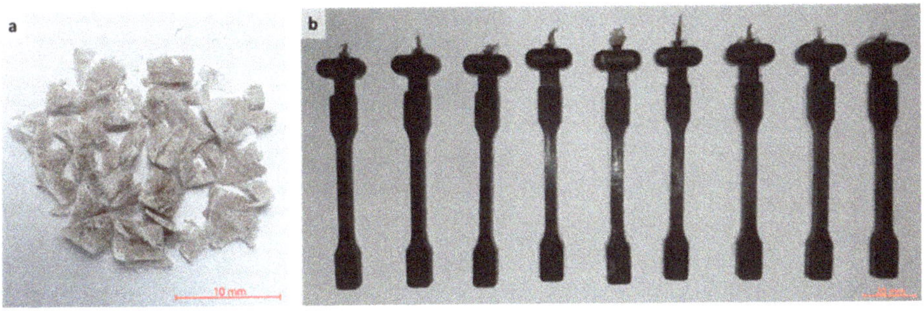

Figure 1. (**a**) Granulate with PLA and flax fibres. (**b**) Specimens of PLA/wool composite after injection moulding.

2.3. Degradation Methods

The composites obtained were exposed to UV-A and UV-B radiation in a specially prepared ageing chamber (12 W bulb). Samples were placed on foil stands in Petri dishes, in which a small layer of water (25 mL) was poured to ensure constant humidity during the ageing process. The samples were exposed to UV radiation for 12 days. The study was conducted at room temperature.

The next portion of the composite samples was composted. Freshly cut grass, forest mulch, wet soil, and flower fertiliser soil were used to prepare the compost. All ingredients were mixed and placed in containers. The composite samples were placed 5 cm deep in such a prepared composter. The soil was watered every 5 days to ensure moisture and covered with a punctured stretch film to ensure air flow. A lamp was directed at the entire set of containers to keep the site warm. The samples remained in the composter for 16 days. The study was conducted at room temperature.

All types of samples prepared, their labels and composition and the degradation process used are summarised in Table 1.

Table 1. List of all the sample types tested, their composition, and the degradation process they were subjected to.

Sample Name	PLA (% wt.)	Flax (% wt.)	Wool (% wt.)	UV Degradation Process	Soil Degradation Process
PLA	100	-	-	-	-
W10	90	-	10	-	-
W20	80	-	20	-	-
L10	90	10	-	-	-
L20	80	20	-	-	-
PLA_UV	100	-	-	+	-
W10_UV	90	-	10	+	-
W20_UV	80	-	20	+	-
L10_UV	90	10	-	+	-
L20_UV	80	20	-	+	-
PLA_CO	100	-	-	-	+
W10_CO	90	-	10	-	+
W20_CO	80	-	20	-	+
L10_CO	90	10	-	-	+
L20_CO	80	20	-	-	+

2.4. Research Methods

Contact angle testing was performed on a Krüss DSA 10 MK2 goniometer to determine the changes in wettability of the samples before and after degradation. Roughness measurement was performed using the Hommelwerke T1000 profilometer. The software calculated R_a

(average roughness of a surface) and R_z (difference between the tallest "peak" and the deepest "valley" in the surface) to characterise the roughness. Mechanical tests were carried out on a Zwick/Roell RetroLine 1445 test machine. The method used was the three-point bending of the specimens (PN-EN ISO 178). Each type of composite tested was subjected to DMA (dynamic mechanical analysis) tests on a DMA 850 (TA Instruments). The tests were carried out at a constant frequency of 10 Hz and in the temperature range of −60 ÷ 100 °C, with a heating rate of 5 °C/min, using three-point bending mode. PLA, L10, and W10 samples (before and after degradation processes) were subjected to thermogravimetric analysis (TG). Measurements were preformed using a TGA 550 (TA Instruments) thermal analyser, in platinum crucibles, under a nitrogen atmosphere. The heating rate was 10 °C/min. The study was conducted at the temperature range 45 ÷ 600 °C. DSC analysis was performed on PLA, L10, and W10 samples (before and after degradation processes), using DSC 1 from Mettler Toledo with STARe software. The samples were subjected to a heating (10 °C/min)/cooling (30 °C/min)/heating (10 °C/min) temperature program in the temperature range of 0–210 °C, under nitrogen atmosphere, with a flow rate of 30 mL/min. The samples were placed in pierced and sealed aluminium pans. The reference sample was an empty crucible. Using data from the DSC analysis, the degree of crystallinity (X_c) was calculated using the following formula (Equation (1)):

$$X_c = \frac{\Delta H_m}{\Delta H_m^0 (1 - a)} \tag{1}$$

where ΔH_m is the heat of melting of studied PLA composite, $\Delta H_m{}^0$ is the heat of melting of 100% crystalline polymer (for PLA 93.7 J/g [25]), and a is the mass fraction of the additive (fibres). The microstructure of the composites was investigated using an FEI Nova NANO SEM 200 scanning electron microscope equipped with an EDS analyser from EDAX and using a Keyence VHX-900F digital microscope.

3. Results and Discussion

The results of the contact angle tests provided information on the surface quality of the manufactured biocomposites before and after degradation. The contact angle results obtained are presented as a graph in Figure 2.

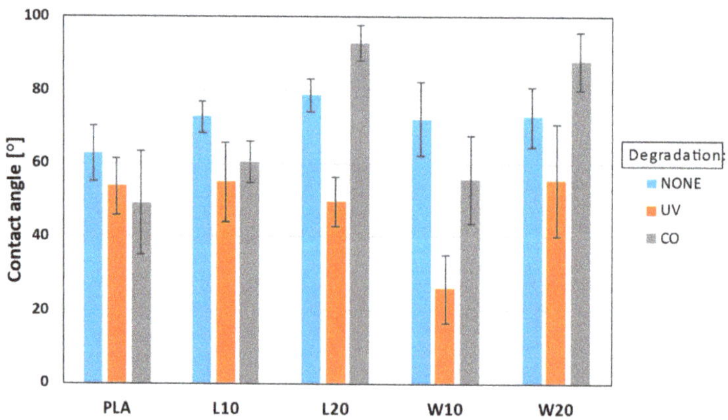

Figure 2. Contact angle of PLA/natural fibre composites before and after degradation process. The error bars represent the standard deviation of measurements made 6 times for each sample.

The incorporation of natural fibres of both wool and flax has a significant effect on the wettability of the PLA-based composites. It was observed that the degradation of pure PLA under UV light and composting conditions causes a decrease in the contact angle. The presence of natural fibres increases the values of contact angles on the surface of samples of nondegraded composites (immediately after their production). However, after degradation, the measured contact angles of the biocomposites are different, probably related to the uneven degradation of PLA and the presence of areas of exposed fibres. Although cellulose is strongly hydrophilic, there are studies that describe its influence on the wetting of the surfaces and thin layers. The high hydrophilicity of cellulose can be reduced with mineral additives such as TiO_2 [26]. Also, studies by Stark and Matuana [27] confirmed an increase in contact angle value when composite samples containing cellulose are subjected to UV ageing processes. Despite the use of a HDPE matrix, significant changes in the surface of the investigated HDPE with woof flour composites were observed. In our studies, it was found, that there is a tendency for the contact angle to decrease after UV degradation and increase under simulated composting degradation for PLA/natural fibre composites with 20 wt.% fibre content.

Roughness measurements were the second test performed to assess the surface quality of the composite samples. The results obtained in the study are presented in Figures 3 and 4.

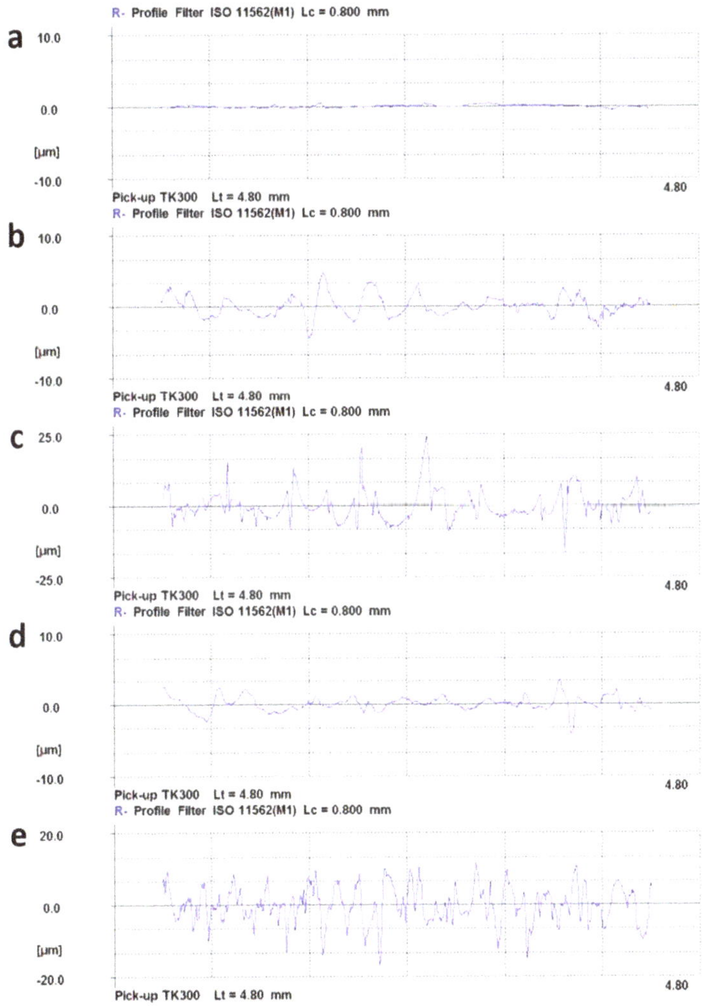

Figure 3. Example plots of sample surface: (**a**) PLA, (**b**) W20, (**c**) W20_UV, (**d**) L20, and (**e**) L20_UV.

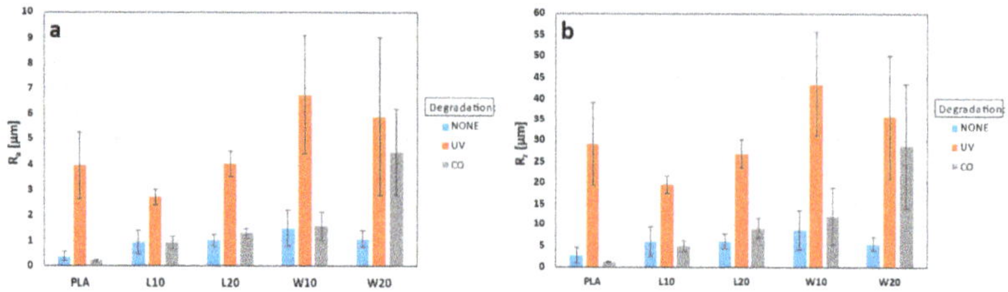

Figure 4. Surface roughness measured for different composites: (**a**) Ra and (**b**) Rz. The error bars represent the standard deviation of measurements made 3 times for each sample.

The evidence of strong UV degradation on the PLA surface was confirmed by the significant increase in roughness after ageing under UV radiation. The presence of natural fibres causes an increased surface roughness in the resulting composite after exposure to ultraviolet (UV) radiation; however, compared to pure PLA, flax fibres slow down degradation, while wool fibres accelerate it. Both linen and wool accelerated degradation in the composter compared to the degradation of pure PLA.

In the case of the addition of wool, an increased roughness was observed compared to pure PLA after both irradiation and composting, while linen delayed the degradation under UV light but accelerated it in the soil, as seen in Figure 4. Other scientists in their research observed similar changes in roughness after ageing processes. The presence of flax fibres in the composite significantly increases the R_a values [18], and the sisal fibre in a PLA matrix, after 15 days in compost, doubles its roughness (from a value of 1.2 μm to values ranging from 2 to 3.5 μm of the R_a index). The pure PLA matrix itself increases its roughness from a value of 0.5 μm to a value of 1 μm. In all studies, it was observed that, with an increase in the content of natural fibres in the composite, the rate of surface roughening also increases [28]. The increase in roughness can be correlated with an increase in the rate of biodegradation because the greater surface area allows easier transport of water deep into the material and the anchoring of fungi and microorganisms in the pores and defects of the composite. The right combination of fibres in the composite will allow the life of the biomaterial to be regulated.

The results of the flexural strength test are summarised in Table 2, and the selected flexural curves are shown in Figure 5. In Figure 5a, it can be seen that the mechanical properties of PLA after composting are slightly altered, while a significantly broken and degraded structure of the material with very poor properties can be seen after UV irradiation. In Figure 5b, flax is observed to have slowed the loss of mechanical properties during UV degradation of the sample, while it significantly accelerated the composting process. For the addition of wool (Figure 5c), there are also similar relationships to those for the addition of flax.

Table 2. Young modulus and tensile strength for PLA and PLA/natural fibre composites before and after degradation.

Sample	E (GPa)	Tensile Strength (MPa)
PLA	2.39 ± 0.08	47.61 ± 0.08
L10	3.22 ± 0.09	60.57 ± 9.73
L20	3.73 ± 0.37	63.59 ± 8.76
W10	2.20 ± 0.28	33.82 ± 6.77
W20	2.04 ± 0.05	23.86 ± 0.82
PLA_UV	1.09 ± 0.12	3.13 ± 1.13
L10_UV	2.68 ± 0.21	28.22 ± 5.94
L20_UV	3.37 ± 0.50	19.71 ± 9.23
W10_UV	1.24 ± 0.17	7.64 ± 3.59
W20_UV	1.08 ± 0.01	13.35 ± 3.35
PLA_CO	2.59 ± 0.07	51.28 ± 5.15

Table 2. *Cont.*

Sample	E (GPa)	Tensile Strength (MPa)
L10_CO	2.46 ± 0.33	39.56 ± 5.70
L20_CO	2.30 ± 0.29	28.05 ± 8.40
W10_CO	1.36 ± 0.52	13.66 ± 10.46
W20_CO	1.06 ± 0.54	13.98 ± 11.12

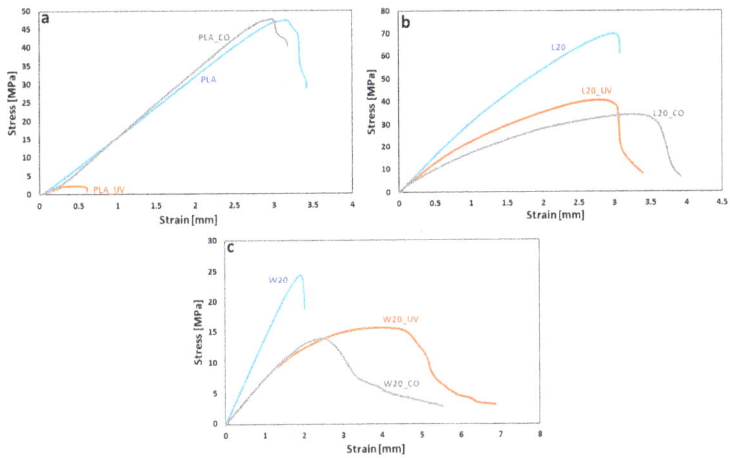

Figure 5. Comparison charts of bending curves: (**a**) PLA samples, (**b**) L20 samples, and (**c**) W20 composites.

PLA has poor resistance to UV radiation (PLA_UV—a 92% drop in transmitted force during testing). Radiative degradation leads to the almost-complete destruction of PLA samples. PLA in damp soil, on the other hand, does not change as much (PLA_CO—a 10% drop in transmitted force during testing). The use of flax as a filler in the PLA matrix reduced the destructive impact of UV radiation on the PLA matrix in the investigated biocomposites. However, it was observed that, in this case, degradation in moist soil (L20_CO) has a more significant effect. However, this degradation is at a level of 63% compared to a sample containing flax and not exposed to degradative factors. Wool as an additive, similar to the addition of flax, slows UV degradation. However, the addition of wool fibre results in a significant decrease in the transmitted force values when exposed to moist soil (compared to a similar sample with flax). PLA undergoes hydrolytic degradation that involves several steps, including the diffusion of water into the amorphous phase of PLA, macromolecules chain scission by hydrolysis of the ester bonds, and the formation of the oligomers and monomers; and, in the next step, we see partial hydrolysis of the crystalline fractions [29]. Since washed wool fibre is hygroscopic and hydrophilic, it facilitates the transport of water deep into the material, leading to the hydrolysis of the PLA matrix and its degradation, thus accelerating the degradation of the entire material. UV degradation was slowed down by each type of fibre, and more force had to be applied to destroy the composite. The more fibres, the greater the difference in destructive forces, and the addition of flax slowed UV degradation significantly more than the addition of wool. For the degradation caused by composting, an acceleration of the degradation rate can be seen under the influence of fibre addition, since the force used to destroy the composite is, in each case, lower than that used with pure PLA. For L10 samples, the difference is very small compared to PLA. In this case, the greater the addition, the faster the degradation, and the wool fibres accelerated the degradation more.

It can be seen (Table 2) that the mechanical properties of the composite increased after the addition of flax fibre in both cases compared to pure PLA. For small amounts of wool fibre, the properties are comparable to those of PLA, but with the large addition of these animal fibres, the mechanical properties decreased. Few studies have focused on the influence of sheep wool fibres on the mechanical properties of polymer matrix composite materials. Wool fibre is characterised by relatively low tensile strength compared to even natural fibres from plants (around 120 MPa). Sharma et al. [22] tested a modified composite with sheep wool fibres in an epoxy matrix and obtained a composite with improved mechanical properties of 45 MPa. Such strength values are typical for pure plastics and even some thermoplastics. Hence, the conclusion was drawn that sheep wool fibres are not suitable as a reinforcing modifier for composites, but rather as a filler or functional additive. The ageing process (degradation of the biocomposite in water) occurs quite rapidly. In other studies, the strength values decreased to a level of 30% (L20 samples experienced a 50% decrease after exposure to a humid environment) [30,31]. Modification with cut, short natural fibres in polymer matrices results in an improvement in mechanical properties by up to 30%. The continued increase in the proportion of natural fibres in composites is associated with a slight increase in conjunction with deterioration of technological conditions (flow decrease and increased porosity of the biocomposite) [32].

The second test used to evaluate the mechanical properties and how they change as a consequence of the conditions and degree of degradation was a dynamic mechanical analysis (DMA), for which some of the curves are shown (Figure 6). The range to which particular attention was paid includes temperatures from −30 °C to 50 °C, as within this range, the material would be mainly used.

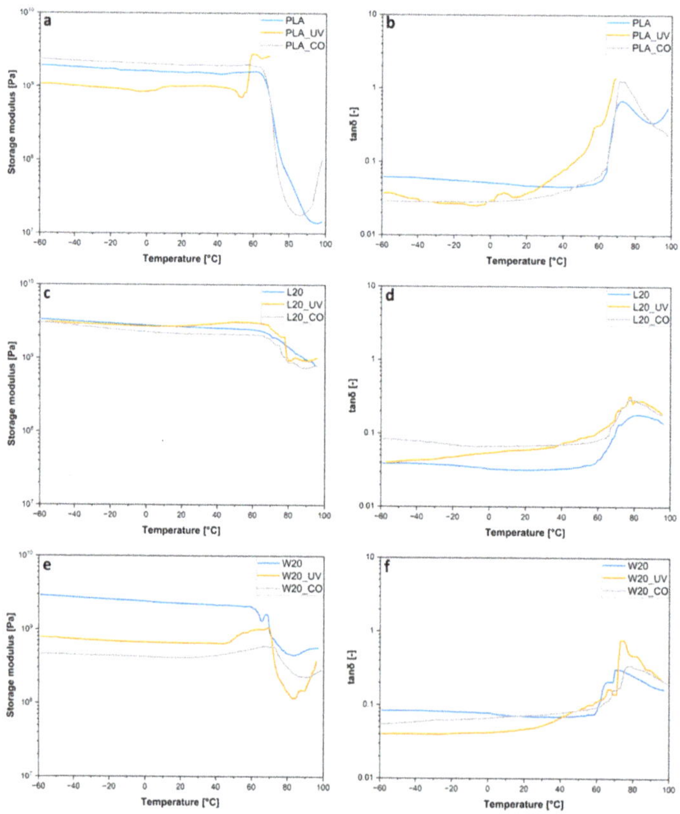

Figure 6. Comparison charts of DMA curves: (**a**) storage modulus of PLA samples, (**b**) tanδ of PLA samples, (**c**) storage modulus of L20 samples, (**d**) tanδ of L20 samples, (**e**) storage modulus of W20 samples, and (**f**) tanδ of W20 samples.

From the graphs shown in Figure 6, it can be seen that a very high storage modulus was obtained for the materials immediately after manufacture. As confirmed by other tests, PLA samples subjected to composting were not significantly different from PLA samples not subjected to degradation. Therefore, the values of the storage modulus between these samples do not have significant differences, and the slightly lower modulus for the PLA sample without degradation may be due to slow degradation and physical ageing processes in the composted sample. As a result, composted PLA is characterised by higher degree of crystallinity and stiffness, and the storage modulus is higher. For each fibre, with the exception of W10, the value of storage modulus improved compared to pure PLA. It can be seen that a large content of wool fibres destabilises the material properties more quickly. In addition, the low content of fibres significantly reduces the difference in the loss modulus responsible for stress dissipation. From the available studies, it can be observed that the addition of flax fibres does not affect the glass transition temperature, but it does alter the ratio of the storage modulus to the loss modulus. Short flax fibres stiffen the composite and decrease the PLA macromolecules' mobility [21,33,34]. Composites with a PLA matrix containing flax fibres increase the value of the storage modulus almost twice [35]. The UV-exposed PLA sample was the only one to deteriorate before reaching 100 °C. When composted samples were tested for pure PLA and the L10 composite, no deterioration in properties was obtained. For the L20_CO sample, both the storage modulus and the loss modulus decreased steadily and dropped by 15% compared to the nondegraded L20. A very high decrease was recorded for the wool composites, indicating a very high degradation of these composites in the ground. For the composted samples, there is a clear difference between the flax and wool composites, which have significantly lower properties.

The TG curves and the corresponding DTG curves are presented in Figure 7. A summary of the most important parameters determined from the TG curves to assess the dynamics of material degradation is included in Table 3.

From Figure 7, it can be seen that the manufactured composites show high thermal stability. The greatest difference can be seen for sample W10_CO, where degradation occurs significantly earlier than for the other samples (Figure 7e). In Table 3, it can be seen that much higher residue values at 600 °C were obtained for the samples with natural fibres. The degradation under UV light accelerated the weight loss in the samples with flax; this is due to the strong destruction of the matrix, which accounted for 90% of the weight of the composites. For wool fibres in the case of composting, no differences can be seen at first; they only appear around 350 °C, where the fibres play a significant role in the acceleration of degradation.

For flax fibres with PLA, rapid changes occur from about 300 °C, where the degradation of the polymer itself begins, and the presence of fibres shifts the temperature of maximum degradation towards higher values [17]. PLA samples reinforced with other plant fibres (coir, pineapple leaves, and bamboo) also showed stability in the same temperature range, and from about 350 °C, they degraded in one step [36,37]. The TG studies showed that the presence of both flax and wool fibres increased the degradation under UV light and during composting.

A DSC study was carried out on PLA, L10, and W10 samples to assess the effect of the presence of the fibre on the processing and degradation properties of the polymer. Plots of the first- and second-cycle heating curves are shown in Figure 8, and a summary of the most important parameters analysed is given in Table 4. Additionally, the glass transition temperature value from the analysis of the DMA curves (Figure 6), calculated as a peak in the tan(δ) plot, is added to the table for comparison.

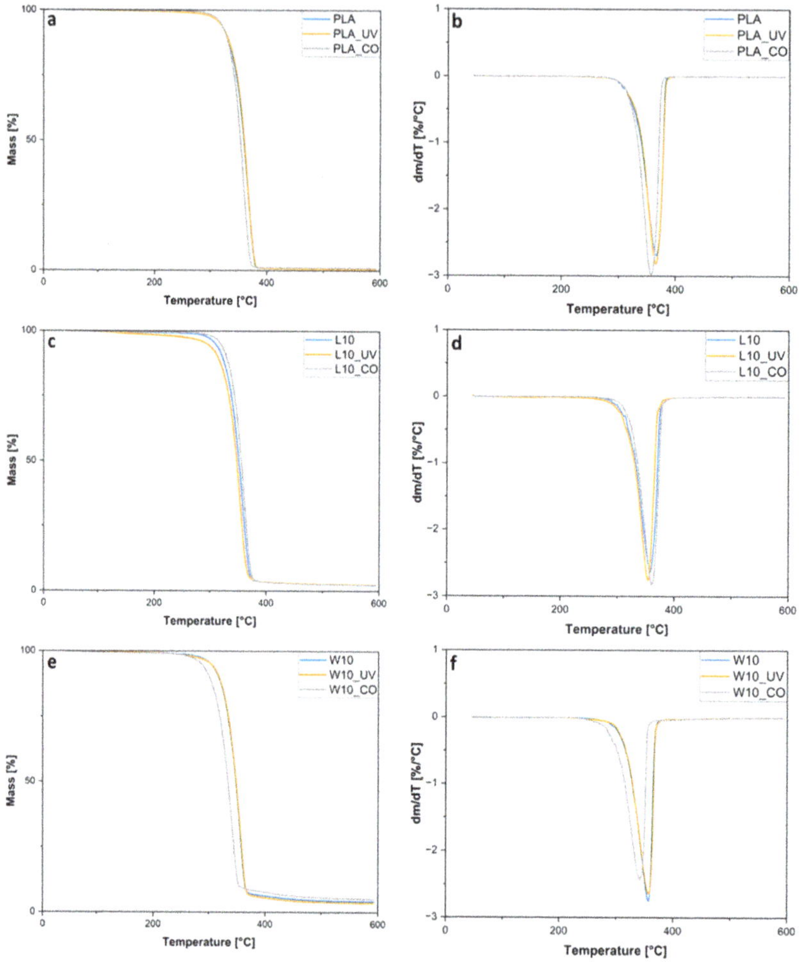

Figure 7. Comparison chart of TG and DTG curves: (**a**) TG curve of PLA samples, (**b**) DTG curve of PLA samples, (**c**) TG curve of L10 samples, (**d**) DTG curve of L10 samples, (**e**) TG curve of W10 samples, and (**f**) DTG curve of W10 samples.

Table 3. Results of TG analysis *.

Sample	$T_{1\%}$ (°C)	$T_{3\%}$ (°C)	$T_{5\%}$ (°C)	$T_{20\%}$ (°C)	$T_{50\%}$ (°C)	T_{DTGmax} (°C)	W_r (%)
PLA	288	309	318	342	359	367	0.53
L10	266	301	310	334	351	358	2.17
W10	239	292	305	330	348	357	4.16
PLA_UV	267	307	317	343	360	367	0.57
L10_UV	155	271	296	329	348	355	2.46
W10_UV	235	289	305	331	349	358	3.58
PLA_CO	290	311	318	338	353	358	0.99
L10_CO	285	311	319	340	356	361	2.31
W10_CO	237	275	287	315	334	341	5.11

* $T_{1(3,5,20,50)\%}$—temperature at 1, 3, 5, 20, and 50% mass loss; T_{DTGmax}—degradation temperature; W_r—char residue at 600 °C.

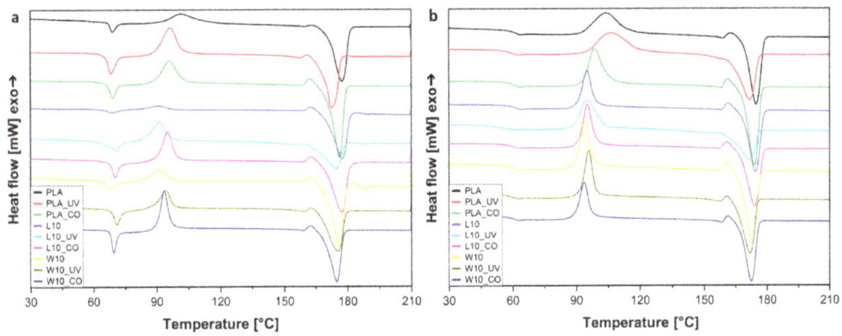

Figure 8. DSC curves of PLA/natural fibre composites: (**a**) first heating cycle and (**b**) second heating cycle.

Table 4. Results * of T_g from DMA, DSC analysis and crystallinity in 1st and 2nd heating cycle.

Sample	$T_{g,DMA}$ (°C)	$T_{g,DSC1}$ (°C)	X_{c1}	T_{m1} (°C)	$T_{g,DSC2}$ (°C)	X_{c2}	T_{m2} (°C)
PLA	74	67	55%	177	60	54%	175
L10	75	64	48%	177	60	49%	174
W10	72	65	50%	176	59	49%	173
PLA_UV	-	64	52%	172	58	55%	172
L10_UV	76	64	31%	175	54	41%	172
W10_UV	80	68	46%	175	59	46%	172
PLA_CO	74	65	57%	176	60	56%	174
L10_CO	73	67	51%	177	59	49%	174
W10_CO	78	67	49%	175	59	48%	173

* $T_{g,DMA}$—glass transition temperature obtained from DMA; $T_{g,DSC1(2)}$—glass transition temperature obtained from DSC (1: 1st heating cycle; 2: 2nd heating cycle); $X_{c1(2)}$—crystallinity degree (1: 1st heating cycle; 2: 2nd heating cycle); $T_{m1(2)}$—melting temperature obtained from DSC (1: 1st heating cycle; 2: 2nd heating cycle).

In the first heating cycle (Figure 8a), strong relaxation of the polymer can be seen after the injection process in each case. Performing the degradation by UV irradiation led to the destruction of the polymer structure, as can be seen when comparing the graphs for PLA and PLA_UV. During the second heating cycle (Figure 8b), a significant reduction in processing-induced relaxation and significant exothermic peaks can be observed from cold crystallisation. These are more pronounced for samples with added natural fibres, which can thus be considered to be the centres of crystallisation.

Due to the different characteristics of the methods, the glass transition temperatures obtained from DMA are higher than the corresponding temperatures from DSC (the PLA_UV sample in DMA was destroyed before the result could be obtained). In the second heating cycle, the glass transition occurred at lower temperatures, and the lowest values were obtained for the samples after UV degradation. The calculated degree of crystallinity of the pure PLA was 54%, and the presence of fibres reduced its value. The melting temperatures are very close to each other, but lower values were obtained for samples with the addition of fibres, especially wool.

A reduction in T_g values due to the addition of natural fibres has also been noted in other studies. Furthermore, the authors observed that the addition of flax increases the enthalpy of the cold crystallisation of PLA [17]. In the study, the addition of kenaf fibres lowered the melting temperatures of PLA-based composites and accelerated the recrystallisation process [38]. For various plant-based natural fibres, e.g., coir and bamboo, an increase in T_g values can be seen in relation to pure PLA [36,37]. The presence of natural fibres in PLA composites mainly affects the degree of crystallinity of the polymer and its properties during processing; furthermore, a significant reduction in the parameters studied after degradation processes confirms its successful course.

Selected SEM microphotographs are shown in Figure 9. The SEM observations for pure PLA show very clear cracks for the exposed composite. The appearance of the sample after composting is not significantly different from that of the sample immediately after fabrication. Figure 9d,g provide a very good comparison of the size and surface area of plant and animal fibres. It can be seen that PLA homogenised much better with the flax fibres, covered them, and formed a physical bond over a significant area of the fibre.

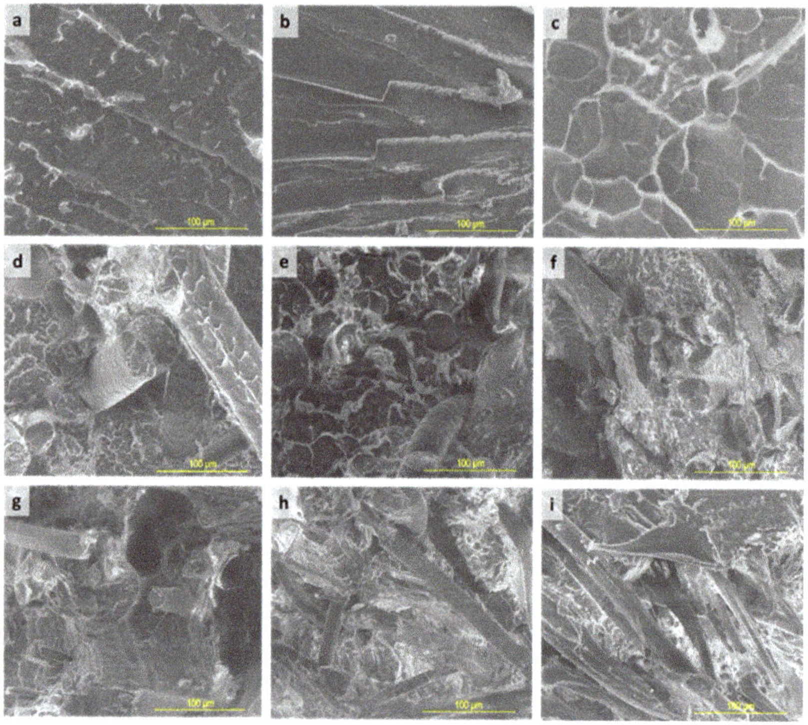

Figure 9. SEM images of samples: (**a**) PLA; (**b**) PLA_UV; (**c**) PLA_CO; (**d**) W20; (**e**) W20_UV; (**f**) W20_CO; (**g**) L20; (**h**) L20_UV; and (**i**) L20_CO.

In Figure 9e,h, it can also be seen that both the fibres and the matrix have been degraded, but in Figure 9f,i, it can be seen that the main degrading component are the fibres. Broken fibres and pulling out of fibres from the matrix (pull-out) have also been observed by other scientists studying similar biocomposites [18,20]. They have reached similar conclusions, noting that the higher the content of natural fibres in the biocomposite, the worse the degradation that occurs [24].

Additionally, EDS studies were performed on the samples, which confirmed the qualitative composition of the composites tested. For the pure PLA and L20 samples, peaks were obtained for carbon and oxygen. For the W20 composites, peaks were obtained for carbon, oxygen, nitrogen, and sulphur.

Surface microscopic observations of the samples after and before degradation were carried out to assess the visual effects of the degree of degradation—Figure 10.

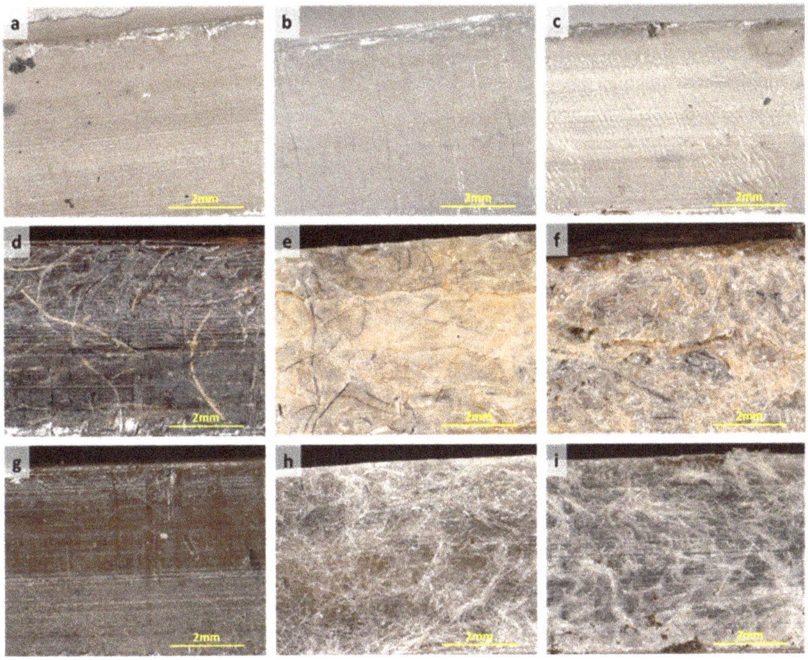

Figure 10. PLA and composite surface observed under a digital microscope (50× magnification): (**a**) PLA; (**b**) PLA_UV; (**c**) PLA_CO; (**d**) W10; (**e**) W10_UV; (**f**) W10_CO; (**g**) L10; (**h**) L10_UV; and (**i**) L10_CO.

When comparing samples made from PLA alone (Figure 10a–c), clear cracks can be seen for the samples degraded under UV radiation. They have also lost their transparency. The composted PLA sample shows slight wrinkling on the surface, which may indicate the beginning of surface degradation, but apart from this, there is little difference from the non-degraded PLA.

For the samples in group W10 (Figure 10d–f), a clear difference can be seen between the degraded samples and the samples after generation. The degraded samples are much lighter in colour, with uneven surfaces and protruding fibre fragments visible. It is caused by the appearance of microcracks in the volume of the PLA sample, on which the passing light disperses, giving the sample a milky discolouration effect. The composted W10 sample shows deep cracks and significant distortion of the whole sample. The brightest (milky) fragments on the edges and around the fibres signal that biodegradation is taking place mainly on the surface and in the adjacent fibres due to their high hygroscopicity. The degraded W20 samples, like the W10 samples, differed significantly in colour from the undegraded sample. Degradation progresses along the fibres. This is likely due to the easy absorption of water in the volume of wool fibres, and in a humid environment, it is easier to degrade the PLA matrix.

The L10 samples (Figure 10g–i), as the wool fibre samples, changed colour after degradation to a lighter colour. In the literature, similar observations can be found; the presence of moisture in the environment accelerates the degradation process of biocomposites. The fibres transport moisture into the interior of the sample, causing them to swell. This phenomenon speeds up the degradation [18,19]. A difference can be seen between the type of degradation of the two samples (Figure 10h,i): the composted sample shows white traces arranged mainly along the fibres and smooth brown areas of the matrix, while on the UV-illuminated sample, the degradation is more uniform throughout the composite. In the

L20 samples, the colour change in the sample during the degradation test was comparable to the samples in the L10 category.

4. Conclusions

The development of composite technology based on natural and biodegradable materials may be one of the answers to the ever-increasing problem of plastic waste disposal. A major challenge is to develop materials with properties similar to those of conventional polymers at a competitive price.

In this work, two types of composites based on a PLA with different contents of flax or wool fibres were obtained. The effect of the type of fibre on PLA degradation was investigated. For samples with flax added, composites with better properties than pure PLA were obtained, the better the greater the fibre addition. Flax fibres accelerated the simulated degradation in the environment and slowed the fast degradation of pure PLA under UV exposure. Wool fibres in small amounts did not significantly change the properties of PLA, and high filler addition resulted in lower mechanical properties. Wool fibres significantly accelerated degradation by composting, mainly due to their high hygroscopicity, which allowed water transport and hydrolysis of the polymer.

Given the latest trends associated with a circular economy, more and more attention is paid to materials that are of natural origin and do not burden the environment. Sheep's wool fibres have been present in our civilisation for thousands of years and are characterised by unique functional properties. The amount of waste, mainly packaging, is constantly increasing and the main problem is the decomposition time, which is inadequate for the duration of use. Wool accelerates the degradation of polymer matrix materials and can serve as a factor in controlling and determining the lifetime of a biocomposite that contains sheep's wool. Due to the hygroscopic nature of washed wool, water can be transported deep into the material, and, thus, hydrolysis of the matrix occurs throughout the composite, accelerating the degradation of the entire material. Natural fibres, through their structure and properties, improved PLA degradation. The use of low-cost natural fibres or even a waste product such as sheep wool further reduces the price and increases the attractiveness and competitiveness of such a material on the market.

Author Contributions: Conceptualisation, E.S. and P.S.; methodology, E.S., P.S. and K.P.; software and validation, E.S. and P.S.; formal analysis, E.M., P.S. and K.P.; investigation, E.S.; resources, E.M., P.S. and K.P.; data curation, E.S. and P.S.; writing—original draft preparation, E.S. and P.S.; writing—review and editing, K.P. and E.M.; visualisation, E.S., E.M. and K.P.; supervision, K.P. and P.S.; project administration and funding acquisition, K.P. All authors have read and agreed to the published version of the manuscript.

Funding: This work was supported from the subsidy of the Ministry of Education and Science for the AGH University of Krakow (Project No. 16.16.160.557).

Institutional Review Board Statement: Not applicable.

Informed Consent Statement: Not applicable.

Data Availability Statement: The data used to support the findings of this study are included within the article. Additional data are available from the corresponding author upon request.

Conflicts of Interest: The authors declare no conflicts of interest.

References

1. Law, K.L.; Narayan, R. Reducing Environmental Plastic Pollution by Designing Polymer Materials for Managed End-of-Life. *Nat. Rev. Mater.* **2022**, *7*, 104–116. [CrossRef]
2. Lau, W.W.Y.; Shiran, Y.; Bailey, R.M.; Cook, E.; Stuchtey, M.R.; Koskella, J.; Velis, C.A.; Godfrey, L.; Boucher, J.; Murphy, M.B.; et al. Evaluating Scenarios toward Zero Plastic Pollution. *Science* **2020**, *369*, 1455–1461. [CrossRef]
3. Motru, S.; Adithyakrishna, V.H.; Bharath, J.; Guruprasad, R. Development and Evaluation of Mechanical Properties of Biodegradable PLA/Flax Fiber Green Composite Laminates. *Mater. Today Proc.* **2020**, *24*, 641–649. [CrossRef]
4. Ao, X.; Vázquez-López, A.; Mocerino, D.; González, C.; Wang, D.-Y. Flame Retardancy and Fire Mechanical Properties for Natural Fiber/Polymer Composite: A Review. *Compos. Part B Eng.* **2024**, *268*, 111069. [CrossRef]

5. Xu, H.; Xie, L.; Chen, Y.-H.; Huang, H.-D.; Xu, J.-Z.; Zhong, G.-J.; Hsiao, B.S.; Li, Z.-M. Strong Shear Flow-Driven Simultaneous Formation of Classic Shish-Kebab, Hybrid Shish-Kebab, and Transcrystallinity in Poly(Lactic Acid)/Natural Fiber Biocomposites. *ACS Sustain. Chem. Eng.* **2013**, *1*, 1619–1629. [CrossRef]
6. Johari, A.P.; Mohanty, S.; Kurmvanshi, S.K.; Nayak, S.K. Influence of Different Treated Cellulose Fibers on the Mechanical and Thermal Properties of Poly(Lactic Acid). *ACS Sustain. Chem. Eng.* **2016**, *4*, 1619–1629. [CrossRef]
7. Yadav, V.; Singh, S. A Comprehensive Review of Natural Fiber Composites: Applications, Processing Techniques and Properties. *Mater. Today Proc.* **2022**, *56*, 2537–2542. [CrossRef]
8. Ejaz, M.; Azad, M.M.; Shah, A.U.R.; Afaq, S.K.; Song, J.-I. Mechanical and Biodegradable Properties of Jute/Flax Reinforced PLA Composites. *Fibers Polym.* **2020**, *21*, 2635–2641. [CrossRef]
9. Molik, E.; Szatkowski, P.; Flis, Z.; Suchorowiec, K.; Szczepanik, E.; Niemiec, M.; Komorowska, M.; Matusevičius, P. Tradition and Innovative in the Protection of the Natural Environment of Mountain Regions. *Acta Sci. Pol. Form. Circumiectus* **2023**, *22*, 33–40. [CrossRef]
10. Altin, M.; Yildirim, G.Ş. Investigation of Usability of Boron Doped Sheep Wool as Insulation Material and Comparison with Existing Insulation Materials. *Constr. Build. Mater.* **2022**, *331*, 127303. [CrossRef]
11. Kandola, B.K.; Mistik, S.I.; Pornwannachai, W.; Horrocks, A.R. Effects of Water and Chemical Solutions Ageing on the Physical, Mechanical, Thermal and Flammability Properties of Natural Fibre-Reinforced Thermoplastic Composites. *Molecules* **2021**, *26*, 4581. [CrossRef]
12. Carrasco, M.F.A.; Rouault, N.J.; Ferri Azor, J.M.; López-Martínez, J.; Samper Madrigal, M.D. A New Bio-Based Fibre-Reinforced Polymer Obtained from Sheep Wool Short Fibres and PLA. *Green Mater.* **2019**, *8*, 79–91. [CrossRef]
13. Szatkowski, P.; Tadla, A.; Flis, Z.; Szatkowska, M.; Suchorowiec, K.; Molik, E. Production of Biodegradable Packaging with Sheep Wool Fibres for Medical Applications and Assessment of the Biodegradation Process. *Anim. Sci. Genet.* **2022**, *18*, 57–67. [CrossRef]
14. Szatkowski, P.; Tadla, A.; Flis, Z.; Szatkowska, M.; Suchorowiec, K.; Molik, E. The Potential Application of Sheep Wool as a Component of Composites. *Rocz. Nauk. Pol. Tow. Zootech.* **2021**, *17*, 1–8. [CrossRef]
15. Kim, N.K.; Lin, R.J.T.; Bhattacharyya, D. Extruded Short Wool Fibre Composites: Mechanical and Fire Retardant Properties. *Compos. B Eng.* **2014**, *67*, 472–480. [CrossRef]
16. Bhiogade, A.; Kannan, M.; Devanathan, S. Degradation Kinetics Study of Poly Lactic Acid(PLA) Based Biodegradable Green Composites. *Mater. Today Proc.* **2020**, *24*, 806–814. [CrossRef]
17. Pawłowska, A.; Stepczyńska, M.; Walczak, M. Flax Fibres Modified with a Natural Plant Agent Used as a Reinforcement for the Polylactide-Based Biocomposites. *Ind. Crops. Prod.* **2022**, *184*, 115061. [CrossRef]
18. Sit, M.; Ling, J.; Jiang, C.; Zhang, Z.; Khalfallah, M.; Ioos, F.; Grossmann, E.; Dhakal, H.N. Influence of Accelerated Weathering on the Properties of Flax Reinforced PLA Biocomposites. *Results Mater.* **2022**, *16*, 100333. [CrossRef]
19. Pantaloni, D.; Shah, D.; Baley, C.; Bourmaud, A. Monitoring of Mechanical Performances of Flax Non-Woven Biocomposites during a Home Compost Degradation. *Polym. Degrad. Stab.* **2020**, *177*, 109166. [CrossRef]
20. Pantaloni, D.; Melelli, A.; Shah, D.U.; Baley, C.; Bourmaud, A. Influence of Water Ageing on the Mechanical Properties of Flax/PLA Non-Woven Composites. *Polym. Degrad. Stab.* **2022**, *200*, 109957. [CrossRef]
21. Nassiopoulos, E.; Njuguna, J. Thermo-Mechanical Performance of Poly(Lactic Acid)/Flax Fibre-Reinforced Biocomposites. *Mater. Des.* **2015**, *66*, 473–485. [CrossRef]
22. Kishor Sharma, Y.; Meena, A.; Sahu, M.; Dalai, A. Experimental Investigation on Mechanical and Thermal Characteristics of Waste Sheep Wool Fiber-Filled Epoxy Composites. *Mater. Today Proc.* **2023**. [CrossRef]
23. Abdellah, M.Y.; Sadek, M.G.; Alharthi, H.; Abdel-Jaber, G.T.; Backar, A.H. Characteristic Properties of Date-Palm Fibre/Sheep Wool Reinforced Polyester Composites. *J. Bioresour. Bioprod.* **2023**, *8*, 430–443. [CrossRef]
24. Haque, A.N.M.A.; Naebe, M.; Mielewski, D.; Kiziltas, A. Waste Wool/Polycaprolactone Filament towards Sustainable Use in 3D Printing. *J. Clean. Prod.* **2023**, *386*, 135781. [CrossRef]
25. Garlotta, D. A Literature Review of Poly(Lactic Acid). *J. Polym. Environ.* **2001**, *9*, 63–84. [CrossRef]
26. Rbihi, S.; Aboulouard, A.; Laallam, L.; Jouaiti, A. Contact Angle Measurements of Cellulose Based Thin Film Composites: Wettability, Surface Free Energy and Surface Hardness. *Surf. Interfaces* **2020**, *21*, 100708. [CrossRef]
27. Stark, N.M.; Matuana, L.M. Characterization of Weathered Wood–Plastic Composite Surfaces Using FTIR Spectroscopy, Contact Angle, and XPS. *Polym. Degrad. Stab.* **2007**, *92*, 1883–1890. [CrossRef]
28. Rajesh, G.; Ratna Prasad, A.V.; Gupta, A.V.S.S.K.S. Soil Degradation Characteristics of Short Sisal/PLA Composites. *Mater. Today Proc.* **2019**, *18*, 1–7. [CrossRef]
29. Limsukon, W.; Rubino, M.; Rabnawaz, M.; Lim, L.-T.; Auras, R. Hydrolytic Degradation of Poly(Lactic Acid): Unraveling Correlations between Temperature and the Three Phase Structures. *Polym. Degrad. Stab.* **2023**, *217*, 110537. [CrossRef]
30. Avci, A.; Eker, A.A.; Bodur, M.S.; Candan, Z. Water Absorption Characterization of Boron Compounds-Reinforced PLA/Flax Fiber Sustainable Composite. *Int. J. Biol. Macromol.* **2023**, *233*, 123546. [CrossRef]
31. Hamdan, M.H.M.; Siregar, J.P.; Cionita, T.; Jaafar, J.; Efriyohadi, A.; Junid, R.; Kholil, A. Water Absorption Behaviour on the Mechanical Properties of Woven Hybrid Reinforced Polyester Composites. *Int. J. Adv. Manuf. Technol.* **2019**, *104*, 1075–1086. [CrossRef]
32. John, M.J.; Anandjiwala, R.D. Chemical Modification of Flax Reinforced Polypropylene Composites. *Compos. Part A Appl. Sci. Manuf.* **2009**, *40*, 442–448. [CrossRef]

33. Du, Y.; Wu, T.; Yan, N.; Kortschot, M.T.; Farnood, R. Fabrication and Characterization of Fully Biodegradable Natural Fiber-Reinforced Poly(Lactic Acid) Composites. *Compos. B Eng.* **2014**, *56*, 717–723. [CrossRef]
34. Adam, J.; Korneliusz, B.A.; Agnieszka, M. Dynamic Mechanical Thermal Analysis of Biocomposites Based on PLA and PHBV—A Comparative Study to PP Counterparts. *J. Appl. Polym. Sci.* **2013**, *130*, 3175–3183. [CrossRef]
35. Agüero, Á.; Garcia-Sanoguera, D.; Lascano, D.; Rojas-Lema, S.; Ivorra-Martinez, J.; Fenollar, O.; Torres-Giner, S. Evaluation of Different Compatibilization Strategies to Improve the Performance of Injection-Molded Green Composite Pieces Made of Polylactide Reinforced with Short Flaxseed Fibers. *Polymers* **2020**, *12*, 821. [CrossRef]
36. Awad, S.; Siakeng, R.; Khalaf, E.M.; Mahmoud, M.H.; Fouad, H.; Jawaid, M.; Sain, M. Evaluation of Characterisation Efficiency of Natural Fibre-Reinforced Polylactic Acid Biocomposites for 3D Printing Applications. *Sustain. Mater. Technol.* **2023**, *36*, e00620. [CrossRef]
37. Inseemeesak, B.; Siripaiboon, C.; Somkeattikul, K.; Attasophonwattana, P.; Kiatiwat, T.; Punsuvon, V.; Areeprasert, C. Biocomposite Fabrication from Pilot-Scale Steam-Exploded Coconut Fiber and PLA/PBS with Mechanical and Thermal Characterizations. *J. Clean. Prod.* **2022**, *379*, 134517. [CrossRef]
38. Chen, P.-Y.; Lian, H.-Y.; Shih, Y.-F.; Chen-Wei, S.-M.; Jeng, R.-J. Preparation, Characterization and Crystallization Kinetics of Kenaf Fiber/Multi-Walled Carbon Nanotube/Polylactic Acid (PLA) Green Composites. *Mater. Chem. Phys.* **2017**, *196*, 249–255. [CrossRef]

applied sciences

MDPI

Review

Upcycling Wine Industry Waste: Dealcoholized Grape Pomace as a Platform for Bio-Based Material Innovation

Jorge Miguel Matias [1], Fernando Braga [2] and Alice Vilela [3,*]

[1] School of Agrarian and Veterinary Sciences, University of Trás-os-Montes and Alto Douro, 5001-801 Vila Real, Portugal; jmmmds@gmail.com
[2] Chemistry Research Centre-Vila Real (CQ-VR), Department of Chemistry, School of Life Sciences and Environment, University of Trás-os-Montes and Alto Douro, 5001-801 Vila Real, Portugal; fbraga@utad.pt
[3] Chemistry Research Centre-Vila Real (CQ-VR), Department of Agronomy, School of Agrarian and Veterinary Sciences, University of Trás-os-Montes and Alto Douro, 5001-801 Vila Real, Portugal
* Correspondence: avimoura@utad.pt

Abstract

The wine industry produces substantial amounts of organic waste, particularly in the form of dealcoholized grape pomace—the primary residual biomass that remains after the fermentation process and the extraction of alcohol from winery by-products. This study explores the potential of upcycling dealcoholized pomace, an often-overlooked by-product, into a sustainable platform for innovative bio-based materials. Using a multidisciplinary approach that combines materials science, biotechnology, and principles of the circular economy, we carefully examine the physical, chemical, and mechanical properties of dealcoholized pomace. Our research includes comprehensive analyses of its structural integrity, biodegradability, and potential applications, including biocomposites, eco-friendly packaging solutions, and other sustainable materials. The results of our study highlight not only the promising performance characteristics of dealcoholized pomace, such as its strength-to-weight ratio and biocompatibility, but also underscore its significant role in advancing waste valorization strategies. By effectively transforming waste into valuable resources, we contribute to the development of sustainable materials, thereby supporting a more circular economy within the wine industry and beyond.

Keywords: waste valorization; grape pomace; dealcoholized grape pomace; spent seedless grape pomace; bio-based materials; circular economy; food loss and waste; bioplastics

1. Overview of Wine Production

Although global wine production and consumption show a declining trend, 237.3 million hectoliters were produced in 2023, with 144.5 million hectoliters originating from the European Union alone [1]. Annual wine industry operations consequently involve not only harvesting, transporting, and processing millions of tonnes of grapes but also managing large amounts of liquid effluents and biowastes, including grape stalks, grape marc (pomace), wine lees, and winery waste-activated sludge. The proper disposal of these by-products represents a technical challenge and a significant financial investment. However, if untreated, these residues pose a serious environmental risk [2].

In the European Union, which for decades has included five of the world's top ten wine-producing countries, regulatory measures promote the valorization of grape pomace (GP) and wine lees [3–5]. These policies offer financial incentives for wineries to deliver such by-products to authorized distilleries, where valuable compounds, such as grape pigments for the food industry, ethanol used as bioethanol or a raw material in fortified wines and spirits, and calcium tartrate for tartaric acid production, can be recovered [6].

Although estimates of the volume of GP generated as a by-product of winemaking vary depending on grape variety, vineyard practices, and processing techniques [7], there is general agreement that it accounts for 15 to 25% of the total processed grape weight [8,9], corresponding to approximately 4.7 to 7.9 million tonnes worldwide in 2023.

1.1. Definition and Composition of GP

Fresh grape pomace (GP) can be broadly defined as the solid residue composed of skins, seeds, and occasionally stems, resulting from grape pressing, which may or may not contain unfermented sugars [7].

This complex and heterogeneous bio-waste consists of approximately 425 kg of grape skins, 225 kg of grape seeds, and 249 kg of grape stalks (GS) per tonne, with a high and variable moisture content ranging from 50% to 72% [7]. Among the structural and bioactive compounds present, lignin, pectic substances, sugars, proteins, minerals, and phenolic compounds were quantified [8,10–14] (Table 1).

Table 1. Representative composition of GP (g/100 g, fresh weight basis).

Component	Amount	References
Moisture content	50–72	[7]
Lignin	16.8–24.2	[10]
Pectic substances	~20	[8]
Soluble sugars	1.5–6.2	[10,11]
Proteins	2.7–3.8	[10,14]
Phenolic compounds	5–10	[12,13]
Minerals	1.8–4.6	[10,11]

Traditionally, GP undergoes two distinct unit operations for recovering valuable substances. The choice of extraction process hinges on the unfermented sugar content and the levels of tartrate and grape pigments.

Fermented pomace from red wine production, which has low residual sugar, can be directly subjected to steam stripping distillation if ethanol recovery is the only desired outcome. The resulting product is a hydroalcoholic solution containing between 20 and 35% alcohol by volume (v/v), along with small amounts of ethyl acetate, methanol, and acetic acid. This solution is subsequently redistilled in appropriate columns to yield an alcohol with a concentration of 52–80% (v/v). This process is known in the wine industry as pomace "dealcoholization" (Figure 1).

Conversely, pomace from white or fortified wine production, still rich in unfermented sugars, is subjected to a countercurrent leaching process. This yields a dilute hydroalcoholic solution (approximately 3% (v/v) alcohol and 30–50 g/L sugars), which is then fermented to increase its alcohol content before distillation. Similarly, if red pomace contains sufficient grape pigments or tartrates to warrant extraction, the same leaching process is applied, followed by concentration of the heat-sensitive dilute pigment solution at low temperatures using a multiple-effect evaporator (Figure 2).

In the case of dealcoholized grape pomace (DGP), large-scale industrial applications typically involve partial dehydration by pressing, followed by drying in rotary furnaces to a final moisture content of approximately 5%. This process stabilizes the material microbiologically, enabling sieving to separate grape skins from seeds. Industrially, the recovered seeds can be used for grape seed oil extraction, as a component in animal feed, or as an efficient solid biomass fuel [15].

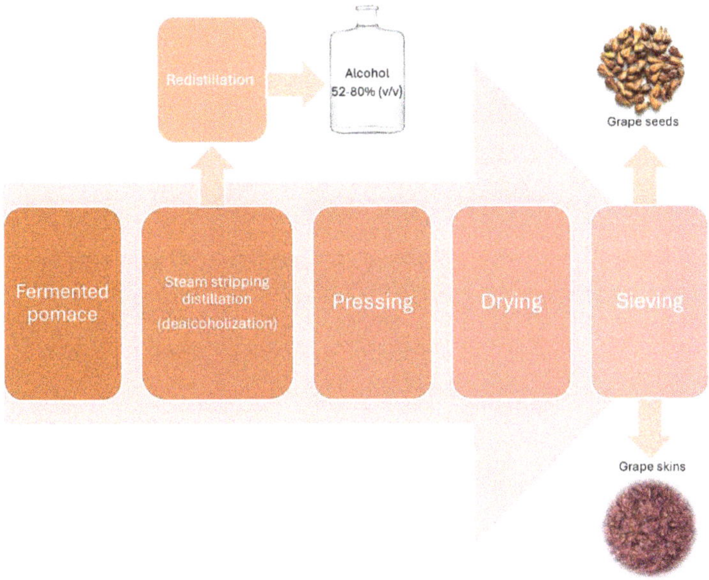

Figure 1. Schematic diagram of the distillation process for fermented pomace. Adapted from Braga et al. [6].

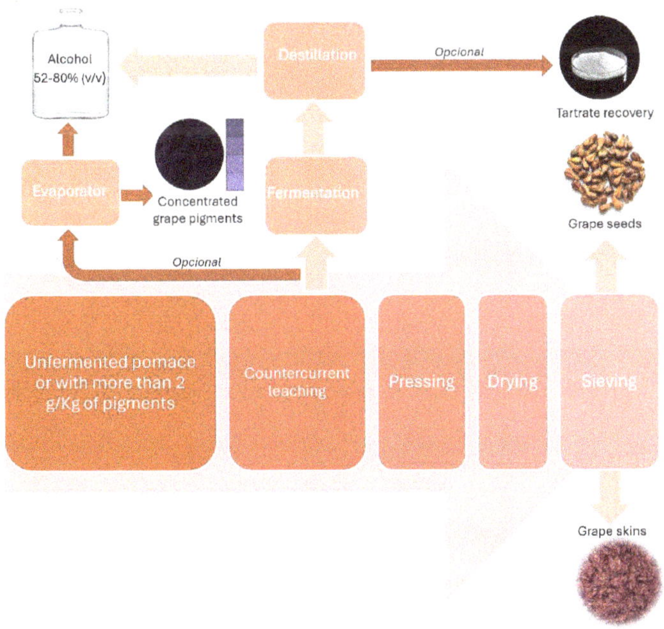

Figure 2. Schematic diagram of the countercurrent leaching process of pomace. Adapted from Braga et al. [6].

If no viable market exists for the seeds, or if the distillery lacks steam generators powered by solid fuels, the DGP may be used as soil mulch, as an organic amendment, or be composted—often in combination with other bio-wastes that offer a complementary C/N ratio [16,17].

While the recovery of specific bioactive compounds is affected by the dealcoholization processes mentioned earlier [18], structural compounds are largely unaffected by the conditions used due to their high thermal stability and low water solubility [19].

1.2. Rationale for Upcycling in the Context of the Circular Economy and Bioeconomy

As a result of seed separation after drying dealcoholized grape pomace, a new lignocellulosic material composed mainly of grape skins is now produced. This material is sometimes known as spent seedless grape pomace (SSGP).

In recent years, beyond its traditional uses, there has been growing interest in applying circular economy principles to enhance the value of SSGP [20]. Initial research has identified several valuable classes of compounds present in SSGP [21,22], including dietary fibers [23] and bioactive compounds with antioxidant, anticancer, cardioprotective, and anti-inflammatory activities [24–26].

Additionally, SSGP contains components with functional properties such as food colorants, texturizing agents, and natural antioxidants [27].

More recent studies have shifted focus toward developing sustainable extraction methods and optimizing conditions to maximize the recovery of these compounds [28,29]. Researchers have also begun exploring the use of SSGP as a second-generation feedstock for synthesizing polymer building blocks or as a reinforcing filler in composite materials [30].

Within the framework of the circular economy, the upcycling of SSGP contributes directly to reducing biomass waste and greenhouse gas emissions, extending material life cycles, and improving resource efficiency across the agri-food sector. Although dealcoholization may slightly reduce the levels of specific bioactive compounds, SSGP still retains a significant portion of its original composition, making it a versatile input for the development of secondary products. This makes it a valuable and versatile input for the development of secondary products. Its composition supports a wide range of applications, including food ingredients, dietary supplements, cosmetics, and biodegradable materials, which reduce our reliance on synthetic additives and non-renewable resources [31]. Furthermore, the removal of alcohol enhances the safety and regulatory compliance of these applications, particularly in food and nutraceutical products, by lowering ethanol content to acceptable or negligible levels [32].

From a bioeconomy perspective, valorizing SSGP supports a regenerative model in which biomass is repurposed efficiently and repeatedly across different sectors. This approach aligns with EU and global bioeconomy strategies by fostering local value chains and promoting rural development, especially in wine-producing regions where large volumes of pomace are generated annually [33,34]. When combined with green extraction technologies, such as supercritical CO_2 or subcritical water extraction, or when integrated into biopolymer systems based on polylactic acid (PLA), poly(butylene succinate), or starch derived from agri-food by-products, SSGP enables the sustainable recovery of high-value compounds. These innovations offer a low environmental impact and reinforce the role of bio-based solutions in accelerating the transition to a more sustainable industry [35].

2. Physicochemical Properties of SSGP

As mentioned earlier, because DGP is inherently prone to microbial instability, whether produced by steam distillation or countercurrent leaching, drying is always required if it is not intended for composting [36–38]. For this reason, the final physicochemical characteristics of SSGP are only marginally influenced by grape variety and winemaking method and are primarily determined by thermal and oxidative effects (Table 2).

Table 2. Representative composition of SSGP (g/100 g, as-is basis).

Component	Amount	References
Moisture content	3–7	[36,37]
Lignin	27–56	[26,37,39]
Proteins	2.8–3.0	[36]
Phenolic compounds	0.5–0.6	[38]
Minerals	1.5–10	[36,37]

A comparison of GP (Table 1) and SSGP (Table 2) reveals significant physicochemical transformations resulting from processing. Most notably, SSGP exhibits a drastically lower moisture content (3–7%) compared to the high moisture levels of fresh GP (50–72%), indicating extensive drying. This drying and leaching process also leads to the complete removal of soluble sugars from SSGP, which are present at 1.5–6.2% in fresh GP. Phenolic compounds are substantially reduced, decreasing from 5–10% in GP to just 0.5–0.6% in SSGP. Conversely, the removal of moisture and soluble components results in a relative concentration of other constituents: lignin content increases from 16.8–24.2% in GP to a much higher 27–56% in SSGP. Protein levels remain relatively consistent (2.7–3.8% in GP vs. 2.8–3.0% in SSGP), while mineral content in SSGP (1.5–10%) exhibits a broader range and a potentially higher upper limit than in GP (1.8–4.6%). Pectic substances, prominent in GP (~20%), are not reported for SSGP, suggesting their removal or significant alteration.

3. Pretreatment and Processing Technologies

Efficient processing and extraction of DGP and SSGP necessitate appropriate pretreatment methods to disrupt the complex plant cell wall matrix and enhance the accessibility and recovery of target molecules such as bioactive polyphenols and dietary fibers. These pretreatments can be broadly categorized into mechanical, chemical, and enzymatic approaches, often used in combination to maximize yields and product quality.

Mechanical pretreatments are fundamental first steps, primarily involving drying, milling, and size reduction. Drying is crucial for reducing moisture content, preventing microbial spoilage, concentrating compounds, and preparing the material for subsequent processing. Methods such as convective air drying, freeze-drying, and microwave-assisted drying are employed, with the choice influencing energy consumption and the preservation of thermolabile compounds, including certain polyphenols [40]. Following drying, milling or grinding reduces particle size, significantly increasing the surface area available for solvent penetration and interaction. This enhanced surface area is directly correlated with improved extraction efficiency of both polyphenols and the subsequent accessibility of dietary fiber components [41].

Chemical pretreatments predominantly involve the use of solvents for extraction, but they can also include treatments to modify fiber structure. For the extraction of phenolic compounds, solvents such as water, ethanol, methanol, acetone, and their aqueous mixtures have been investigated. Ethanol–water mixtures are particularly favored due to their Generally Recognized as Safe (GRAS) status and their ability to efficiently extract a broad range of substances. The choice of solvent, temperature, and extraction time is a critical parameter influencing yield and selectivity [42]. Emerging green solvents, such as deep eutectic solvents (DES), are also gaining attention for their efficacy and environmental benefits. For dietary fiber modification or isolation, mild acid or alkaline treatments can be used to hydrolyze hemicellulose or delignify the material, making cellulose more accessible. However, this process must be carefully controlled to preserve other valuable components.

Enzymatic pretreatments provide a more targeted and gentle approach to deconstructing the plant cell wall, thereby releasing entrapped bioactive compounds and modifying the characteristics of dietary fiber. Enzymes such as cellulases, hemicellulases, pectinases, and tannases can be used individually or in combinations. These enzymes catalyze the hydrolysis of specific polysaccharides, including cellulose, hemicellulose, and pectin, which form the structural backbone of the cell wall. This process improves the release of polyphenols and increases the soluble dietary fiber fraction. Enzyme-assisted extraction (EAE) often yields higher polyphenol concentrations under gentler conditions, such as lower temperatures and less organic solvent, compared to conventional methods, thereby preserving their bioactivity [43]. Furthermore, enzymatic treatment can enhance the functional properties of dietary fibers, such as their water-holding capacity and fermentability.

4. Dealcoholized Grape Pomace Applications in Bio-Based Materials

4.1. Bio-Based Materials Used in Our Daily Life

Bio-based materials are increasingly used in daily life as sustainable alternatives to petroleum-based products. These materials are derived from renewable biological resources and are found in a wide range of everyday applications, helping to reduce our environmental impact and support a circular economy.

Made from natural resources such as corn starch, cellulose, sugarcane, and agri-food by-products, bioplastics are used in packaging, cutlery, bowls, straws, and other applications. They are biodegradable or partially biodegradable, making them less harmful to the environment compared to conventional plastics [44–48], as shown in Table 3.

Table 3. Summary of some bio-based materials used in our daily life, including their source and composition.

Bio-Based Material	Daily Life Application(s)	Source/Composition	Reference(s)
Fungal-based biomaterials	Packaging, textiles, leather alternatives, automotive, insulation, fire protection	Fungal mycelium, lignin/cellulose waste	[49]
Bio-polyethylene (Bio-PE), Bio-polypropylene (Bio-PP), Bio-PET	Packaging, durable goods, and electronics	Biomass-derived monomers	[44]
Biofibers, biopolymers, biocomposites	Construction, engineering, biomedical, and commercial products	Cellulose, starch, chitosan, natural fibers	[50,51]
Bioplastics from fruit waste	Packaging, cutlery, bowls, straws	Orange peel (cellulose), corn starch, sugarcane bagasse	[45]
Polylactic acid (PLA), Polyhydroxyalkanoate (PHA), cellulose, starch, proteins, lipids, waxes	Packaging, textiles	Plant-based polymers, natural resources	[46]
Biodegradable nanofiber materials	Energy, environmental, and biomedical applications	Starch, cellulose, chitosan, PLA, PCL	[51]

Materials such as cellulose, starch, and chitosan are processed into fibers, films, and composites for use in construction, textiles, and engineering, offering renewable and sustainable alternatives to synthetic materials [50,51]. Fungal mycelia are utilized to produce biomaterials for packaging, textiles, leather alternatives, automotive components, insulation, and fire protection. These materials are renewable, degradable, and have significant innovation potential [49].

Bio-polyethylene (Bio-PE), bio-polypropylene (Bio-PP), and bio-poly(ethylene terephthalate) (Bio-PET) are developed for packaging, durable goods, and electronics, mimicking the properties of petroleum-based plastics while reducing their carbon footprint [44].

Bio-based materials are widely used in food packaging, shopping bags, and containers, offering improved sustainability and options for recycling or composting [44–48].

Bio-based fibers and polymers are utilized in clothing and textiles, offering biodegradable, renewable alternatives to synthetic fabrics [46,48–50]. Items such as disposable cutlery, bowls, and straws, as well as insulation and automotive components, are increasingly made from biobased materials [45,46,49].

Bio-based materials can also significantly reduce greenhouse gas emissions compared to their fossil-based counterparts, particularly in targeted applications such as packaging and textiles [47,48,52]. These materials are suitable for recycling, composting, and energy recovery, which supports a circular economy and reduces landfill waste [46,47,52].

4.2. Dealcoholized Grape Pomace as an Enhancer or Filler in Bio-Based Materials

Despite its high availability and industrial potential, only a limited number of studies over the past decade have investigated the use of DGP or grape pomace-derived materials as fillers or enhancers in bio-based materials. However, all available studies report significant improvements in properties such as thermal stability, heat resistance, and overall performance of the resulting biocomposites.

Auriemma et al. [53] examined the incorporation of a grape pomace extract (GPE), along with other phenol-based additives such as tannic acid (TA) and lignocellulosic biomass (LC), into poly(3-hydroxybutyrate) (PHB), a biodegradable polymer known for its limited mechanical performance and narrow processing window. Their results demonstrated that these natural additives significantly improved the thermal resistance of PHB, reduced its degradation during melt processing, and preserved its molecular weight. Rheological analysis revealed a slower viscosity decay in doped samples, indicating that LC functioned as a heterogeneous nucleating agent, which promoted crystallization during cooling and potentially mitigated physical aging. This study highlights the potential of grape pomace-derived additives as sustainable and functional stabilizers for bioplastics, such as PHB, offering an eco-friendly alternative to synthetic polymer stabilizers.

Expanding on these findings, similar improvements in thermal and mechanical performance have been observed with other agro-industrial residues, such as sugarcane bagasse and spent grain from beer production, incorporated into polymer matrices like PLA, polypropylene (PP), and high-density polyethylene (HDPE) [54–58]. These natural fillers typically enhance thermal stability, promote crystallinity, and improve resistance to thermal degradation and aging. Such improvements are often attributed to enhanced interfacial adhesion between the fibers and the polymer matrix, particularly when compatibilizers or chemical modifications are employed to improve compatibility. DGP exhibits similar reinforcing behavior, and its wide availability contributes to a reduction in the required polymer content, thereby increasing the sustainability and cost-effectiveness of the resulting biocomposites.

In terms of direct applications, Gowman et al. [59] explored the development of biocomposites using bio-based poly(butylene succinate) (BioPBS) reinforced with DGP via melt extrusion and injection molding. Composites containing up to 50 wt% DGP were produced, with specific formulations compatibilized using in situ synthesized maleic anhydride-grafted BioPBS (MA-g-BioPBS). The incorporation of DGP significantly improved both flexural and impact strength, with the 57:40:3 BioPBS/GP/MA-g-BioPBS blend showing the best overall performance, achieving increases of 28.4% and 59% in flexural and impact strength, respectively, compared to neat BioPBS. Additionally, the heat distortion temperature increased by 14.3%, and DGP acted as a reinforcing phase that enhanced the material's mechanical integrity. Scanning electron microscopy revealed improved interfacial adhesion in compatibilized samples, while thermogravimetric analysis confirmed the thermal stability of DGP under processing conditions.

More recently, interest has emerged in the development of bioactive laminated composite films incorporating graphene-based polyethylene (GPE). Vázquez et al. [60] developed bioactive laminated films based on bacterial cellulose (BC) and chitosan, enriched with GPE as an antioxidant and glycerol as a plasticizer, targeting applications in sustainable food packaging. Formulations were systematically varied to evaluate their mechanical and thermal properties. The results indicated strong interactions between components, leading to improved strength and flexibility. Tensile strength values ranged from 16.92 to 32.71 MPa, and films with GPE showed up to a fourfold increase in elongation (from 2.33% to 8.63%) compared to control samples. Puncture resistance also improved with the addition of GPE and glycerol. When applied as separators for Havarti cheese, the films effectively reduced lipid oxidation by 67.3% after 60 days of storage while maintaining antioxidant activity. These findings support the use of GP-derived components in biodegradable, active packaging solutions that align with the principles of the circular economy.

Several authors have also concluded that the GPE imparts antioxidant properties to films, which is beneficial for food packaging applications by delaying oxidation and extending shelf life [61]. The integration of pomace results in more compact and well-bonded film structures, as confirmed by microscopy and spectroscopy analyses [62].

4.3. *Incorporation of Dealcoholized Grape Pomace in Biodegradable Plastics*

Incorporating DGP or materials derived from grape pomace into biodegradable plastics represents an emerging strategy to improve the sustainability and functionality of packaging materials.

In 2023, the Science Learning Hub [63] reported on the development of biodegradable vine net clips made from DGP, proposed as a sustainable alternative to conventional plastic clips used in vineyards. Conducted by Scion's Biopolymers and Chemicals division, the project addressed a primary environmental concern, i.e., the estimated 16.8 million plastic clips used annually in New Zealand vineyards, which are rarely collected and often accumulate as persistent microplastic waste. The bioplastic composite was formulated by blending PLA with dried grape pomace, resulting in a material with sufficient mechanical strength for vineyard use and enhanced biodegradation compared to PLA alone. According to the authors, the vine clips biodegrade more rapidly, offering an eco-friendly solution for single-use applications in viticulture and demonstrating a tangible step forward in achieving circular bioeconomy objectives.

This study demonstrates the feasibility of using DGP as a functional filler to enhance the mechanical and thermal performance of biodegradable polymer composites, supporting the design of more sustainable materials.

More recently, Titone et al. [64] explored the incorporation of dried and ground grape pomace into the commercial biodegradable polymer blend Mater-Bi (MB)—a mixture of PLA and poly(butylene adipate-co-butylene terephthalate) (PBAT)—to improve its properties and sustainability. The study developed and characterized biocomposites containing 10% and 20% DGP, assessing their mechanical, chemical, and degradation behavior compared to neat MB. The addition of DGP enhanced the antioxidant properties of the biocomposites and modified their rheological behavior, particularly increasing the complex viscosity at low frequencies. A modest improvement in elastic modulus and crystallinity was also observed. Soil degradation tests indicated an accelerated weight loss at higher GP content, suggesting enhanced biodegradability. NMR analysis further revealed compositional changes in PBAT during degradation, including an increase in terephthalic acid content. This work highlights the potential of grape pomace as a functional filler that can boost both the performance and environmental sustainability of biodegradable plastics.

Beyond DGP, other by-products obtained after dealcoholization, drying, and screening—such as GS, grape seeds, and grape seed flour—can also be used as additives or reinforcements in biopolymer films and foams. These materials are intended to enhance mechanical, antioxidant, and barrier properties while promoting environmental compatibility, and they have been investigated in the development of biodegradable plastics (Table 4). GS can be effectively used as fillers in cassava starch-based foams, integrating well without causing agglomeration or impairing foam structure. These foams are fully biodegradable within seven weeks and are suitable for packaging low-moisture food products [65].

Table 4. Additional examples of DGP-derived components incorporated in biodegradable plastics.

Example/Material Type	DGP Component	Biodegradable Matrix/Polymer	Key Properties/Findings	References
Starch-based foams with grape stalks	Grape stalks (from pomace)	Cassava starch	Good morphology, complete biodegradation in 7 weeks, suitable for food storage with low moisture content	[65]
Biodegradable spoons with grape flour	Grape flour (from grape pomace)	Proso millet, wheat, xanthan, palm oil	High strength, improved antioxidant activity, enhanced nutritional profile	[66]
Packaging films with grape seed lignin	Grape seed lignin (from pomace)	PHB/PHA blend	Improved gas barrier, antioxidant activity, enhanced biodegradability, non-toxic degradation products	[67]

Grape seed flour has been applied in the production of biodegradable cutlery. The resulting spoons provide a sustainable alternative to plastic utensils, combining structural strength with added nutritional and health benefits. Specifically, the flour enhances the antioxidant activity and improves the nutritional profile of the final product [66]. In another application, grape seed lignin has been incorporated into polyhydroxyalkanoate (PHA/PHB) films to improve their antioxidant capacity, gas barrier performance, and biodegradability. These films also show improved decomposition rates and yield non-toxic degradation by-products [67].

Grape pomace-derived materials consistently demonstrate rapid and complete biodegradation in soil or compost, with no formation of recalcitrant or toxic residues. Additionally, components such as lignin contribute to increased tensile strength, improved gas barrier properties, and overall durability of biopolymer films and foams [65,67]. They also exhibit significant antioxidant activity, which can extend the shelf life of packaged foods and enhance their nutritional value [66].

Starch-based foams made with GS and PHA/PHB films containing grape seed lignin have shown promise for packaging foods—especially those with low moisture content—and for extending the shelf life of fruits such as grapes. While these positive results are promising, large-scale adoption remains limited due to production costs, variability in agro-waste availability, and the need for further optimization of mechanical and functional properties [68].

Continued research is essential to refine formulations and processing techniques, thereby maximizing the benefits of incorporating grape pomace while ensuring economic feasibility and scalability.

5. Environmental Benefits of Incorporating Dealcoholized Grape Pomace in Bio-Based Materials

Incorporating DGP into bio-based materials offers several environmental benefits, primarily by transforming agricultural waste into valuable, sustainable products. This approach reduces reliance on fossil-based plastics, lowers greenhouse gas emissions, and supports the development of biodegradable materials. Using grape pomace, a by-product of the wine industry, also helps manage large volumes of seasonal agricultural waste, converting it into a functional material rather than waste destined for landfill or incineration [69]. Grape pomace can be converted into porous carbons and fibers for use in biocomposites, superhydrophobic coatings, and biodegradable films, adding value to this waste stream [70].

Replacing petroleum-based plastics with grape pomace-based biocomposites significantly reduces the global warming potential, fossil resource use, and terrestrial acidification, primarily when clean energy is utilized in the processing [68].

Life cycle assessments show that incorporating DGP or DGP-derived materials into biocomposites lowers greenhouse gas emissions and fossil resource scarcity compared to both sugarcane-based and fossil-based polyethylene [71]. Bio-based polymers derived from grape pomace and other agro-wastes have a lower environmental footprint compared to synthetic plastics, as they are renewable, biodegradable, and contribute to reducing plastic pollution.

6. Challenges and Limitations

A primary challenge in valorizing DGP or SSGP lies in its inherent variability in composition. This variability stems from diverse factors, including the grape variety, specific viticultural practices, the climatic conditions of the vintage year, the winemaking process itself, and the subsequent dealcoholization method (e.g., steam distillation vs. leaching). As a result, general extraction or processing techniques often yield inconsistent outcomes in terms of polyphenol profiles, fiber characteristics, or other target compounds. This necessitates dedicated studies to adapt and optimize these techniques for specific streams, moving beyond one-size-fits-all approaches to ensure consistent product quality and yield. The scaling up of valorization processes and the management of supply chain logistics present considerable practical challenges, particularly for companies that are typically distilleries. These facilities are often not equipped or designed for the sophisticated downstream processing required to produce pharmaceutical raw materials or high-purity food ingredients. Investment in new equipment, specialized personnel, and robust quality control systems is substantial. Moreover, distilleries may lack the necessary certifications (e.g., Good Manufacturing Practice, or GMP) to access high-value markets, posing a significant barrier to realizing the full economic potential of GP-derived products. Ensuring economic viability and cost competitiveness is another critical hurdle, even though distilleries are often strategically located in large wine-growing regions, which can minimize their carbon footprint and reduce the costs associated with collecting fresh grape pomace. The processes involved in drying, extracting, purifying, and stabilizing valuable compounds from DGP or SSGP can be energy-intensive and require significant capital investment. The final products must then compete in markets where established alternatives already exist, meaning valorization pathways must be not only technically feasible but also economically viable to justify the investment and operational costs. Finally, regulatory and standardization hurdles significantly impact the commercialization of SSGP-derived products. There is often a lack of specific regulations or established quality standards for novel ingredients extracted from SSGP, such as defined polyphenol extracts or specialized dietary fibers. This regulatory ambiguity creates uncertainty for both producers and consumers, complicates international trade, and slows market acceptance. Navigating the complex approval processes for food additives, novel foods, or active pharmaceutical ingredients requires substantial effort and investment, often acting as a deterrent for smaller enterprises or those new to these regulated sectors [72–74].

7. Emerging Research and Innovations

7.1. Nanocellulose Extraction from Grape Pomace

Cellulose is a natural, biodegradable biopolymer made up of β-anhydro-D-glucose units. While traditionally sourced from forest wood, agro-industrial by-products like fruit peels and husks are now recognized as valuable sources of cellulose. Recycling these materials can significantly reduce the environmental impact of food waste in agricultural and industrial activities [75].

Agro-industrial wastes offer low-cost and sustainable options for nanocellulose due to their consistent availability. This approach not only helps minimize disposal pollution but also promotes economic viability [75].

The extraction of cellulose and nanocellulose typically involves several steps, including pre-treatment, purification, acid hydrolysis of amorphous regions, and mechanical treatment to achieve nanoscale sizes. Acid hydrolysis and mechanical methods are widely used in industrial recovery [75].

Extraction techniques range from conventional methods with toxic chemicals like sodium hydroxide and sulfuric acid to newer approaches that aim for higher purity cellulose. Pre-treatment steps, such as hot water washing, are essential for quality recovery but are often neglected in reports [75].

Newer, more environmentally friendly techniques are being developed and discussed. In Table 5, we summarize some of them.

Table 5. Newer and environmentally friendly extraction techniques. Adapted from Picot-Allain and Emmambux [75].

Extraction Method	Description	Advantages	Limitations
Enzyme-assisted extraction	Uses enzymes to produce cellulose nanofibrils from sources like sugar beet pulp.	Reduces water consumption; avoids chemical pre-treatments.	Enzyme activity can be influenced by temperature, pH, and secondary metabolites present in agro-industrial waste.
High-pressure homogenization	Forces cellulose slurry through a nozzle under high pressure to isolate nanostructures.	Eco-friendly, efficient, no need for organic solvents.	Clogging due to insoluble cellulose requires pre-treatment, as well as increased investment and training.
Ultrasonication	Applies ultrasonic waves to modify the morphology of nanocellulose.	Higher power leads to individualized nanofibrils with uniform width.	Not explicitly mentioned, but results can vary with power settings.
Steam explosion	Combines mild acid treatment, steam explosion, and ultrasonication to produce fibrillar nanocellulose.	Low capital cost, low energy demand, and environmentally friendly.	Limited selectivity—cannot specifically recover nanocrystals or nanofibers.
Supercritical CO_2 and cold plasma extraction	Emerging green technologies for cellulose extraction.	Potentially green alternatives.	Cold plasma lacks sufficient scientific validation for the extraction of agro-industrial nanocellulose.

The physico-chemical properties of nanocellulose from different sources need to be characterized. Nanocellulose exhibits enhanced properties compared to micro- and macrocellulose, including an improved fiber–matrix interface and good dispersion in matrices. Nanocellulose suspensions are transparent due to their smaller and homogeneous fiber size. Important properties evaluated include the crystallinity index and thermal stability. Techniques such as X-ray diffraction (XRD), ^{13}C nuclear magnetic resonance (NMR) spectroscopy, and infrared (IR) spectroscopy are used to assess crystallinity, with XRD being the most commonly employed method. Recovered nanocellulose typically exhibits an increase in crystallinity index and a higher thermal decomposition temperature compared to the raw source material [75].

Nanocellulose's abundance, non-toxic nature, mechanical properties, and biodegradability make it attractive for various applications. Some of the applications are summarized in Table 6.

Table 6. Nanocellulose applications. Adapted from [75].

Application Area	Nanocellulose Use
Cosmetics	Nanocellulose suspensions exhibit optical properties similar to those of nano-TiO_2 particles, which are used in concealers and sunscreens.
Drug Delivery/Medical Applications	Nanocellulose films serve as wound dressings, inhibiting bacterial growth, particularly when combined with honey. Nanocellulose-alginate is utilized in 3D printing for personalized dressings and serves as a potential carrier for ions such as calcium.
Food/Biodegradable Plastic	Acetylated nanocellulose films exhibit reduced solubility in non-aqueous food systems or during dry storage. Starch–nanocellulose biocomposites are biodegradable and suitable for use in food packaging. Chitosan-oxidized nanocellulose films enhance strength, moisture barrier properties, and thermal stability. Nanofibrillated cellulose biocomposites provide high tensile strength and modulus for short-life packaging applications.
Green Tire Tech/Rubber Compounding	Rice husk nanocellulose composites reduce rolling resistance, resulting in lower fuel use and emissions. Nanocellulose crystals improve rubber's tensile strength, stress, and elongation at break.

Extensive grape pomace produced from red wine-making can seriously pollute the environment and represents a waste of resources. Although active constituents, such as polyphenols or vitamins, are sometimes extracted, much of the grape pomace, rich in cellulose, is often treated as waste [76].

An eco-friendly method for extracting Cellulose Nanocrystals (CNCs) from grape pomace can be achieved using a simple and environmentally friendly deep eutectic solvent (DES). The green DES used had a composition of lactic acid and choline chloride in a 2:1 molar ratio. The extraction process involved pretreating crushed grape pomace to remove impurities, lignin, and hemicellulose, resulting in cellulose. This cellulose is then hydrolyzed using the DES. A key advantage of DES is its eco-friendliness and ease of preparation. The components of the DES (lactic acid and choline chloride) and ethyl acetate used for separation can be recycled, offering a sustainable and straightforward preparation method [77]. The extracted CNCs successfully obtained are rod-like crystals with an average length of 241.5 ± 45.3 nm and a diameter of 22.0 ± 3.9 nm. They show higher crystallinity (95.2%) than the original cellulose (89.6%), indicating the successful removal of amorphous regions. FT-IR and electric conductivity titration confirmed the presence of carboxylic acid groups (0.63 mmol/g) on the CNC surface, resulting from esterification with lactic acid. The surface-modified CNCs demonstrated good thermal stability and excellent dispersibility in water, remaining stable for at least 24 h [76].

A study was conducted to evaluate different methods for extracting nanocellulose, specifically cellulose nanofibers (CNFs), from yerba mate sticks (YMSs). The study successfully extracted nanocellulose (CNF) from YMSs using mild chemical treatments and steam explosion. The results demonstrated that the treatments effectively removed amorphous components, such as lignin and hemicellulose, leading to an increase in the crystallinity index of the samples. The use of both alkaline treatment and steam explosion favored the isolation of cellulose crystals, resulting in the highest crystallinity in the YMS-ASBHS sample. TEM and AFM analyses confirmed the isolation of nanocellulose in all treated samples, with the YMS-ASBHS and YMS-SBHS samples exhibiting better morphology and size characteristics, suggesting that steam explosion facilitates the removal of amorphous components. Notably, the study found that CNF could be obtained from YMS even without the alkaline treatment step (YMS-SBHS), suggesting a potential route for reducing processing steps and minimizing liquid waste generation [77].

Another angle is the possibility of utilizing wine industry wastes as a nutrient source for producing bacterial cellulose (BC) instead of using a costly commercial medium, such as the standard Hestrin–Schramm (HS) medium [78]. In the study by Ogrizek et al. [78], it was concluded that white grape pomace is a more suitable substitute for expensive commercial glucose than red grape pomace for producing bacterial cellulose. Using white pomace as a substitute carbon source resulted in a significantly higher yield and almost five times higher water-holding capacity, along with greater flexibility compared to BC produced in standard HS medium. While red grape pomace did produce BC, the resulting films were brittle. BC produced from red pomace media could be valuable for applications requiring fragmented, highly porous particles, such as in the food industry. However, white pomace BC shows greater potential for broader applications in sectors such as textiles and biomedicine, including uses such as wound dressings, hygienic face masks, and carriers for active ingredients [77].

7.2. Biochar and Activated Carbon from Residuals

As mentioned before, grape production and wine industry operations generate significant biomass residues, which are often treated as waste due to high transportation costs and low demand for by-products. Open burning, a standard disposal method, harms the environment [79].

Thermochemical conversion technologies can process biomass closer to harvest sites, improving transportation efficiency and producing valuable bio-products. Distributed-scale systems are being developed for this purpose; however, knowledge gaps exist regarding their outputs and integration into existing supply chains, posing investment risks [79].

There are few studies on thermochemical conversion of grape and wine by-products, making it necessary to compare these systems with those for forest biomass and other sources.

The study conducted by Anderson et al. [79] characterizes the producer gas, biochar, and activated carbon produced by a 700 kg h^{-1} prototype gasification system (TEA) and a 225 kg h^{-1} pyrolysis system (BSI) using coniferous sawmill and forest residues. The goal was to understand the products from distributed-scale thermochemical conversion systems for forest biomass, which are not widely deployed or well-characterized in this sector [79]. (i) The gas from the TEA gasifier had a higher energy content than the gas from the BSI pyrolysis system. For the TEA system, gas from sawmill residues averaged 12.4 MJ m^{-3}, and gas from forest residues averaged 9.8 MJ m^{-3}. Gas from the BSI system averaged 1.3 MJ m^{-3} for mill residues and 2.5 MJ m^{-3} for forest residues. (ii) Biochars from both systems showed similar particle size distributions and bulk density but varied in pH and carbon content. (iii) Biochars from both systems were successfully activated using steam activation, achieving BET surface areas within the range of commercial activated carbon [79].

Distributed thermochemical conversion systems can help meet operational needs in wood product manufacturing, including waste disposal, heat production, power generation, and the creation of value-added products. The gas and biochar outputs from the studied systems can produce heat, electricity, and various marketable products. However, uncertainties remain [79].

Gale et al. [80] investigated how hydrothermal carbonization (HTC) and slow pyrolysis (SP) produce distinct biochar precursors from corn stover, with HTC facilitating more efficient biomass breakdown at lower temperatures. HTC biochars contain more oxygen-based surface functional groups. The preparation method significantly influences the properties of activated carbon (AC); AC from corn stover and HTC biochars exhibit higher surface areas and better vanillin adsorption compared to AC from SP biochars. While surface area is important, pore structure and surface functional groups also impact adsorption performance. Gale et al. [80] also suggest that AC from corn stover could be as effective as that from HTC and SP biochars, indicating that the additional biochar synthesis step may not be necessary for producing AC adsorbents for phenolic compounds.

The study by Gonçalves et al. [81] demonstrates that Brewer's Spent Grain (BSG) and Brewer's Surplus Yeast (BSY) can be effectively utilized as raw materials to produce value-added products, such as granular activated carbon (GAC) and bio-oil, via pyrolysis and CO_2 activation. The implementation of this reuse strategy can provide significant benefits by generating valuable products, creating jobs, and addressing pollution issues associated with brewery residues.

7.3. Integration with Other Agri-Food Waste Streams

The agri-food sector generates significant waste due to inefficiencies in the supply chain, contributing to environmental degradation and greenhouse gas emissions, which account for 19–29% of total emissions. About 25% of resources are used to produce food that is lost or wasted. The improper disposal of organic waste, particularly from processed fruits and vegetables, poses substantial economic and environmental challenges [82].

According to the same author [82], utilizing food waste through a biorefinery approach can help achieve sustainable development. This concept integrates biomass conversion processes to produce fuels, power, and other valuable chemicals. By refining agro-industrial by-products, essential nutrients can be extracted, enabling market flexibility. By-products from the processing of tomatoes, olives, grapes, and apples are rich in health-promoting bioactives. For example, tomato waste, which accounts for 5 to 19% of the processed mass, contains carotenoids, polyunsaturated fatty acids, and pectin. Olive oil production can result in 30% waste, with olive pomace still containing valuable nutrients. Grape pomace also provide antioxidant-rich residues.

An Integrated Biorefinery Model could merge olive and grape pomace, utilizing supercritical CO_2 extraction to obtain oils, which can be used in nutraceuticals and biodiesel production. Subsequently, the solid residue can then be utilized for biogas production through anaerobic digestion. Overall, integrated biorefineries can mitigate the environmental impact of food waste while generating high-value products and energy, promoting more efficient use of diverse agri-food processing wastes [82].

7.4. Smart Materials and Responsive Composites

Smart materials can undergo significant changes in their properties under controlled conditions, enabling them to detect and respond to stimuli for beneficial effects. They demonstrate "smart behavior" by reacting reliably to physical or chemical inputs. Recent advancements include self-healing components, sensing materials, and shape-changing substances, with a focus on intelligent bio-based composites utilizing natural biopolymers, such as cellulose, which can be enhanced with inorganic elements to achieve dynamic responses. Applications span biomaterials, robotics, and artificial muscles.

According to Ramakrishnan et al. [83], two-dimensional smart membranes are gaining traction for their roles in separation processes, healthcare, and environmental remediation, showcasing capabilities such as pressure transduction and energy harvesting. Advancements in smart biodegradable food packaging utilize plant extracts, such as anthocyanins, for quality monitoring and shelf-life enhancement, alongside innovative photoactivated materials. Smart materials are paving the way for improvements in health and well-being, facilitating innovations in medical applications, from physiological monitoring to minimally invasive surgery. Their development is crucial for advancements in wireless communication, smart buildings, and sustainable design, using materials such as piezoelectric and thermochromic options that enhance energy efficiency.

Smart hybrid Nickel–Titanium (NiTi) materials are increasingly used due to their smart properties. Combining NiTi alloys creates a "smart material" with smart alloy properties. NiTi alloys benefit various sectors, with a primary focus on automotive applications, including suspension systems, wing mirrors, and safety lock units. NiTi can also be used in shape-memory alloys for the automotive sector.

8. Industrial and Market Perspectives

8.1. Current Commercial Initiatives and Case Studies

The valorization of winemaking by-products and waste presents significant economic, social, and environmental advantages, often yielding higher financial returns than wine production. Grape pomace, in particular, has proven economically efficient for valorization. Rising waste disposal costs in the EU emphasize the need for viable recycling solutions. Economically, waste valorization can reduce management costs, sustain jobs, and enhance productivity by leveraging by-products like anthocyanins for use in food, pharmaceuticals, and cosmetics. This boosts income for wine producers and promotes high-performance products. Socially, benefits include improved agricultural management, adherence to European labor standards, job creation, and enhanced quality of life in rural areas. Environmentally, it promotes responsible management of emissions and resource use, reduces waste, and enables the utilization of by-products as fertilizers, which in turn protects the environment and creates job opportunities [84].

In the article "The Exploitation of By-Products in a Waste Management Process", the authors address the universal contemporary challenge of sustainability within the global economic, social, cultural, and physical environment. They highlight the critical role of interdisciplinary Supply Chain Management (SCM) and its evolution into Sustainable SCM (SSCM) and Green SCM. The wine industry is identified as a fertile ground for developing multidisciplinary collaborations and applying innovative entrepreneurship practices in both forward and reverse agrifood chains [85]. The paper explicitly studies the exploitation of opportunities arising from wine production and waste management in this new environment. It presents a start-up business plan developed through a collaboration between the Agricultural University and the Harokopio University of Athens, Greece. The focus is on a wine waste management company located on the island of Crete, Greece, demonstrating how challenges can be transformed into opportunities for innovation and more efficient use of by-products and wastes.

The study's methodology involved an action research framework and a business plan goal. The location, Kissamos province in Crete, was chosen due to its tradition in viticultural farming, wine production, and historical innovative capability. Data were collected through interviews with managers of potential supplier wineries and potential customer pharmaceutical and food companies. Secondary data provided technical information for the company's setup and costs. Analytical tools used included PESTEL and SWOT analyses. A technical and financial analysis, including profit and loss (P/L) evaluation and economic performance indicators, was also conducted.

The start-up company in Kissamos, West Crete, will focus on elaborating wineries' by-products, specifically pomace, to produce polyphenols. Pomace is a significant by-product of Greek wineries. The company's potential suppliers are seven large wineries identified as having medium-level sustainable orientation ("unexploiters") with potential for future adoption of sustainable practices. Potential customers are from the food, cosmetics, and pharmaceutical industries, with an interest in polyphenols derived from grapes. The SWOT analysis revealed key factors (Table 7).

Table 7. SWOT analysis, retrieved from Malindretos et al. [85].

Strengths	Weaknesses
- Crete's tradition in grape farming	- High dependence on supplier wineries and grape production volumes
- Educated managers	- Absence of institutional support for green initiatives
- Differentiated final product (polyphenols) aligned with sustainability trends	- High investment costs
- Potential long-term agreements with customers	- Lack of specialized employees
Opportunities	**Threats**
- Growing global awareness of qualitative, nutritious food and environmental issues	- Global economic crisis
- No current competitors in this niche	
- High entry barriers due to required expertise and investment	
- Potential future financial support from EU funds	

The financial evaluation of the start-up project yields positive results, with initial infrastructure costs estimated at over EUR 1.9 million and annual operational costs projected to exceed EUR 1.58 million. By utilizing pomace from suppliers, the company can produce 656.25 tons of polyphenols annually, with a forecasted first-year revenue of EUR 1,968,750, potentially increasing to over EUR 2 million within five years. Key financial indicators suggest that the investment is sustainable. The study highlights that entrepreneurship is driven by innovative behavior, appropriate infrastructure, and collaboration between academia and industry, which can enhance sustainability in the wine sector. Partnerships among local stakeholders are essential for developing sustainable solutions and creating job opportunities. The role of universities and authorities in promoting sustainable practices is crucial for overall well-being.

8.2. Market Potential for GP-Derived Materials

Although most grape pomace could be composted and returned to vineyards to help close the carbon cycle, this practice is not widely implemented. At the same time, there is growing consumer demand for natural compounds over synthetic alternatives, as well as an increasing focus on the sustainability of agricultural practices. This context highlights a vast array of potential applications for the bioactives found in grape pomace. Despite this potential, no prior assessment of the market potential for value-added uses of grape pomace had been conducted (Table 8) [86,87].

Table 8. Alternative bioactive applications and market potential. Adapted from Dwyer et al. [86].

Application	Key Details/Benefits
Composting	Currently, the majority of grape pomace used in Ontario and B.C. is composted on-site and recycled into the vineyard. It has an initial carbon-to-nitrogen (C: N) ratio of approximately **27:1**, which is within the optimal range (25:1 to 35:1) for composting. **Composting increases the nitrogen content** (e.g., to 2.35 wt%), which is **beneficial for vineyard growth**. Using only composted grape pomace as the sole nitrogen source is neither feasible nor economical; supplemental fertilizer is required. Nitrogen from compost is considered "slow-releasing", and improper distribution year-to-year could lead to detrimental nitrogen overabundance.
Functional Foods	It can be used as a functional food ingredient due to its high fiber and phenolic content. **Consumer acceptance studies** have been conducted for products incorporating grape seed flour or grape pomace skins.
Food Processing	Grape pomace can be used for **biosurfactant production**, utilizing glucose for lactic acid. It can be **added to food products for consistency and texture**. Bioactives, such as antioxidants from pomace, can be added internally to preserve food and are particularly preferred over synthetic antioxidants in meat products. Glucose can be used to produce **Pullulan**, which can be used as a food additive for **texture and low-calorie bulk**.
Cosmetics	**Antioxidants and oils** from grape pomace can be used topically. **Grape seed oil**, rich in linoleic acid, is beneficial for maintaining skin moisture, aiding in the healing of sunburns, and potentially reducing the appearance of acne. **Vitamin E** extracted from grape seeds is also effective when applied topically.
Pharmaceutical/Biomedical	**Pullulan** has various applications in this field.
Supplements	Products like **Bioflavia**, an organic dried red grape skin powder, are already being sold as supplements. **Grape seed oil** is also available as a supplement.

The article by Dwyer et al. [86] assesses the market potential of red grape pomace alternatives, highlighting existing products. The estimated market potential for red grape skins produced in Ontario and British Columbia (B.C.) in 2011 was approximately CAD 499 million for Ontario and CAD 185 million for B.C., assuming 100% utilization as supplements. For grape seed oil, the potential was over CAD 5 million combined from both provinces. These numbers reflect theoretical market values based on complete utilization and do not account for factors such as composting or extraction costs, meaning the actual potential would likely be lower. Despite this, they indicate significant annual production of pomace and potential commercial uses. Current primary uses involve composting, but there are opportunities in functional foods and supplements. Further research is necessary to assess economic viability and explore alternative applications, as grape pomace is expected to remain primarily composted until proven otherwise.

Recently, innovative uses have been found for grape pomace, indicating its increasing market potential. Dos Santos Silva et al. [87] discuss the technological potential of grape pomace for use in the meat industry. The text addresses the compounds of interest within grape pomace and elucidates the challenges associated with their application in meat products. The adoption of nano- and microencapsulation technologies, along with the implementation of active packaging as carriers for grape pomace extracts, significantly enhances their application in meat and meat products. These technologies, whether applied individually or in combination, facilitate the controlled release of bioactive compounds during storage while also masking undesirable flavors associated with phenolic compounds.

In addition to the technological advancements, the incorporation of grape pomace within the meat industry aligns with the growing demand for natural additives. It aligns with the United Nations' Sustainable Development Goals. This practice mitigates environmental impacts and enables the extraction of high-value molecules from industrial waste.

The work of Manso et al. [88] examines an integrated biorefinery approach that aims to maximize the utilization of white grape pomace (WGP) through a combination of hydrothermal and organosolv treatments, followed by solid-state fermentation (SSF). The hydrothermal pretreatment process was optimized to enhance sugar extraction, achieving a concentration of up to 80.94 g L^{-1} of reducing sugars, which can be utilized as a substrate for microalgae cultivation. The organosolv treatment facilitated the selective recovery of polyphenols and lignin, with the highest yield of polyphenols (2.27 g kg^{-1} dry weight) being obtained at a temperature of 150 °C. Additionally, a lignin yield of 34.72% was achieved at 200 °C. The residual solid fractions were subsequently subjected to solid-state fermentation using the fungi *Aspergillus oryzae*, *Neurospora intermedia*, and *Rhizopus oryzae*, resulting in a noteworthy increase in protein content, with *A. oryzae* achieving a protein content of 17.6% on a dry weight basis. This multi-step integrated approach demonstrates a scalable and sustainable strategy for converting WGP into multiple high-value bio-products, thereby advancing the principles of the circular economy within the food sector.

As previously mentioned, the main solid by-products and waste from winemaking are grape pomace (GP), wine lees, and grape solids (GSs). Regulations in Spain require industrial waste to be recycled, valorized, or properly disposed of to prevent environmental contamination. Traditionally, wineries have opted for disposal and paid a "polluter pays" fee. However, a significant rise in disposal fees—along with fines for unauthorized discharges that can reach EUR 30,000 to EUR 40,000—has made waste valorization a more attractive and economically viable alternative. Penalties can also include mandatory decontamination of affected areas or even prison sentences.

In this context, valorization emerges as both an environmentally responsible and cost-effective strategy. Although it mentions data from 2011, the work of Devesa-Rey et al. [89], which highlights several approaches for valorizing different winery wastes, including trimming residues, grape pomace, and wine lees, remains relevant. Their findings suggest that these strategies not only reduce the costs associated with biotechnological processes but also contribute to sustainable waste management.

Table 9 summarizes the valorization strategies proposed by Devesa-Rey et al. [89] for DGP.

Table 9. Strategies for valorizing grape pomace.

Material	Description Composition	Potential Uses and Valorisation Methods	Specific Process Notes
Grape Pomace	Lignocellulosic material is generated after pressing. Distilled grape pomace is also produced.	Source of hemicellulosic sugars (xylose and glucose).	Can be hydrolysed. Usable as a substrate for lactic acid and biosurfactant production by *Lactobacillus pentosus*. Mineral and organic nitrogen enhance xylose consumption.
		Production of biosurfactants and extracellular bioemulsifiers.	*Lactobacillus pentosus* can produce cell-bound biosurfactants and extracellular bioemulsifiers from hydrolysates. These can reduce surface tension and effectively stabilize kerosene/water emulsions. Other microorganisms can also utilize grape seed oil for the production of biosurfactants.
		Extraction of phenolic compounds with antioxidant properties.	Can use organic solvents or supercritical carbon dioxide.
		Extraction of condensed tannins.	For enology and as wood adhesives.
		Extraction of oil from grape seeds.	For the food industry.
		Production of anti-allergic substances.	Achieved through lactic acid fermentation with *Lactobacillus*.
		Obtaining hydrolytic enzymes and bioethanol.	Achieved through solid-state fermentation.
		Component for plant growing media.	Used after stabilization treatments like composting. Composting with animal manure can reduce phytotoxicity. Winery waste compost, including grape marc, was successfully evaluated for use as a fertilizer for crops such as tomatoes.
		Component for a high-quality, low-cost substrate for cultivating edible mushrooms.	Part of the vineyard waste is used for this.
		Improved agronomic value through vermicomposting.	Reduces C:N ratio, conductivity, and phytotoxicity. Increases humic materials, nutrients, and pH. Performs better than compost as an amendment for mining wastes.
		Natural adsorbent for metals (e.g., copper and lead).	In winery effluent pretreatment.
		A valuable source of energy and protein for ruminants.	Can be combined with vine shoots and wine lees.
		Source of microbial and human food.	Can be bioconverted by *Pleurotus* spp. and through solid-state fermentation, potentially combined with vineyard pruning and wine lees.

9. Industry–Academia Partnerships

Policies, Subsidies, and Incentives Promoting Bio-Based Innovation

The Bio-based Industries Joint Undertaking (BBI JU) catalyzes sustainable bio-based economic growth in Europe. Established in 2014 as a public–private partnership between the European Commission and the Bio-based Industries Consortium, it supports research, innovation, and investment in bio-based industries. BBI JU aims to overcome market barriers to private investment, promoting a bio-based economy that utilizes domestic renewable raw materials for producing food, feed, chemicals, materials, and fuels. It embodies the principles of the circular economy under the Horizon 2020 initiative.

Mengal et al. [90] identify several challenges facing the European bio-based industry: (i) lack of critical mass at the European level in terms of scale, excellence, and innovation potential; (ii) the sector is fragmented, with diverse organizations and companies that have not collaborated across industries and regions before; (iii) geographical issues related to the supply of biomass feedstock; (iv) difficulty in linking advancements in bioeconomy research and innovation with commercialization and market uptake; (v) complex and substantial technology and innovation challenges in bio-based industries and value chains; (vi) struggling to set up new value chains due to the diverse range of involved players who have not worked together; (vii) potential investments and resulting knowledge are being implemented outside of Europe; (viii) the need to bridge the "valley of death" between proof-of-concept and scaling up to commercial levels.

The BBI JU aims to establish a competitive and sustainable Europe by transitioning to a low-carbon society using renewable raw materials. This initiative focuses on the local production of chemicals, materials, and fuels, fostering job creation and rural development while emphasizing sustainability and resource efficiency. Funded with EUR 975 million in public money, it aims to leverage at least EUR 2.73 billion in private investment between 2014 and 2024. It provides a stable framework for strategic collaboration between industry partners, encouraging long-term private sector investment in demonstration and large-scale projects. Its operations are managed by the BBI JU Program Office, which is overseen by a Governing Board comprising representatives from the EC and BIC. The Strategic Innovation and Research Agenda (SIRA) outlines the priority topics for annual work programs. BBI JU funds collaborative research and innovation projects, typically requiring participants from at least three different Member States or Associated Countries, across various Technology Readiness Levels (TRLs), generally ranging from TRL 3 to TRL 8 (Table 10).

Table 10. Project type, acronyms, and focus from several projects [90].

Project Type	Acronym	Focus/Scope
Coordination and Support	CSAs	Address cross-sectoral challenges and support value chains through studies and networking, such as standardization and regulatory issues.
Research and Innovation	RIAs	Validate specific technologies at the laboratory and pilot levels.
Innovation	IAs	Address the entire value chain from feedstock to market applications.
Demonstration Actions	IA.DEMO	**Scale up** technologies to prove readiness for commercial production and demonstrate the business case.
Flagship Actions	IA.FLAG	Establish **first-of-a-kind industrial-scale biorefineries** by scaling up demonstrated technologies.

Funding levels differ by organization and project type. Universities, academic bodies, and non-profits can receive 100% funding for eligible costs in Research and Innovation Activities (RIAs), Coordination and Support Actions (CSAs), and Innovation Actions (IAs).

Large industries receive up to 70% funding for Flagship and Demonstration projects, but not for Research, Innovation, and Advanced (RIA) or Commercialization and Small Business (CSA) projects. SMEs receive 100% for RIAs and CSAs and 70% for Flagship and Demonstration projects.

The benefits of a strong bio-based industrial sector for European citizens include a reduced reliance on fossil fuels, contributions to climate targets, environmentally friendly growth, greener products, significant reductions in greenhouse gas emissions, waste utilization in line with the circular economy, optimized land use, job creation, and support for regional development. Key performance indicators (KPIs) for monitoring include cross-sector connections, new bio-based value chains, industrial investments, collaborative projects, emission reductions, and the establishment of biorefineries and flagship projects. The overall impact is assessed based on socio-economic and environmental benefits. The BBI initiative aligns with the circular economy and the 2012 European Bioeconomy Strategy, focusing on sustainable biomass supply, efficient processing, innovative bio-based products, and market growth. It engages various feedstocks, including agri-based, forest-based, aquatic, bio-waste, and CO_2. Additionally, the BBI JU promotes synergies with national strategies and other EU funding programs and has launched a plan to enhance participation from underrepresented regions and stakeholders [89].

10. Final Remarks

The incorporation of DGP as a filler or additive in polymer matrices provides a substantial enhancement of mechanical characteristics, including increased tensile strength, elasticity, and flexibility, along with notable antioxidant properties. These enhancements render the resultant composites particularly appealing for sustainable packaging applications and other functional uses.

Integrating grape pomace into polymer systems has proven to be a proficient strategy for elevating the thermal properties of biocomposites. This method not only enhances heat resistance and thermal stability but also promotes sustainability by valorizing agricultural by-products, thereby reducing overall polymer consumption.

DGP in its various processed forms can be effectively utilized in the development of biodegradable plastics, enhancing their environmental credentials, mechanical robustness, and antioxidant efficacy. Such biocomposite materials show significant promise for applications in single-use items and food packaging; however, further research and development efforts are necessary to overcome commercial production hurdles and optimize the performance characteristics of these materials.

The employment of DGP in bio-based materials presents distinct environmental advantages by minimizing waste, reducing carbon emissions, and facilitating the transition to biodegradable and renewable products. This approach not only addresses the challenges associated with waste management but also contributes to the establishment of more sustainable material cycles.

Valorizing FLW presents a significant opportunity to mitigate environmental impacts. However, standardization and quality control are essential for scalable processes, which combine FLW valorization with efficient food production and emerging technologies to meet global needs.

Sustainable wine production is vital for the industry's future, benefiting environmental, social, and economic sustainability. The sector can significantly contribute to various Sustainable Development Goals (SDGs), such as SDG 2 (Zero Hunger) through sustainable farming practices, SDG 3 (Good Health) by promoting responsible drinking, and SDG 12 (Responsible Consumption) through sustainable production and consumption. Academic literature can guide the industry in addressing sustainability challenges effectively.

11. Future Directions and Opportunities

11.1. Multidisciplinary Approaches for Material Development

Effective waste management faces challenges like inadequate funding, poor infrastructure, and limited public awareness. Traditional landfills harm the environment, while advanced thermal conversion technologies, such as gasification and pyrolysis, can be costly. Circular economy principles prioritize composting and anaerobic digestion over incineration and landfilling, as they enhance soil properties and support sustainable food systems [90].

Proposals suggest integrating multiple disciplines to tackle global food waste challenges, specifically through anaerobic digestion. Key areas include (i) techniques such as metagenomics and proteomics, which enhance the understanding of microbial processes in biogas production; (ii) utilization of nanomaterials to enhance waste treatment efficiency, reduce costs, and improve biogas output through conductive materials; (iii) optimizes anaerobic digestion processes for better efficiency; and (iv) analyzing data and predict digester performance using methods like artificial neural networks [91].

A collaborative and multidisciplinary approach, supported by robust policies, is essential for effective and sustainable food waste management worldwide. More integrated research is needed to explore these methodologies.

11.2. Advanced Functionalization and Hybrid Materials

The article by Otoni et al. [92] explores the potential of using agri-food residues, specifically food losses and waste (FLW), to develop next-generation bioplastics and advanced materials within the context of a circular bioeconomy. It highlights the environmental impacts of food side streams, emphasizing the importance of reducing food loss and waste (FLW). The distinction is made between food loss, which occurs before retail, and food waste, often resulting from consumer behavior. Approximately 14% of food is lost before retail, particularly in cereals and pulses. The valorization of FLW involves transforming these residues into high-value materials, rather than returning them to food, by utilizing methods such as polymerization and biosynthesis to produce bioplastics and other materials.

Various building blocks derived from food loss and waste (FLW) are explored: (i) Monomers and modified molecules, which include hydroxy acids, epoxides, fatty acids, and lactic acid sourced from fruits, vegetables, and oils, are usable for producing polymers such as polyesters and polyurethanes. D-Limonene from citrus peels is a notable monomer, although challenges remain in its polymerization. (ii) Soluble biopolymers (pectin, starch) and insoluble ones (lignin) can be extracted for applications in films, coatings, and medical devices. Lignin enhances cellulose's wet strength, enabling the creation of bioplastics, while proteins like gelatin are functional in medical wearables. Additionally, nanocelluloses from plant waste and chitin nanocrystals from crustacean shells can be utilized, along with biogenic silica microparticles obtained from pineapple peels.

FLW composition influences film formation, with each component contributing unique properties. Films made from carrots, parsley, and cauliflower show potential but may require reinforcement to achieve optimal strength.

Manufacturing processes significantly affect material properties. Utilizing FLW can lower production costs through methods such as spinning, casting, and 3D printing. Mycelium-based products are also emerging.

11.3. Integration into Local Circular Economy Models

The growing interest in sustainability within the wine industry highlights the increasing consensus that circular economy solutions are essential for achieving the United

Nations' Sustainable Development Goals (SDGs). By implementing circular economy strategies, wine production processes can improve both environmental and economic sustainability. Key components of these strategies include waste valorization and industrial symbiosis.

However, a gap exists between theoretical sustainability strategies and their practical implementation in the wine industry. This gap can be attributed to several factors, including the substantial investments of time, resources, and technology required, financial constraints faced by smaller producers, a lack of knowledge, and difficulties in coordinating practices throughout the entire supply chain.

The review by Abbate et al. [20] identifies two main thematic areas: (i) Attitudes towards sustainability in the wine industry include drivers and barriers to adopting sustainable practices, the development of circular business models, the integration of technologies, and performance indicators. Drivers include internal values, stakeholder expectations, marketing, innovation, and competitiveness. Consumers are increasingly willing to pay more for sustainably produced wines. (ii) Wine waste valorization focuses on transforming by-products into valuable resources. This includes bioenergy from grape marc and cheese whey, biorefineries, and the extraction of bioactives from vine pruning. Other applications include grape seed oil for packaging and anthocyanins for food, cosmetics, and pharmaceuticals. Wine waste can also be processed into compost and organic fertilizer for functional foods.

Practical applications include integrating grape residues into consumer products, new extraction methods, and innovations such as WineLeather and antioxidant-rich asphalt. Economic and environmental benefits include reduced disposal costs and new revenue streams. Future research should explore technologies in vineyards, partnerships for by-product exchange, waste prevention, and the evaluation of sustainability through cost–benefit and life cycle analyses.

11.4. Alignment with UN Sustainable Development Goals (SDGs)

Sustainability is vital for the long-term survival of the wine industry. Embracing sustainable practices protects natural resources, ensures fair employment, and leverages technology to improve production quality while minimizing environmental impacts. These practices enhance wine quality by reducing chemical use, conserving water, and improving waste management. Additionally, sustainability enhances the industry's image and tourism while lowering production costs through improved resource management.

The wine industry makes a significant contribution to the economy, culture, and local employment, while also having a positive impact on ecosystems and promoting the use of renewable energy. It engages in practices that reduce pesticide and fertilizer use, improve water quality, and enhance biodiversity.

Recognized as a key partner in the United Nations' Sustainable Development Goals (SDGs), the wine industry supports economic growth, promotes employment and social justice, and fosters environmental preservation. It creates jobs, invests in education, and removes gender barriers, ultimately improving farmers' incomes and quality of life. However, unsustainable wine production can harm development goals, including the use of pesticides, loss of biodiversity, excessive water use, and greenhouse gas emissions from transportation, packaging, and processing. Vineyard expansion can also lead to deforestation, soil erosion, and reduced land availability.

The study by Martínez-Falcó et al. [93] used a bibliometric analysis of the Web of Science (WoS) database. A specific search equation, incorporating Boolean operators and wildcards, identified relevant articles published between 1997 and 2022. The PRISMA statement guided article selection, narrowing 114 initial records to 107 after removing duplicates and irrelevant items. Bibliometric reviews provide a comprehensive overview, highlighting research trends, priorities, and connections between authors.

Key findings include (i) a 633.33% increase in SDG-related wine research from 2015 (3 articles) to 2022 (22 articles), mainly due to the UN's 2015 adoption of the SDGs; (ii) main research areas such as Environmental Sciences, Green Sustainable Science, and Food Science Technology, with contributions from Horticulture and Agriculture, underscoring the multidisciplinary focus.

11.5. A Roadmap for the Valorization of Dealcoholized Grape Pomace in Bio-Based Materials: Opportunities, Challenges, and Pathways for Industrial Uptake

DGP has emerged as a promising bio-based material for biocomposites and biodegradable plastics, particularly in applications such as food packaging, agricultural tools, and single-use items. Existing studies consistently demonstrate that DGP improves mechanical performance, thermal stability, and antioxidant activity when used as a filler or functional additive. These performance gains, along with DGP's abundant availability and compatibility with circular economy principles, position it as an attractive candidate for developing sustainable materials.

Nevertheless, despite encouraging results, the field remains in a relatively early stage and would benefit from a more application-oriented research strategy. To support industrial uptake and advance scientific development, the following roadmap identifies key research priorities.

First, DGP's role as a sustainable filler has been demonstrated, but its practical relevance could be significantly enhanced by focusing on application-specific performance metrics. For example, researchers should evaluate barrier properties for food packaging applications, investigate aging resistance for outdoor or agricultural uses, assess biodegradability under both industrial and home composting conditions, and examine shelf-life stability for active packaging systems.

Second, greater attention must be given to the optimization of processing parameters. This includes understanding the influence of drying methods and particle size on DGP's reinforcing capabilities, fine-tuning compatibilizer formulations to enhance interfacial adhesion within different polymer matrices, and examining how extraction or purification methods affect the final material properties.

Third, while the environmental advantages of DGP valorization are well-articulated, the economic dimension requires stronger support. We consider that conducting a robust economic feasibility analysis—factoring in drying costs, energy demands, and market competitiveness—is essential to substantiate claims of sustainable valorization and to guide investment decisions.

Fourth, the path to industrial scale-up will benefit from standardization and integration into existing manufacturing frameworks. Establishing common protocols for the collection, drying, and storage of DGP will help ensure uniform quality. Moreover, defining key performance indicators and developing scalable, industry-aligned processing techniques are critical for consistent and widespread application.

Despite its potential, several scaling challenges must be addressed to ensure the successful commercial deployment of DGP-based materials. One significant barrier is supply chain variability, as the volume and composition of DGP can vary considerably depending on grape variety, regional practices, and vintage conditions, thereby complicating efforts toward consistent sourcing and standardization. Another significant hurdle is the energy-intensive and logistically complex process of drying wet pomace, especially for small and medium-sized wineries that may lack the infrastructure for in situ valorization. Regulatory issues also pose obstacles, particularly concerning approvals for food-contact materials or classification of DGP composites under national biodegradability standards.

Ultimately, a significant gap remains in market readiness and industry awareness. Many potential users are still unaware of DGP's functional benefits, highlighting the need for demonstration projects, stakeholder outreach, and more transparent communication of performance advantages.

DGP offers strong potential as a high-performing, sustainable, and circular filler for bio-based materials. To fully realize its value, the field must evolve beyond proof-of-concept work and pursue targeted application testing, economic and environmental validation, and the resolution of technical and regulatory bottlenecks. Coordinated efforts among researchers, industry partners, and policymakers will be essential. With the proper support, DGP-based materials can make a significant contribution to the development of a circular and bio-based economy, particularly in wine-producing regions where this resource is both abundant and underutilized.

Author Contributions: Conceptualization, A.V.; writing—original draft preparation, J.M.M., F.B., and A.V.; writing—review and editing, A.V. and F.B.; supervision, A.V. All authors have read and agreed to the published version of the manuscript.

Funding: This study was funded by the Vine & Wine Portugal-Driving Sustainable Growth Through Smart Innovation Project, number 67, AAC: 02/C05-i01/2022, sub-project B1.5.1.—Alcohol a la carte: Reducing alcohol in the wine after fermentation, without any loss of aromas. Financed by the NextGenerationEU "Programa de Recuperação e Resiliência (PRR)/Alianças Mobilizadora". The CQ-VR [grant number UIDB/00616/2020 and UIDP/00616/2020—https://doi.org/10.54499/UIDB/00616/2020], FCT—Portugal, and COMPETE also funded this study. National funds from the FCT also supported this work as part of the Portuguese Foundation for Science and Technology's projects UI/04033 and LA/P/0126/2020 (https://doi.org/10.54499/LA/P/0126/2020). This work is supported by National Funds by FCT—Portuguese Foundation for Science and Technology, under the projects UI/04033 and LA/P/0126/2020 (https://doi.org/10.54499/LA/P/0126/2020).

Institutional Review Board Statement: Not applicable.

Informed Consent Statement: Not applicable.

Data Availability Statement: No new data were created or analyzed in this study.

Acknowledgments: The authors thank the Chemistry Research Center (CQ-VR) and the project PRR—Vine & Wine Portugal—Driving Sustainable Growth Through Smart Innovation for their financial support.

Conflicts of Interest: The authors declare that they have no conflicts of interest.

References

1. OIV. State of the World Vine and Wine Sector in 2023. Available online: https://www.oiv.int/sites/default/files/2024-04/OIV_STATE_OF_THE_WORLD_VINE_AND_WINE_SECTOR_IN_2023.pdf (accessed on 5 May 2025).
2. Matos, C.; Castro, M.; Baptista, J.; Valente, A.; Briga-Sá, A. The use of water in wineries: A review. *Sci. Total Environ.* **2024**, *951*, 175198. [CrossRef] [PubMed]
3. Regulation (EU) 2019/787 of the European Parliament and of the Council of 17 April 2019 on the Definition, Description, Presentation and Labelling of Spirit Drinks, the Use of the Names of Spirit Drinks in the Presentation and Labelling of Other Foodstuffs, the Protection of Geographical Indications for Spirit Drinks, the Use of Ethyl Alcohol and Distillates of Agricultural Origin in Alcoholic Beverages, and Repealing Regulation (EC) No 110/2008. Available online: https://eur-lex.europa.eu/eli/reg/2019/787/2024-05-13 (accessed on 29 April 2025).
4. Commission Delegated Regulation (EU) No 1143/2014 of the European Parliament and of the Council of 12 March 2019 Supplementing Regulation (EU) No 1308/2013 of the European Parliament and the Council as Regards Wine-Growing Areas Where the Alcoholic Strength May Be Increased, Authorised Oenological Practices and Restrictions Applicable to the Production and Conservation of Grapevine Products, the Minimum Percentage of Alcohol for By-Products and Their Disposal, and Publication of OIV Files. Available online: https://eur-lex.europa.eu/legal-content/en/TXT/?uri=CELEX:32019R0934 (accessed on 29 April 2025).

5. Regulation (EU) 2021/2117 of the European Parliament and of the Council of 2 December 2021 Amending Regulations (EU) No 1308/2013 Establishing a Common Organisation of the Markets in Agricultural Products, (EU) No 1151/2012 on Quality Schemes for Agricultural Products and Foodstuffs, (EU) No 251/2014 on the Definition, Description, Presentation, Labelling and the Protection of Geographical Indications of Aromatised Wine Products and (EU) No 228/2013 Laying Down Specific Measures for Agriculture in the Outermost Regions of the Union. Available online: https://eur-lex.europa.eu/legal-content/EN/TXT/?uri=CELEX:32021R2117 (accessed on 29 April 2025).
6. Braga, F.G.; Lencart e Silva, F.A.; Alves, A. Recovery of Winery By-products in the Douro Demarcated Region: Production of Calcium Tartrate and Grape Pigments. *Am. J. Enol. Vitic.* **2002**, *53*, 41–45. [CrossRef]
7. Nerantzis, E.T.; Tataridis, P. Integrated enology-utilization of winery by-products into high added value products. *J. Sci. Technol.* **2006**, *1*, 79–89. [CrossRef]
8. Chakka, A.K.; Babu, A.S. Bioactive Compounds of Winery by-products: Extraction Techniques and their Potential Health Benefits. *Appl. Food Res.* **2022**, *2*, 100058. [CrossRef]
9. Dávila, I.; Robles, E.; Egüés, I.; Labidi, J.; Gullón, P. The biorefinery concept for the industrial valorization of grape processing by-products. In *Handbook of Grape Processing By-Products—Suitable Solutions*; Galanakis, C.M., Ed.; Academic Press: Cambridge, MA, USA, 2017; pp. 29–53.
10. González-Centeno, M.R.; Rosselló, C.; Simal, S.; Garau, M.C.; López, F.; Femenia, A. Physico-chemical properties of cell wall materials obtained from ten grape varieties and their byproducts: Grape pomaces and stems. *LWT Food Sci. Technol.* **2010**, *43*, 1580–1586. [CrossRef]
11. Muhlack, R.A.; Potumarthi, R.; Jeffery, D.W. Sustainable wineries through waste valorisation: A review of grape marc utilisation for value-added products. *Waste Manag.* **2018**, *72*, 99–118. [CrossRef]
12. Telini, B.P.; Villa, L.C.; Vainstein, M.H.; Lopes, F.C. From Vineyard to Brewery: A Review of Grape Pomace Characterization and Its Potential Use to Produce Low-Alcohol Beverages. *Fermentation* **2025**, *11*, 57. [CrossRef]
13. Karastergiou, A.; Gancel, A.L.; Jourdes, M.; Teissedre, P.L. Valorization of Grape Pomace: A Review of Phenolic Composition, Bioactivity, and Therapeutic Potential. *Antioxidants* **2024**, *13*, 1131. [CrossRef]
14. Ramírez, R.; Delgado, J.; Rocha-Pimienta, J.; Valdésa, M.E.; Martín-Mateosa, M.J.; Ayuso-Yuste, M.C. Preservation of white wine pomace by high hydrostatic pressure. *Heliyon* **2023**, *9*, e21199. [CrossRef]
15. Beres, C.; Costa, G.N.S.; Cabezudo, I.; da Silva-James, N.K.; Teles, A.S.C.; Cruz, A.P.G.; Mellinger-Silva, C.; Tonon, R.V.; Cabral, L.M.C.; Freitas, S.P. Towards integral utilization of grape pomace from winemaking process: A review. *Waste Manag.* **2017**, *68*, 581–594. [CrossRef]
16. Arvanitoyannis, I.S.; Ladas, D.; Mavromatis, A. Wine waste treatment methodology. *Int. J. Food Sci. Technol.* **2006**, *41*, 1117–1151. [CrossRef]
17. Semitela, S.; Pirra, A.; Braga, F.G. Impact of mesophilic co-composting conditions on the quality of substrates produced from winery waste activated sludge and grape stalks: Lab-scale and pilot-scale studies. *Bioresour. Technol.* **2019**, *289*, 121622. [CrossRef] [PubMed]
18. Silva, A.; Silva, V.; Igrejas, G.; Gaivão, I.; Aires, A.; Klibi, N.; Enes Dapkevicius, M.L.; Valentão, P.; Falco, V.; Poeta, P. Valorization of Winemaking By-Products as a Novel Source of Antibacterial Properties: New Strategies to Fight Antibiotic Resistance. *Molecules* **2021**, *26*, 2331. [CrossRef]
19. Mohan, D.; Charles, U.; Pittman, C.U., Jr.; Steele, P.H. Pyrolysis of Wood/Biomass for Bio-oil: A Critical Review. *Energy Fuels* **2006**, *20*, 848–889. [CrossRef]
20. Abbate, S.; Centobelli, P.; Di Gregorio, M. Wine waste valorisation: Crushing the research domain. *Rev. Manag. Sci.* **2025**, *19*, 963–998. [CrossRef]
21. Bustamante, M.A.; Moral, R.; Paredes, C.; Pérez-Espinosa, A.; Moreno-Caselles, J.; Pérez-Murcia, M.D. Agrochemical characterisation of the solid byproducts and residues from the winery and distillery industry. *Waste Manag.* **2008**, *28*, 372–380. [CrossRef]
22. Charalampia, D.; Koutelidakis, A. Grape pomace: A challenging renewable resource of bioactive phenolic compounds with diversified health benefits. *MOJ Food Process Technol.* **2016**, *3*, 262–265. [CrossRef]
23. Beres, C.; Freitas, S.P.; Godoy, R.L.O.; Oliveira, D.C.R.; Deliza, R.; Iacomini, M.; Mellinger-Silva, C.; Cabral, L.M.C. Antioxidant dietary fiber from grape pomace flour or extract: Does it make any difference on the nutritional and functional value? *J. Funct. Foods* **2019**, *56*, 276–285. [CrossRef]
24. Ferrer-Gallego, R.; Silva, P. The Wine Industry By-Products: Applications for Food Industry and Health Benefits. *Antioxidants* **2022**, *11*, 2025. [CrossRef]
25. Peixoto, C.M.; Dias, M.I.; Alves, M.J.; Calhelha, R.C.; Barros, L.; Pinho, S.P.; Ferreira, I.C.F.R. Grape pomace as a source of phenolic compounds and diverse bioactive properties. *Food Chem.* **2018**, *253*, 132–138. [CrossRef]
26. Yu, J.; Ahmedna, M. Functional Components of Grape Pomace: Their Composition, Biological Properties and Potential Applications. *Int. J. Food Sci. Technol.* **2013**, *48*, 221–237. [CrossRef]

27. Djilas, S.; Canadanovic-Brunet, J.; Cetkovic, G. By-products of fruits processing as a source of phytochemicals. *Chem. Ind. Chem. Eng. Quart.* **2009**, *15*, 191–202. [CrossRef]
28. Rondeau, P.; Gambier, F.; Jolibert, F.; Brosse, N. Compositions and chemical variability of grape pomaces from French vineyard. *Ind. Crop. Prod.* **2013**, *43*, 251–254. [CrossRef]
29. Otero-Pareja, M.J.; Casas, L.; Fernández-Ponce, M.T.; Mantell, C.; Ossa, E.J.M. Green Extraction of Antioxidants from Different Varieties of Red Grape Pomace. *Molecules* **2015**, *20*, 9686–9702. [CrossRef] [PubMed]
30. Nanni, A.; Parisi, M.; Colonna, M. Wine By-Products as Raw Materials for the Production of Biopolymers and of Natural Reinforcing Fillers: A Critical Review. *Polymers* **2021**, *13*, 381. [CrossRef]
31. Rezzadori, K.; Benedetti, S.; Amante, E.R. Proposals for the valorization of winery wastes: Part 1: Technological potential for production of high added-value products. *Waste Manag.* **2012**, *32*, 2039–2053. [CrossRef]
32. European Commission. *Communication from the Commission to the European Parliament, the Council, the European Economic and Social Committee and the Committee of the Regions. A New Circular Economy Action Plan for a Cleaner and More Competitive Europe*; COM(2020) 98 Final; European Commission: Brussels, Belgium, 2020.
33. García-Lomillo, J.; González-SanJosé, M.L. Applications of wine pomace in the food industry: A review. *Trends Food Sci. Technol.* **2017**, *68*, 191–201. [CrossRef]
34. Antonic, B.; Jancikova, S.; Dordevic, D.; Tremlova, B. Grape Pomace Valorization: A Systematic Review of the Existing Valorization Concepts. *Foods* **2020**, *9*, 39. [CrossRef]
35. Bordiga, M.; Travaglia, F.; Locatelli, M. Valorisation of grape pomace: An overview of the present and future biotechnological applications. *Food Bioprod. Process.* **2019**, *117*, 106–123. [CrossRef]
36. Spinei, M.; Oroian, M. The Potential of Grape Pomace Varieties as a Dietary Source of Pectic Substances. *Foods* **2021**, *10*, 867. [CrossRef]
37. Amendola, D.; De Faveri, D.M.; Egües, I.; Serrano, L.; Labidi, J.; Spigno, G. Autohydrolysis and organosolv process for recovery of hemicelluloses, phenolic compounds and lignin from grape stalks. *Bioresour. Technol.* **2012**, *107*, 267–274. [CrossRef] [PubMed]
38. Frincu, M.; Teodorescu, R.I.; Barbulescu, D.I.; Tudor, V.; Dumitrache, C.; Hirjoaba, L.N.; Matei, F. The grape as potential source of biologically active compounds. *Sci. Papers Ser. B Hortic.* **2019**, *63*, 103–107.
39. Bassani, A.; Alberici, N.; Fiorentini, C.; Giuberti, G.; Dordoni, R.; Spigno, G. Hydrothermal Treatment of Grape Skins for Sugars, Antioxidants and Soluble Fibers Production. *Chem. Eng. Trans.* **2020**, *79*, 127–132. [CrossRef]
40. Orsat, V.; Chang, M.W. Chapter 7—Drying of Fruits and Vegetables for Functional Food Ingredients. In *Prebiotics and Probiotics in Food, Nutrition and Health*; Siddiqui, M.W., Wickramasinghe, W.A.K.G., Eds.; Academic Press: Cambridge, MA, USA, 2020; pp. 111–133.
41. Peternel, L.; Sokač Cvetnić, T.; Gajdoš Kljusurić, J.; Jurina, T.; Benković, M.; Radojčić Redovniković, I.; Jurinjak Tušek, A.; Valinger, D. The Effects of Drying and Grinding on the Extraction Efficiency of Polyphenols from Grape Skin: Process Optimization. *Processes* **2024**, *12*, 1100. [CrossRef]
42. Azmir, J.; Zaidul, I.S.M.; Rahman, M.M.; Sharif, K.M.; Mohamed, A.; Sahena, F.; Jahurul, M.H.A.; Ghafoor, K.; Norulaini, N.A.N.; Omar, A.K.M. Techniques for extraction of bioactive compounds from plant materials: A review. *J. Food Eng.* **2013**, *117*, 426–436. [CrossRef]
43. Domínguez-Rodríguez, G.; Marina, M.L.; Plaza, M. Enzyme-assisted extraction of bioactive non-extractable polyphenols from sweet cherry (*Prunus avium* L.) pomace. *Food Chem.* **2021**, *339*, 128086. [CrossRef] [PubMed]
44. Siracusa, V.; Blanco, I. Bio-Polyethylene (Bio-PE), Bio-Polypropylene (Bio-PP) and Bio-Poly(ethylene terephthalate) (Bio-PET): Recent Developments in Bio-Based Polymers Analogous to Petroleum-Derived Ones for Packaging and Engineering Applications. *Polymers* **2020**, *12*, 1641. [CrossRef]
45. Yaradoddi, J.S.; Banapurmath, N.R.; Ganachari, S.V.; Soudagar, M.E.M.; Sajjan, A.M.; Kamat, S.; Mujtaba, M.A.; Shettar, A.S.; Anqi, A.E.; Safaei, M.R.; et al. Bio-based material from fruit waste of orange peel for industrial applications. *J. Mater. Res. Technol.* **2021**, *17*, 3186–3197. [CrossRef]
46. Reichert, C.L.; Bugnicourt, E.; Coltelli, M.B.; Cinelli, P.; Lazzeri, A.; Canesi, I.; Braca, F.; Martínez, B.M.; Alonso, R.; Agostinis, L.; et al. Bio-Based Packaging: Materials, Modifications, Industrial Applications and Sustainability. *Polymers* **2020**, *12*, 1558. [CrossRef]
47. Wojnowska-Baryła, I.; Kulikowska, D.; Bernat, K. Effect of Bio-Based Products on Waste Management. *Sustainability* **2020**, *12*, 2088. [CrossRef]
48. Suarez, A.; Ford, E.; Venditti, R.; Kelley, S.; Saloni, D.; Gonzalez, R. Rethinking the use of bio-based plastics to accelerate the decarbonization of our society. *Resour. Conserv. Recycl.* **2022**, *186*, 106593. [CrossRef]
49. Cerimi, K.; Akkaya, K.; Pohl, C.; Schmidt, B.; Neubauer, P. Fungi as source for new bio-based materials: A patent review. *Fungal Biol. Biotechnol.* **2019**, *6*, 17. [CrossRef]
50. Vinod, A.; Sanjay, M.; Suchart, S.; Jyotishkumar, P. Renewable and sustainable biobased materials: An assessment on biofibers, biofilms, biopolymers and biocomposites. *J. Clean. Prod.* **2020**, *258*, 120978. [CrossRef]

51. Zhou, J.; Li, X.; Zhang, Z.; Hou, T.; Xu, J.; Wang, Y.; Ye, H.; Yang, B. Bio-based and bio-degradable nanofiber materials: A sustainable platform for energy, environmental, and biomedical applications. *Chem. Eng. J.* **2024**, *491*, 152105. [CrossRef]
52. Gheewala, S. Life cycle assessment for sustainability assessment of biofuels and bioproducts. *Biofuel Res. J.* **2023**, *10*, 1810–1815. [CrossRef]
53. Auriemma, M.; Piscitelli, A.; Pasquino, R.; Cerruti, P.; Angelini, S.; Scarinzi, G.; Malinconico, M.; Grizzuti, N. Effect of natural phenolics on the thermal and processing behaviour of poly(3-hydroxybutyrate). *AIP Conf. Proc.* **2015**, *1695*, 020016. [CrossRef]
54. Lila, M.; Shukla, K.; Komal, U.; Singh, I. Accelerated thermal ageing behaviour of bagasse fibers reinforced Poly (Lactic Acid) based biocomposites. *Compos. Part B Eng.* **2019**, *156*, 121–127. [CrossRef]
55. Lila, M.; Komal, U.; Singh, I. Thermal post-processing of bagasse fiber reinforced polypropylene composites. *Compos. Commun.* **2020**, *23*, 100546. [CrossRef]
56. Jordá-Reolid, M.; Martínez-García, A.; Ibáñez-García, A.; León-Cabezas, M.; Galvañ-Gisbert, J. Study of Mechanical and Thermal Properties of Environmentally Friendly Composites from Beer Bagasse. *Polymers* **2024**, *16*, 2916. [CrossRef]
57. Balaji, A.; Karthikeyan, B.; Swaminathan, J.; Raj, S. Effect of Filler Content of Chemically Treated Short Bagasse Fiber-Reinforced Cardanol Polymer Composites. *J. Nat. Fibers* **2019**, *16*, 613–627. [CrossRef]
58. Mulinari, D.; Voorwald, H.; Cioffi, M.; Da Silva, M. Cellulose fiber-reinforced high-density polyethylene composites—Mechanical and thermal properties. *J. Compos. Mater.* **2017**, *51*, 1807–1815. [CrossRef]
59. Gowman, A.; Wang, T.; Rodriguez-Uribe, A.; Mohanty, A.; Misra, M. Bio-poly(butylene succinate) and Its Composites with Grape Pomace: Mechanical Performance and Thermal Properties. *ACS Omega* **2018**, *3*, 15205–15216. [CrossRef] [PubMed]
60. Vázquez, M.; Flórez, M.; Cazón, P. A Strategy to Prolong Cheese Shelf-Life: Laminated Films of Bacterial Cellulose and Chitosan Loaded with Grape Bagasse Antioxidant Extract for Effective Lipid Oxidation Delay. *Food Hydrocoll.* **2024**, *156*, 110232. [CrossRef]
61. Silva, R.; De Aguiar, A.; Simões, B.; Da Silva, S.; Higuchi, M.; Roberto, S.; Yamashita, F. Biodegradable Polymer Packaging System for 'Benitaka' Table Grapes during Cold Storage. *Polymers* **2024**, *16*, 274. [CrossRef]
62. Cazón, P.; Puertas, G.; Vázquez, M. Characterization of multilayer bacterial cellulose-chitosan films loaded with grape bagasse antioxidant extract: Insights into spectral and water properties, microstructure, and antioxidant activity. *Int. J. Biol. Macromol.* **2024**, *268*, 131774. [CrossRef]
63. Science Learning Hub—Pokapū Akoranga Pūtaiao. Making Bioplastic Clips from Wine Industry Waste. 2023. Available online: https://www.sciencelearn.org.nz/resources/3061-making-bioplastic-clips-from-wine-industry-waste (accessed on 17 April 2025).
64. Titone, V.; Rapisarda, M.; Pulvirenti, L.; Napoli, E.; Impallomeni, G.; Botta, L.; Mistretta, M.C.; Rizzarelli, P. Sustainable biocomposites based on Mater-Bi and grape pomace for a circular economy: Performance evaluation and degradation in soil. *Polym. Degrad. Stab.* **2025**, *231*, 111091. [CrossRef]
65. Engel, J.; Ambrosi, A.; Tessaro, I. Development of biodegradable starch-based foams incorporated with grape stalks for food packaging. *Carbohydr. Polym.* **2019**, *225*, 115234. [CrossRef]
66. Dordević, D.; Necasova, L.; Antonić, B.; Jančíková, S.; Tremlová, B. Plastic Cutlery Alternative: Case Study with Biodegradable Spoons. *Foods* **2021**, *10*, 1612. [CrossRef]
67. Vostrejš, P.; Adamcová, D.; Vaverková, M.; Enev, V.; Kalina, M.; Machovský, M.; Šourková, M.; Márová, I.; Kovalcik, A. Active biodegradable packaging films modified with grape seeds lignin. *RSC Adv.* **2020**, *10*, 29202–29213. [CrossRef]
68. Maraveas, C. Production of Sustainable and Biodegradable Polymers from Agricultural Waste. *Polymers* **2020**, *12*, 1127. [CrossRef]
69. Guardia, L.; Suárez, L.; Querejeta, N.; Pevida, C.; Centeno, T. Winery wastes as precursors of sustainable porous carbons for environmental applications. *J. Clean. Prod.* **2018**, *193*, 614–624. [CrossRef]
70. Qin, C.; Wang, W.; Li, W.; Zhang, S.; Li, Z. Developing bagasse towards superhydrophobic coatings. *Cellulose* **2021**, *28*, 3617–3630. [CrossRef]
71. Danielski, R. Challenges hindering the commercialization of nutraceuticals derived from agri-food by-products. *J. Food Bioact.* **2023**, *24*, 1–5. [CrossRef]
72. De Oliveira, F.; Takeshita, V.; Mendes, K.; Tornisielo, V.; Alonso, F.; Junqueira, L.; Neto, M.; Lins, H.; Silva, D. Addition of raw feedstocks and biochars to the soil on the sorption–desorption and biodegradation of ^{14}C-saflufenacil. *Int. J. Environ. Sci. Technol.* **2022**, *20*, 1259–1276. [CrossRef]
73. Antonić, B.; Jančíková, S.; Dordević, D.; Tremlová, B. Grape Pomace Valorization: A Systematic Review and Meta-Analysis. *Foods* **2020**, *9*, 1627. [CrossRef] [PubMed]
74. Karastergiou, A.; Gancel, A.-L.; Jourdes, M.; Teissedre, P.-L. Transforming winemaking waste: Grape pomace as a sustainable source of bioactive compounds: This is an original research article submitted in cooperation with Macrowine 2025. *OENO One* **2025**, *59*. [CrossRef]
75. Picot-Allain, M.C.N.; Emmambux, M.N. Isolation, Characterization, and Application of Nanocellulose from Agroindustrial By-products: A Review. *Food Rev. Int.* **2023**, *39*, 941–969. [CrossRef]

76. Fan, Q.; Jiang, C.; Wang, W.; Bai, L.; Chen, H.; Yang, H.; Wei, D.; Yang, L. Eco-friendly extraction of cellulose nanocrystals from grape pomace and construction of self-healing nanocomposite hydrogels. *Cellulose* **2020**, *27*, 2541–2553. [CrossRef]
77. Júnior, M.A.D.; Borsoi, C.; Hansen, B.; Catto, A.L. Evaluation of different methods for extraction of nanocellulose from yerba mate residues. *Carbohydr. Polym.* **2019**, *218*, 78–86. [CrossRef]
78. Ogrizek, L.; Lamovšek, J.; Cuš, F.; Leskovšek, M.; Gorjanc, M. Properties of Bacterial Cellulose Produced Using White and Red Grape Bagasse as a Nutrient Source. *Processes* **2021**, *9*, 1088. [CrossRef]
79. Anderson, N.; Jones, J.G.; Page-Dumroese, D.; McCollum, D.; Baker, S.; Loeffler, D.; Chung, W. A comparison of producer gas, biochar, and activated carbon from two distributed scale thermochemical conversion systems used to process forest biomass. *Energies* **2013**, *6*, 164–183. [CrossRef]
80. Gale, M.; Nguyen, T.; Moreno, M.; Gilliard-AbdulAziz, K.L. Physiochemical properties of biochar and activated carbon from biomass residue: Influence of process conditions on adsorbent properties. *ACS Omega* **2021**, *6*, 10224–10233. [CrossRef]
81. Gonçalves, G.D.C.; Nakamura, P.K.; Furtado, D.F.; Veit, M.T. Utilization of brewery residues to produces granular activated carbon and bio-oil. *J. Clean. Prod.* **2017**, *168*, 908–916. [CrossRef]
82. Paini, J.; Benedetti, V.; Ail, S.S.; Castaldi, M.J.; Baratieri, M.; Patuzzi, F. Valorization of wastes from the food production industry: A review towards an integrated agri-food processing biorefinery. *Waste Biomass Valorization* **2022**, *13*, 31–50. [CrossRef]
83. Ramakrishnan, T.; Kumar, S.S.; Chelladurai, S.J.S.; Gnanasekaran, S.; Sivananthan, S.; Geetha, N.K.; Arthanari, R.; Assefa, G.B. Recent developments in stimuli responsive smart materials and applications: An overview. *J. Nanomater.* **2022**, *1*, 4031059. [CrossRef]
84. Soceanu, A.; Dobrinas, S.; Sirbu, A.; Manea, N.; Popescu, V. Economic aspects of waste recovery in the wine industry. A multidisciplinary approach. *Sci. Total Environ.* **2021**, *759*, 143543. [CrossRef] [PubMed]
85. Malindretos, G.; Tsiboukas, K.; Argyropoulou-Konstantaki, S. Sustainable wine supply chain and entrepreneurship: The exploitation of by-products in a waste management process. *Int. J. Bus. Sci. Appl. Manag. (IJBSAM)* **2016**, *11*, 34–46. [CrossRef]
86. Dwyer, K.; Hosseinian, F.; Rod, M. The market potential of grape waste alternatives. *J. Food Res.* **2014**, *3*, 91–106. [CrossRef]
87. dos Santos Silva, M.E.; Grisi, C.V.B.; da Silva, S.P.; Madruga, M.S.; da Silva, F.A.P. The technological potential of agro-industrial residue from grape pulping (*Vitis* spp.) for application in meat products: A review. *Food Biosci.* **2022**, *49*, 101877. [CrossRef]
88. Manso, M.; Agnihotri, S.; Lennartsson, P.R.; de Marañón, I.M.; Cebrián, M.; Ibarruri, J. From waste to value: Upcycling white grape pomace through a multiproduct biorefinery approach. *J. Environ. Chem. Eng.* **2025**, *13*, 117249. [CrossRef]
89. Devesa-Rey, R.; Vecino, X.; Varela-Alende, J.L.; Barral, M.T.; Cruz, J.M.; Moldes, A.B. Valorization of winery waste vs. the costs of not recycling. *Waste Manag.* **2011**, *31*, 2327–2335. [CrossRef] [PubMed]
90. Mengal, P.; Wubbolts, M.; Zika, E.; Ruiz, A.; Brigitta, D.; Pieniadz, A.; Black, S. Bio-based Industries Joint Undertaking: The catalyst for sustainable bio-based economic growth in Europe. *New Biotechnol.* **2018**, *40*, 31–39. [CrossRef] [PubMed]
91. Chidi, B.S.; Okudoh, V.I.; Hutchinson, U.F.; Ngongang, M.M.; Maphanga, T.; Madonsela, B.S.; Shale, K.; Lim, J.W.; Ntwampe, S.K.O. A Perspective on Emerging Inter-Disciplinary Solutions for the Sustainable Management of Food Waste. *Appl. Sci.* **2022**, *12*, 11399. [CrossRef]
92. Otoni, C.G.; Azeredo, H.M.; Mattos, B.D.; Beaumont, M.; Correa, D.S.; Rojas, O.J. The food–materials nexus: Next generation bioplastics and advanced materials from agri-food residues. *Adv. Mater.* **2021**, *33*, 2102520. [CrossRef]
93. Martínez-Falcó, J.; Martínez-Falcó, J.; Marco-Lajara, B.; Sánchez-García, E.; Visser, G. Aligning the Sustainable Development Goals in the wine industry: A bibliometric analysis. *Sustainability* **2023**, *15*, 8172. [CrossRef]

applied sciences

MDPI

Review

Banana Peels: A Genuine Waste or a Wonderful Opportunity?

Mursleen Yasin, Shreya Gangan and Sunil K. Panchal *

School of Science, Western Sydney University, Hawkesbury Campus, Richmond, NSW 2753, Australia; 20485678@student.westernsydney.edu.au (M.Y.); sagangan21@gmail.com (S.G.)
* Correspondence: s.panchal@westernsydney.edu.au; Tel.: +61-7-4570-1932

Abstract

Banana is the second-highest fruit produced in the world and is a staple food for over 400 million people. Only 40% of the crop is utilised, leading to 114 million tonnes of banana waste annually. Banana peel constitutes about ~40% of the whole fruit, and it is considered a domestic and food industry waste. It is enriched with macronutrients, micronutrients and bioactive compounds, which can provide antioxidant, anti-microbial, antibiotic, pharmaceutical and nutraceutical properties. Banana peels also have higher nutrient value than banana pulp, and they are used in traditional medicines to treat diabetes, diarrhoea, inflammation, ulcers, burns and cough. Given its diverse bioactive properties, banana peel waste is now being explored within the framework of a circular economy to promote waste recycling and reduce environmental impact. This review highlights the nutritional and health properties of banana peel while providing opportunities for waste reduction. Potential applications of banana peels include anti-cancer and anti-fungal agents, biosorbents, natural emulsifiers, reducing agents, biofertilisers, food industry ingredients for bakery products, natural preservatives and food fortification. Exploring banana peel waste potential not only contributes to sustainable waste management but also enhances environmentally friendly innovation for the benefit of human health and the environment.

Keywords: banana waste; banana peel; bioactive compounds; phytochemicals; circular economy; waste recycle; food fortification

1. Introduction

Fruit and vegetables are essential parts of the human diet. The recommended daily intake of vegetables is five servings, and for fruit, it is two servings. The Banana (genus *Musa*) is one of the most important fruits and belongs to the Musaceae family. The genus *Musa* has over 50 species, and its varieties are differentiated based on the way of consumption (cooking, dessert or fresh) [1]. After citrus, banana is the highest-produced fruit globally [2] and ranked fourth in agricultural commodity production after rice, wheat and milk [3,4]. It is usually grown in 150 countries of tropical and sub-tropical regions of the world [5]. India is the highest banana-producing country, with ~34.5 million tonnes of bananas in 2022, which is one-third of the total global production [6,7]. The production in India was almost three times that in China (~11.8 million tonnes), which is the second largest producer of banana, followed by Indonesia (~9.2 million tonnes) (Figure 1) [7]. Overall, Asia contributes 51% towards banana production, followed by the Americas (24%). Oceania only covers 1.2% of the total banana production in the world. According to the Australian Banana Growers Council, Australia produced 362,000 tonnes of bananas in 2020/2021. The most common varieties grown in India, the Philippines and Ecuador belong to the sub-groups Cavendish or Gros Michel [8,9].

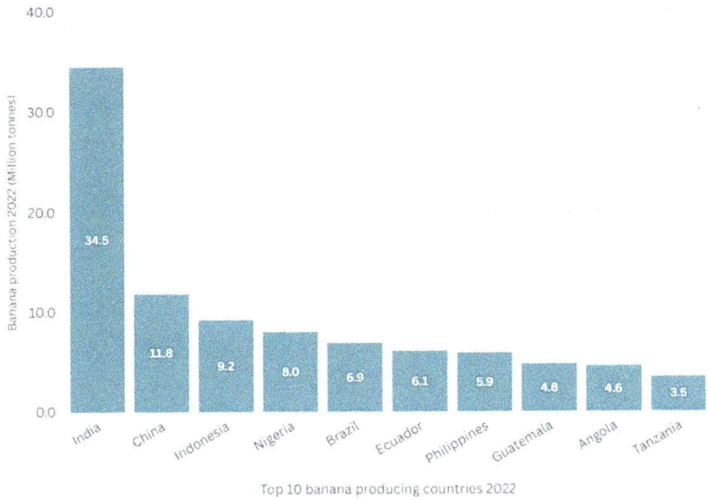

Figure 1. The top ten banana-producing countries in 2022 [7].

In tropical and subtropical regions, banana cultivation is a source of nutrition and a major contributor to the national and local economy in the producing countries. Globally, 15% of the bananas are exported to non-producing countries to provide nutritional diversity in those areas [10,11]. For instance, in Ecuador and Costa Rica, banana and banana products are the second largest agricultural commodities [10]. Banana is evaluated on different external and internal factors for export. The outer appearance is based on the peel colour, fruit shape and size, as well as the defects on the peel. The fruit quality parameters are taste, sweetness, texture, aroma, shelf-life and nutritional values of the fruit [12]. Therefore, banana export is difficult and is limited. Moreover, removal of debris or dust from the fruit, blowing of adjacent leaves and rubbing of petioles also lead to high surface scarring and physical damage to the fruit [12]. This ultimately leads to a reduced export of banana.

The utilisation of banana peel waste is becoming popular to progress toward sustainable development goals. Recently, banana fibre has been used by the textile industry to promote natural fibre used in fabric production [13]. Moreover, banana fibre is also used in the paper, cardboard and packaging industries, along with novel banana leather. Banana fibre has been used to make paper, which is then used to make wallets and paintings [14]. Some applications of banana peel waste include use in agriculture, food and pharmaceutical industries, such as biodegradable pots, biofertilisers, food preservatives, functional foods, antioxidants and anti-microbial drug ingredients [4,15–21]. Innovative extraction techniques such as microwave-assisted extraction, ultrasound-assisted extraction, and pulse-electric-field extraction using deep eutectic and compressed solvents are now suggested for the isolation of bioactive compounds from peels. These novel techniques provide time- and energy-efficient extraction and a higher yield of bioactive ingredients [22]. For example, pulse-electric-field extraction had a higher phenolic compound concentration extracted from Kepok banana peel than the solvent extraction method [23]. However, more research is needed to test novel technologies to evaluate the yield efficiency of bioactive compounds from banana peel and their impact on the environment. Further studies are required to understand the full potential of banana peel and increase the efficacy of utilising this waste for practical application in the circular economy.

The concept of circular economy has gained popularity with the increased amount of food and agro-waste and its harmful disposal to the land. It is defined as reusing or reducing waste by recycling it into useful products and valuable resources. The eco-innovation strategy aims at a 'zero waste' economy [24]. The horticulture products, including mango, apple, banana, citrus, grapes, watermelon, onion, tomatoes, potatoes, cabbage, broccoli, eggplant and cucumber, are sources of agro-waste as seeds, shells, leaves, pseudo stem, pomace, rind and pulp [25]. The agro-waste is a source of many valuable biochemicals and natural resources such as polyphenols, carbohydrates, fibre, prebiotics, protein, pigments, vitamins and lipids. Innovative strategies are now being explored to valorise waste. For example, polyphenols, flavonoids and vitamin C from citrus peels are used as antioxidants, anti-bacterial agents, food packaging and food fortification. Apple pomace is used in the food industry to make ice cream, biscuits, muffins, mayonnaise and crackers and in the pharmaceutical industry as a stabilising agent and nanoparticles [26]. Banana peel is also used for waste valorisation in a similar way to promote the circular economy strategy. This review highlights the potential of banana peel waste as a valuable resource, emphasising its diverse applications across industries while contributing to sustainability, waste reduction and innovative product development.

2. Dessert Bananas and Cooking Bananas

Banana is a perennial crop that is produced around the year. On average, a fruit weighs 125 g, which constitutes ~75% water and the rest is dry matter [27]. The colour of the fruits varies from red to purple and yellow depending on the ripening stages [27].

On a culinary basis, banana production is divided into two categories: dessert bananas and cooked bananas (plantains). The pulp of dessert bananas is firm when unripe and soft during maturation [28]. Therefore, these fruits are eaten raw after ripening or cooked, baked, dried and ground for flour. Moreover, they are also processed to make vinegar, starch and chips. This group belongs to *M. acuminata* or hybrid *Musa* × *paradisiaca* or *M. sapientum* (*M. acuminata* × *M. balbisiana*) Morton. Almost 56% of global banana production is for the dessert category, and ~97% of the banana export is for this variety [29]. Cavendish is the most popular cultivar of this group [30]. Cooking bananas are harvested at the green stage and can be used in different dishes and to make starch. The consumption and popularity of banana sub-groups are differentiated by region, cultivar and sugar concentration. Dessert banana has higher demand worldwide whereas cooking banana is mostly consumed in developing countries such as Africa and Southeast Asia [31]. In Asia, cooking varieties are usually from sub-groups of the cultivar "pisang awak" [31]. Cooking bananas are usually larger and angular, with more starch than dessert bananas.

On the basis of ploidy and genome, bananas are grouped as diploid, triploid or tetraploid. Triploid bananas are cultivated the most in the world. The culinary groups are developed by intra- or inter-hybridisation between two ancestors, i.e., *M. acuminata* Colla (A genome) and *M. balbisiana* Colla (B genome) [32]. The genomic groups of dessert, plantains and cooking bananas are AAA, AAB and ABB, respectively [31,33].

3. Banana Composition and Waste

Banana fruit is important economically and nutritionally for food security and it is a staple food for 400 million people around the world [34]. Banana ranks second highest in the fruit category after citrus and fourth most important crop after rice, wheat and corn [3,4]. The consumption of banana has increased with the changes in dietary patterns, especially for increasing fibre intake, which is an essential nutrient. Uganda was the country with the highest per capita banana consumption in 2023. The average intake per person was ~300 kg per year in Uganda, followed by Papua New Guinea, with ~120 kg per person consumption every year. Other higher consumers of the banana per capita are Samoa, Rwanda, Laos, Burundi, Dominica and Peru [7]. The vast popularity of the banana is due to their high nutritional value and soft texture. Banana is enriched with fibre, sugars, vitamin B_6, vitamin

C, manganese, amino acids, and phenolics such as gallic acid, catechin, epicatechin, tannins and anthocyanins [35]. It is a low-fat food with a high energy content [35].

The waste produced from banana includes deformed, small size and rotten fruit, peel, leaves, pseudo stem parts such as flowers and underground stem (rhizome). Banana peel constitutes ~40% of the fruit and pseudo stem parts make up 60% of the total plant biomass. Harvesting of each tonne of banana fruit generates about 3 tonnes of pseudo stem, 150 kg of rachis, 480 kg of leaves and 100 kg of rejected fruit waste [36]. Some on-farm ways to reuse the waste include putting waste fruit in feedstock with rice bran, molasses, legumes and grasses. The stalk can be chopped or fed directly to the animals [37]. Banana peel is also used as animal food and added to the feedstock of monkeys, poultry, goats, livestock, poultry, etc. [27]. Pseudo stem, leaves and fruit stalks are also used as mulch or composted for manure [37]. However, the conventional ways of banana waste management are now discouraged because of increased environmental concerns and the spread of plant diseases. One of the main reasons for banana wilt spread is the application of manure or mulch from infected banana plants into the banana field [38]. Over the last decade, increased consumer demand for banana has also increased waste production to ~114 million tonnes every year. Only about 40% of the total banana crop is used and the rest is treated as waste biomass [2]. The percentage of losses and waste in the supply chain can reach up to 35% [39]. The banana supply chain is affected by several challenges at each stage that ultimately contribute to waste and loss. Different challenges in each stage are outlined in Figure 2 [40].

Apart from losses at plantation and harvest stages, the banana is prone to many postharvest losses because of its climacteric nature. The metabolic processes, such as respiration and transpiration, continue after harvesting, making it more perishable [41]. Therefore, temperature plays a key role during the storage of banana which is suggested to be kept at 13 °C at both green and ripen stages. Banana is prone to chilling injury at lower temperatures (below 10 °C) and rapid ripening above 14 °C [41,42]. It is physiologically sensitive and gets easily damaged due to mishandling during harvesting and postharvest stages. Moreover, mechanical damage can cause bruises and injuries on the peel, which appear as brownish-black spots on the skin [43]. The fruit might be unacceptable for the market due to the discolouration of the skin. The mechanical damage to the fruit can further cause changes in chemical composition, sugar content, weight loss, colour change, softening and browning of fruit, weight loss, and ultimately, economic loss.

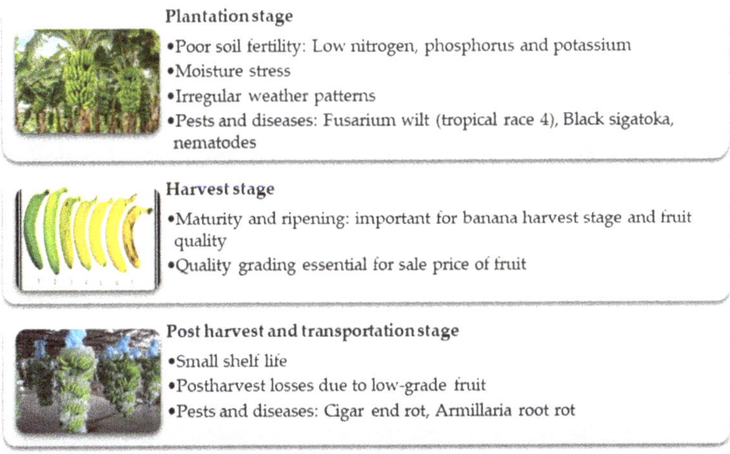

Figure 2. Challenges at each stage of banana production lead to waste production [44].

3.1. Carbon Footprint and Environmental Impacts of Banana Waste

With the increased awareness of climate change and global warming, consumers are now demanding products with the least carbon footprint or global impact. Environmental indicators such as life cycle assessment and carbon footprint are used to evaluate the overall environmental impact of the products. Life cycle assessment helps to understand the overall strategies of the product development to keep it environmentally friendly by analysing the whole product cycle from production to consumption [45]. According to a study, the Ecuadorian banana had a carbon footprint of 1.28 tonnes of CO_2 per tonne of banana [46]. The life cycle assessment of Ecuadorian banana from different stages of the supply chain indicated that the growing stage is the major contributor towards the carbon emission (193.94–220.37 kg CO_2 equivalent/tonnes banana) [47]. In Turkey, the highest impact is at the plantation stage, followed by the transportation stage. The waste disposal of the stems is also one of the major contributors to global warming by banana production [5]. On average, ~114 million tonnes of banana waste is produced globally every year [2]. Natural decomposition of this banana waste releases greenhouse gases when it is dumped in landfills [48].

3.2. Banana Peel

Banana peel is a by-product of the food processing industry and household consumption. Peel is the outer layer or covering that protects the soft, edible part or pulp of the fruit. It consists of ~40% of the total fruit weight, thus banana consumption or use leads to the production of large quantities of banana peel waste every year [49]. It is estimated that around 36 million tonnes of banana peels are produced every year [49]. Traditionally, it has been used for cooking and medicinal purposes, as well as in animal feed. However, a large portion is treated as general waste and discarded in landfills [50]. A study revealed that in Costa Rica, the disposal of rejected bananas in the rivers has caused an imbalance in the aquatic ecosystem. The oxygen demand in the rivers increased because of the higher carbohydrate content in banana biomass, leading to the reduction of aquatic animals [51].

Due to the environmental impact, banana peel waste is now being explored for its potential antioxidant, anti-microbial, antibiotic, pharmaceutical and nutraceutical properties [52]. Previously, it has been used to treat ulcers, burns, diarrhoea, inflammation, cough and diabetes [53].

3.2.1. Nutritional Composition of Banana Peel

Cultivar, variety and ripening stage of the fruit define the nutrient composition of banana peel. Banana peel composition is variable based on these above factors and contains crude protein (2–8%), crude fat (3–11%), carbohydrate (3%) and polyunsaturated fatty acids such as linoleic acid [54]. Table 1 represents the comparison of the proximate composition of peels between the two categories (dessert and cooking) and different cultivars. Banana peels from various cultivars and groups were analysed to compare their nutrient composition. As shown in Table 1, the ripening stage and banana type significantly influence carbohydrate content, including starch and sugar. In the unripe stage, carbohydrates primarily exist as starch. However, as ripening progresses, starch molecules undergo hydrolysis, converting into soluble sugars and leading to an increased total soluble sugar content in ripe banana peels [55]. Moreover, plantain peel has an amylose percentage of starch ranging from 22% to 38%, which makes it less sweet and more resistant to starch hydrolysis [56].

Table 1. Differences between the proximate composition of banana peel from dessert banana and cooked banana.

Component (% of Dry Matter)	Dessert Banana Peel			Plantain Peel (Cooking Banana)		
	Cavendish [57]	*M. sinensis* [58]	Yangambi Km5 [59]	French Clair [57]	*M. paradisiaca* [58]	Big Ebanga [59]
Total dietary fibre	43.2 (green) 49.7 (ripe)	4.17	49.9 (green) 47.9 (ripe)	32.9 (green) 46.9 (ripe)	4.06	35.9 (green) 37.3 (ripe)
Crude protein	6.3 (green) 8.1 (ripe)	2.48	6.9 (green) 7.9 (ripe)	8.3 (green) 9.1 (ripe)	2.23	8.1 (green) 8.6 (ripe)
Starch	11.1 (green) 3.3 (ripe)	-	14 (green) 2.6 (ripe)	35.4 (green) 3.2 (ripe)	-	39.3(green) 0.1 (ripe)
Other soluble sugars	3.7 (green) 32 (ripe)	67.29 (total sugar)	1.4 (green) 33.2 (ripe)	1.7 (green) 42.2 (ripe)	68.57 (total sugar)	4.3 (green) 38.3 (ripe)
Cellulose	-	-	15.2 (green) 15.6 (ripe)	-	-	6.4 (green) 6.1 (ripe)
Crude fat	3.8 (green) 5.7 (ripe)	1.24	6.3 (green) 10.9 (ripe)	4.6 (green) 5.9 (ripe)	1.10	2.2 (green) 3.6 (ripe)
Lignin	-	-	7.3 (green) 13.3 (ripe)	-	-	7.9 (green) 15 (ripe)
Ash	9.6 (green) 12.8 (ripe)	3.95	10.4 (green) 10.7 (ripe)	8.8 (green) 8.2 (ripe)	2.23	6.4 (green) 7.4 (ripe)
Amino acids	4.3 (green) 4.6 (ripe)	-	-	5.8 (green) 5.6 (ripe)	-	-

Banana peel is enriched with essential macronutrients and micronutrients. It comprises Ca, K, Mg, P, Zn, Fe, Cu, Mn and amino acids such as leucine, valine, phenylalanine and threonine. Banana peel is also enriched with pectin (10–21%), lignin (6–12%), cellulose (~6–15%) and hemicellulose (6.4–20%) [60]. The most abundant mineral is P, followed by Na, Ca, Fe and Mg in the peel of *Musa sapientum* [61]. Their concentrations (mg/100 g) in the peel were found to be 211.3, 115.1, 59.1, 47.0 and 45.5, respectively [30,61]. In another study, K content was highest in the peel of *M. sinensis,* followed by Ca and P (1708.6, 40.9, 28.62 and 27.84 mg/100 g, respectively) [58]. Thus, the abundant minerals found in the peel are P and K. However, the results vary across the studies depending on the type of banana, its maturation stage and the method used for respective analysis [44,58,61]. The major minerals found in banana pulp are K, Mg, P and Na, while peel has higher concentrations of P, Ca, Na, Mg and K, as shown in Table 2.

However, the concentrations of these elements were observed to change in different studies depending on the species, cultivar and maturation stage. A study revealed a lower Fe concentration in three banana varieties: Pachabale (10 mg/100 g), Nendranbale (4 mg/100 g) and Yelakkibale (3.33 mg/100 g). However, they observed a Ca concentration of 44.68 mg/100 g in Yelakkiale banana peels. Vitamin C (17.83 mg/100 g), tannins (1073 mg/100 g), phytic acid and oxalates were also higher in the Yelakkiale variety [62]. Table 2 describes the difference in mineral composition between banana pulp and banana peel.

Table 2. Differences between mineral composition of banana pulp and peel.

Minerals	Banana Pulp (*M. sinensis*) mg/100 g [44]	Banana Peel (*M. sinensis*) mg/100 g [58]	Banana Peel (*M. sapientum*) mg/100 g [61]	Banana Peel (*Musa* ABB) mg/100 g [44]
P	15.47	27.84	211.3	283.1
Fe	0.06	0.07	47	2.72
Mg	29.39	28.62	44.5	138.12
K	350.39	1708.6	4.39	3502.5
Ca	4.64	40.9	59.1	142.9
Na	7.73	9.28	115.1	43.1
Zn	0.15	0.39	0.033	1.64
Cu	0.10	0.06	0.51	0.007
N	273.88	344.79	397.5	-
Mn	0.18	0.35	0.702	0.85

3.2.2. Traditional Uses of Peel and Pulp

Banana pulp, peel and plant are being used as traditional and local medicine in many cultures in Asia, Africa, Oceania and America [63]. The peel of ripened bananas can be used as an antiseptic for wounds to release pain and swelling [64]. It can be wrapped directly on the injury site, with the inner layering touching the skin. In Brazil, banana peels have traditionally been used to treat burn injuries [65]. Peel can also be used to soothe warts and mosquito bites by rubbing it onto the skin [53]. Banana tree is also used for the treatment of anaemia. Psudo-stem is used to treat diarrhoea in children in the form of sweetened juice. The flower of banana is used to treat eye problems and pulmonary diseases [65]. In Asian countries such as India and Pakistan, bananas have been used to treat inflammation, gastrointestinal tract disorders, gastritis, diarrhoea, diabetes and cough. A study reported that a banana diet is used subsequently for children during diarrhoeal episodes [66]. In the Philippines, different parts of the plant are used to treat different ailments. For example, young leaves are used for fever and headache, fruit is used for diarrhoea and juice is used for abdominal pain. Banana is also used for depression, anaemia and blood pressure because of their higher content of omega-3/omega-6 polyunsaturated fatty acids, Fe and K, respectively [53,67,68]. The vitamins B_6 and B_{12}, Mg and K in bananas help the body in nicotine withdrawal and hence can help to quit smoking. Moreover, the high fibre content helps to treat constipation, and high sugar helps to treat hangovers, heartburn and ulcers because of its antacid potential [53].

3.3. *Bioactive Compounds in Banana Peel*

The composition of banana peel is a complex matrix of many important primary and secondary metabolites, such as phytochemicals, antioxidants, anthocyanins, amines, carotenoids and catecholamines (Figure 3) [69,70].

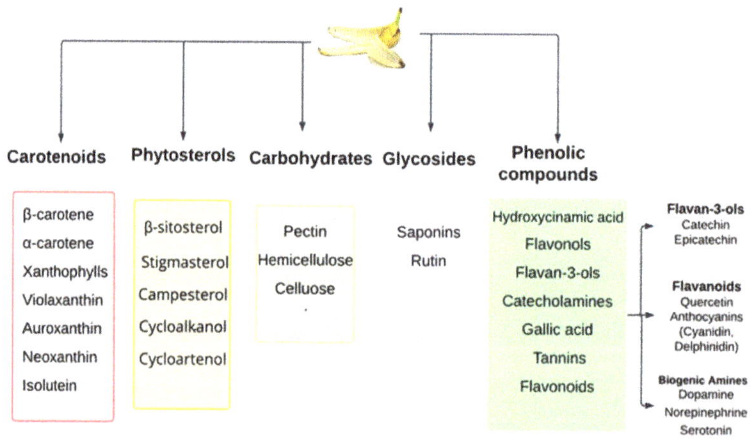

Figure 3. Phytochemicals present in banana peel.

Bioactive compounds in banana peels are secondary metabolites with therapeutic potentials, such as antioxidant effects. Phenolic compounds and carotenoids are the main phytochemicals found in banana peels [71]. Health benefits associated with these compounds are highlighted below. Phenolics, carotenoids, flavonoids and biogenic amines found in ripened and raw bananas are valuable components that can support health benefits [72]. Banana pulp also contains catecholamines (such as serotonin and dopamine), tryptophan and indole compounds [73].

3.3.1. Phenolic Compounds

Phenolic compounds, such as gallic acid, catechin, epicatechin, tannins and anthocyanins, are present in the pulp and peel of banana. Flavonols are the main flavonoids found in the banana, including quercetin, myricetin, kaempferol and cyanidin [74]. These flavonoids have shown many health benefits in several research studies [75–77]. Flavonoids are associated with antioxidant, anti-microbial and anti-inflammatory properties, which are important in preventing cardiovascular diseases, cancer, diabetes and obesity. Flavonoids serve as protective scavengers against the reactive oxygen species and free radicals formed from oxygen that cause ageing and several illnesses [74]. The most abundant flavonoids present in the banana peel are quercetin, myricetin, kaempferol, rutin and cyanidin. Galocatechin was found to be 158 mg/100 g in banana peel, which is the most potent component for antioxidant activity in the peel [78]. A study identified rutin as the most abundant flavonol in the banana peel of the red Yade variety, followed by kaempferol-3-rutinoside and quercetin-deoxyhexose-hexoside (~500 μg/g dry matter, ~175 μg/g dry matter and ~75 μg/g dry matter, respectively) [79].

3.3.2. Carotenoids

Due to their distinctive physiological properties, such as their role as pro-vitamins and as antioxidants, particularly in the scavenging of singlet oxygen, carotenoids have positive effects on health [80]. These are important for performing activities such as pro-vitamin A, antioxidants, cell communication, improved immunity, UV skin protection and macula protection [80]. Carotenoids have been extensively researched for their potential to reduce the risk of illnesses, notably several cancers and eye conditions that were becoming increasingly prevalent across the world. Carotenoids are helpful in preventing cancer, cardiovascular diseases, cataracts and macular degeneration [81]. These are one of the most significant groups of plant pigments, which are categorised as oxygenated hydrocarbon derivatives and pure hydrocarbons (carotenes, xanthophylls). These are by-products of the isoprenoid biosynthesis pathway, and their primary functions in plants are as antioxidants and auxiliary pigments for photosynthesis [80]. Major carotenoids identified in banana peel were lutein, α-carotene, β-carotene, β-cryptoxanthin and lycopene, ranging between 159.66 to 2553.51 μg/100 g among 14 different banana cultivars. Lutein concentration was highest in the unripe peel [82]. A study observed that banana peel is enriched with pro-vitamin A carotenoids such as *trans*-β-carotene, *trans*-α-carotene and *cis*-β-carotene with ~175, 165 and ~92 μg/g dry weight, respectively [83].

3.3.3. Biogenic Amines

Biogenic amines are nitrogenous substances created by the amination of aldehydes and ketones or the decarboxylation of amino acids. It has been demonstrated that banana peel and pulp contain biogenic amines such as serotonin, dopamine and norepinephrine [70]. Biogenic amines are essential for cell survival and the proper functioning of metabolic pathways, such as the synthesis of proteins and hormones and cell replication [84]. However, the excessive accumulation of biogenic amines can lead to cell toxicity and cause food poisoning, diarrhoea, vomiting, headaches, hypertension and carcinogenesis [85].

Banana peel is also enriched with tyrosinase, which is a biocatalyst that produces 3,4-dihydroxyphenyl-l-alanine (L-DOPA). L-DOPA is important as it is the precursor of dopamine and helps in the treatment of Parkinson's disease, a neurodegenerative disorder with reduced dopamine in brain cells. The use of L-DOPA alleviates dopamine by taking the precursor tyrosine across the blood-brain barrier and helps to manage the symptoms of Parkinson's disease [86]. Other bioactive amines present are spermidine, serotonin and putrescine. The biogenic amines make banana peel a useful source for maintaining cholesterol metabolism and cardiovascular diseases [15].

3.3.4. Phytosterol

In functional foods, phytosterols are naturally occurring plant sterols that have a number of favourable health-promoting benefits, including decreasing blood cholesterol and reducing intestinal absorption of dietary cholesterol. They also have anti-cancer effects and modulate the immune system [87].

Other bioactive compounds found in the banana peel are sterols and triterpenes, such as β-sitosterol, stigmasterol, campesterol, cycloalkanol, cycloartenol and 24-methylenecy cloartanol [88]. Some studies observed the total phenolic content in banana peels of 15 varieties at different ripening stages. It ranged from 29.02 to 61.00 mg gallic acid equivalent/100 g for unripe banana peels and 60.39 to 115.70 mg gallic acid equivalent/100 g for ripe banana peels [89]. Another study identified total phenols and total flavonoids to be highest with 80% methanol extraction with concentration at 17.89 mg/g dry weight and 21.04 mg/g dry weight of banana, respectively [90]. The total tannin concentration was found to be 24.2 mg/g. Tannins are important for many medicinal traits such as anti-cancer, antioxidant, anti-inflammatory and anti-bacterial properties.

4. Potentials of Banana Peels

4.1. Antioxidant Activity

Antioxidants are substances that remove free radicals from the body. One of the finest sources of antioxidants is banana peel. It has high concentrations of bioactive substances such as phenols, carotenoids, flavonoids, saponins and tannins [91]. The presence of antioxidant enzymes, including ascorbate peroxidase, catalase, peroxidase and superoxide dismutase, as well as estragole, hexadecenoic acid, ethyl ester, epicatechin and gallocatechin were detected using gas chromatography and mass spectroscopy in banana peels [92]. The banana peel's high concentrations of dopamine, L-DOPA and catecholamines promote antioxidant action [93]. Using the peels from *Musa acuminata*, a semisolid jelly was created to investigate and assess the antioxidant activity of the banana peel. Because the peel itself serves as the gelling component, commercial pectin was omitted from the jelly's formulation. The commercial jelly only has a 65% free radical scavenging activity, but the jelly with a 7% concentration of banana peel has an 84% free radical scavenging activity [94].

4.2. Anti-Bacterial Activity

Because germs are becoming more resistant to the present anti-bacterial treatments, new anti-bacterial agents are now needed. To create novel anti-bacterial agents, researchers are looking at plant materials [95]. The banana peel has anti-microbial effects against both Gram-positive and Gram-negative bacteria such as *Escherichia coli*, *Salmonella enteritidis* and *Pseudomonas aeruginosa* [91]. Gram-positive species such as *Staphylococcus aureus*, *Bacillus subtilis* and *Bacillus cereus* are impacted by banana peels. Peels are also active against *Salmonella typhi*, *Micrococcus luteus* and *Klebsiella pneumonia* [90,96]. *Aggregatibacter actinomycetemcomitans* and *Porphyromonas gingivalis* are two periodontal infections that the banana peel effectively inhibits. Compared to dried peels, fresh banana peels exhibit the largest zone of inhibition against these microorganisms [97]. Moreover, unripe bananas were observed to have higher anti-bacterial activity due to the content of bioactive compounds such as polyphenols, flavonoids, tannins, alkaloids, terpenoids and glycosides [97]. These anti-bacterial metabolites impact the cell membrane of bacteria, which leads to irregular cell shapes and, ultimately, cell death [98]. A study observed the constricted nature of banana peel tannins, which are effective on bacterial peptidoglycan and proteins, thus causing reduced bacterial activity [99].

4.3. Anti-Fungal Activity

Even the synthetic antibiotic agent falls short over the entire spectrum of anti-microbial action. Therefore, the creation of natural antibiotics with broad-spectrum action is necessary [100]. Banana peels can provide a suitable solution as they can inhibit the growth of a large variety of bacteria and fungi. Tannins, alkaloids, saponins and flavonoids from banana peels can provide these anti-microbial actions [101]. Some of the fungi impacted by banana peels are *Aspergillus niger*, *Aspergillus flavus*, *Penicillium digitatum*, *Fusarium oxysporum*, *Candida albicans*, *Saccharomyces cerevisiae* and *Penicillium citrinum*. Due to their effects against *Aspergillus oryzae* and *Rhizopus stolonifer*, banana peels can also be used as a preservative to prevent food spoilage [91].

4.4. Anti-Cancer Activity

Cancer is the leading cause of death worldwide. The efficacy of commercially available anti-cancer medications is declining as a result of rising resistance and excessive cytotoxicity. To treat and prevent cancer, new medicines are investigated [102]. The development of new treatment strategies will benefit from the examination of plant secondary metabolites for their anti-cancer capabilities [102]. Plant-based anti-cancer interventions have been investigated with positive impacts in reducing cancer-related pathologies. Flavonoids and phenols from plant sources are beneficial for the treatment of cancer, as shown in in vitro studies [102]. Due to the abundance of phytochemicals in banana peels, it has shown positive responses as an anti-cancer agent [102].

Tannins are also an important bioactive group of compounds that are enriched with anti-cancer properties. Banana peel contains tannins and studies have identified up to ~24 mg/g concentration in the peel [90]. Banana peel extracts have been used to study anti-cancer properties in vitro. Some of the cell cultures used include breast cancer cell line (MCF-7), human colorectal carcinoma (HCT-116) tumour cell lines and the human umbilical vein endothelial cells (HUVEC cell) [103]. The hexane extract of banana peel and pulp showed the highest efficacy while showing the greatest toxicity against HCT-116 (62%) and MCF-7 (61%) cell lines. In contrast, aqueous and ethanolic extracts exhibited minimal anti-proliferative activities against these cell lines. Notably, all extracts showed negligible cytotoxicity toward normal cell lines, indicating selective action against cancerous cells [103].

The banana is a rich source of carotenoids, and some members of this group, such as lycopene, have anti-cancer activity against prostate cancer in men [83]. It is suggested that higher carotenoid concentration in the diet is effective in preventing cancer, diabetes and cardiovascular diseases [104]. Phytosterols are also effective anti-cancer agents against colon, prostate and breast cancer [105]. The *n*-hexane extract of banana peel effectively inhibited the growth of colon cancer cells due to its potent anti-angiogenic properties [103].

4.5. Other Properties

Banana peel has been identified to have anti-diabetic properties. In a study, rats were orally treated with banana peel extracts from *M. sapientum*, *M. paradisiaca*, *M. cavendish* and *M. acuminata* to test glucose tolerance [106]. The results showed that peels have the capacity to lower blood glucose concentrations. Moreover, it has been suggested that banana peel has enhanced α-amylase inhibitory activity (~81% at 1000 µg/mL) when compared with lemon and pomegranate peels [107].

The anti-ulcer impacts of banana peels have also been observed. The extract of tepal and peel was tested against ulcer index, gastric wall mucus, and pH and volume of gastric juice in ulcer-induced animals. Banana tepal and peel were subjected to a cold extraction technique using methanol and concentration of phytochemicals, along with the determination of anti-ulcer activity on Swiss albino mice. The extracts of banana tepal were found to be enriched with glycosides, phenols and saponins, while the peel had higher flavonoids. Therefore, preventive effects against indomethacin- and pylorus ligation-induced ulcers by ~69 and ~43% were observed, respectively [108]. Moreover, the extract reduces the gastric juice volume, strengthens the gastric wall mucous and decreases the acidity of gastric juice [108]. Leucocyanidin, a compound present in banana peel, was found to be preventive against ulcers [74].

4.6. Pectin Extraction from Banana Peel

Components sourced from natural sources are preferred over synthetic ones for various formulations due to their non-toxic, non-irritating and inexpensive nature [109]. Pectin is a polymer that occurs in nature and includes α-(1,4)-linked D-galacturonic acid units. Pectin interacts with other cell constituents to give plants their mechanical strength [110]. Pectin is most frequently used as an emulsifier, stabiliser and gelling agent. It is seen as a very important component due to its outstanding stabilising and emulsifying abilities [111].

4.7. Banana Peel Extract as a Reducing Agent in the Formulation of Silver Nanoparticles

Silver has a crucial function in metallic nanoparticles as an inorganic anti-bacterial component. Silver nanoparticles may be made for various purposes. The preparation techniques include electrochemical, green chemistry, simple procedure and sono-chemical breakdown of silver complexes [112]. The most common and simple method for producing silver nanoparticles is the reduction of metal salts. Several chemical reducing agents are used, such as hydrazine, sodium citrate, sodium borohydride, ascorbic acid, ammonium formate and dimethylformamide, to produce stable and variably structured silver nanoparticles in an aqueous solution [113,114].

Silver nanoparticles are produced using expensive, time-consuming, energy-intensive and hazardous physical and chemical procedures. Therefore, there is an unmet need for a practical alternative to the plant extract-based method of manufacturing silver nanoparticles [115]. Because it tends to change silver ions (Ag^+ to Ag^0), banana peel extract can be utilised as a reducing agent to produce silver nanoparticles. Silver nitrate ($AgNO_3$) was reduced using banana peel extract to create silver nanoparticles [116]. Surface plasmon resonance (SPR) bands, a unique visual property of noble metals, are present. Banana peel extract is reportedly utilised to biosynthesise silver nanoparticles, which have a clearly discernible SPR band. These biosynthesised nanoparticles have high anti-bacterial activity and successfully help against clinically isolated multidrug-resistant human infections [117]. Compared to Gram-positive bacteria, Gram-negative bacteria are more resistant to them. These nanoparticles also possess powerful anti-fungal and anti-bacterial capabilities [118].

5. Other Uses of Banana Waste

5.1. Lignocellulosic Biomass for Ethanol Production

Agricultural waste from crops such as wheat straw, rice husk and corn stover is used for ethanol production from lignocellulosic waste. The banana is among the top-produced fruit [119] and researchers are now investigating the bioconversion of their waste into ethanol. Almost 2.13 tonnes of fresh lignocellulosic waste is generated with every tonne of banana production [120]. The waste parts include pseudo stem, leaves and rachis. The ethanol production from waste depends on different factors such as time, temperature, pH, particle size and fermentation process [121]. The optimum conditions for enzymatic conversion of banana waste into ethanol were identified in a study [122].

5.2. A Biosorbent of Water Pollutants

With the industrial revolution, water pollution has also increased and many harmful pollutants are released into water without proper treatment [123]. Industries release heavy metals, dyes, pesticides, chiral chemicals, pharmaceuticals and oil particles into the aquatic environment [124]. These pollutants are causing severe threats to human and aquatic life because of their carcinogenic properties. The reproductivity of fish is affected by some elements, even at very low concentrations (1.0 ng/L) [125]. There are several treatment methods being used to decontaminate the water, such as reverse osmosis, filtration, solvent extraction, chelation, ozonation, skimming, adsorption and electrophoresis. Adsorption by activated carbon is one of the most effective methods, but its application is limited due to its high cost [126]. Natural and synthetic adsorbents can be used as alternatives, such as biochar, clay, inorganic materials, chitosan and carbon nanotubes [127,128]. Although efficient, these materials mostly have complicated and expensive procedures for preparation, enhancements and regeneration. Moreover, these processes release secondary pollutants and are non-biodegradable [128]. Hence, biomass materials are now being explored for use in the adsorption of water contaminants as they are easily available, eco-friendly, biodegradable and less expensive [129,130]. The biosorbents effectively being tested are rice husk, banana peel, corn cob, potato peel, coconut husk, cocoa pod, orange peel and sugarcane bagasse [131,132].

Banana peel waste is getting attention to be explored as a biosorbent because it is available all over the world. The overall efficiency of banana peel as an adsorbent depends on factors such as water pH, temperature, contamination level, time of treatment and dosage, reusability, regeneration, thermodynamics and kinetics [133–135]. It has been observed that lower pH values are suitable for decontaminating anionic pollutants and high pH solutions for cationic compounds. The optimum time was 1 h of treatment, and banana peel is an effective biosorbent over a range of temperatures. It can undergo both exothermic and endothermic reactions, making it suitable for use in both tropical and temperate regions [136]. Biosorbent made from banana peel was tested for oil spill removal. The peel biosorbent can be reused up to 10 times after treating it with mechanical action and *n*-hexane to remove the absorbed oil [137]. Banana peel has been used to decontaminate many metals such as Cd, Cu, Pb, Fe, Sr, Zn, Ni, Mn and many dyes such as methylene blue, amido black, Congo red, etc. [128]. Further studies are required to make banana peel a more efficient biosorbent, focusing on the reusability of biomass, effectiveness on more pollutants such as Hg and As, and commercialising it on higher industrial scales [128].

5.3. Use of Banana Peel in Food Sector

The utilisation of agricultural by-products in functional food production has gained significant attention. Banana peel, rich in fibre, essential nutrients, minerals and bioactive compounds, holds promising potential for use as a value-added ingredient in functional foods [138]. Banana peel from ripe and unripe fruits are used to improve the nutritional value of food products such as noodles, jelly, meat and bakery products. Banana peel flour is used as a food fortification ingredient as it is enriched with antioxidants and dietary fibre [139].

Steamed bread was prepared by using 30% green banana pulp flour as a wheat flour substitution. The mineral comparison showed higher Mg, K, Na and Ca content than normal wheat bread, along with higher total dietary fibre and resistant starch. However, the elasticity and chewiness were negatively impacted because of the inconsistency of gluten in the pulp flour [140]. Moreover, the study also used 10% yellow banana pseudo-stem flour and observed an increase in the crust lightness, crumb, total phenolic content, ash and dietary fibre of the banana bread [140]. In another study, 10% yellow banana peel flour was used as a substitute for wheat flour to make chapati bread. The peel chapati bread was softer because of the reduced tear force and high water absorption. The total phenols and flavonoids were also enhanced in peel flour bread compared to the control. The sensory evaluation did not show a significant difference between the acceptance of 10% banana peel flour bread and wheat bread [140]. Another study observed the physiochemical and sensory properties of Egyptian Baladi flatbread made with 5% and 10% banana peel flour as a substitute for wheat flour. It reported higher protein, ash and fat content in banana peel flour bread than in normal wheat flour bread [141]. Moreover, the carbohydrate content decreased from 75% to 50.5% and 44.34%, respectively [141]. The acceptance of banana peel bread was also found to be similar to wheat bread. So, bread with low carbohydrate and enhanced fibre and phenolic content is also proposed as a better alternative for blood glucose control, insulin resistance and obesity [142]. Functional biscuits prepared with 5%, 10% and 15% banana peel flour had improved dietary fibre (~16–37%), ash, moisture content and phenolic composition. The biscuits with 10% peel flour were more accepted by the sensory panelists because of the better colour, taste and texture [143].

Banana peel is a good source of xylitol, which is a naturally produced 5C sugar in plants and is used as a sweetening agent in food products. It has 40% less calories and 75% less carbohydrate than sucrose [4]. Xylitol obtained from banana peel was used in rusk preparation, and it decreased the water activity of the rusk, thus prolonging its shelf life. It also had a good acceptance score [144]. A study used banana peel powder at 3, 6, 9, 12 and 15% in cakes and compared the nutritional, organoleptic, anti-microbial and antioxidant properties. The results indicated that cake was fortified with enhanced nutrients and fat, ash, protein and amino acids. The texture and taste analysis also showed no significant difference, which means the banana peel cake was accepted as much as the regular cake [145]. Further, the addition of 10% peel powder to the cake mix increased the foam stability and acceptance [146]. Another study proposed adding 15% peel flour for the best organoleptic qualities and 30% for fortification purposes in the cake [147]. Banana peels can also be a good ingredient in gluten-free products.

Banana peel powder was used to enhance the fibre content of the chicken sausage. The results showed positive improvement in the fibre and mineral composition of the sausage, with a decrease in total fat from 9.18% (control chicken sausage) to 7.6% in banana peel sausage [148]. Banana peel water extract (2%) was used to produce ground chicken patties and it improved the free radical scavenging activity in the patties [149].

Banana peel is also used as an edible packaging material. It was observed to have similar tensile strength as polyethylene because of its higher fibre content with inorganic nutrients, which improved its mechanical properties. Moreover, these films are degradable and can be composted, offering an environmentally friendly solution [150].

6. Limitations and Future Opportunities

Banana peel has been studied extensively for its potential antioxidant and anti-microbial properties because it has high concentrations of bioactive compounds. However, most of the studies are conducted in vitro, and hence, there is a lack of practical application. The storage, handling, transportation and processing of banana peels also present challenges due to their high moisture content and organic molecules. These factors cause degradation and instability of the peel [151]. Hence, research is required to address a modern and sustainable technology for enhancing the efficiency of banana peel processing and its use in food, pharmaceuticals, agriculture, bioprocessing, paper and textile industries. Utilising banana peel waste as a natural resource contributes to environmental sustainability and human health while supporting the principles of a circular economy.

The addition of waste value also plays an integral role in the Sustainable Development Goals (SDG), especially SDG 2 Zero Hunger, SDG 9 Industry, innovation and Infrastructure, SDG 12 Responsible Consumption and Production, and SDG 13 Climate Action. The use of banana peel as a functional food ingredient for preservation and fortification contributes towards the zero hunger, good health and well-being goals. Technological advancements in processing and using peels for biorefinery, bioprocess, bioethanol production and bioplastics promote sustainability and environmentally friendly innovations. The value addition to waste and the utilisation of its bioactive potential to maximum capacity supports SDG 12 and, ultimately, SDG 13, as these actions will reduce greenhouse gas emissions from waste in landfills. Therefore, integrating banana peels into various industries promotes economic growth, environmental sustainability and public health (Figure 4).

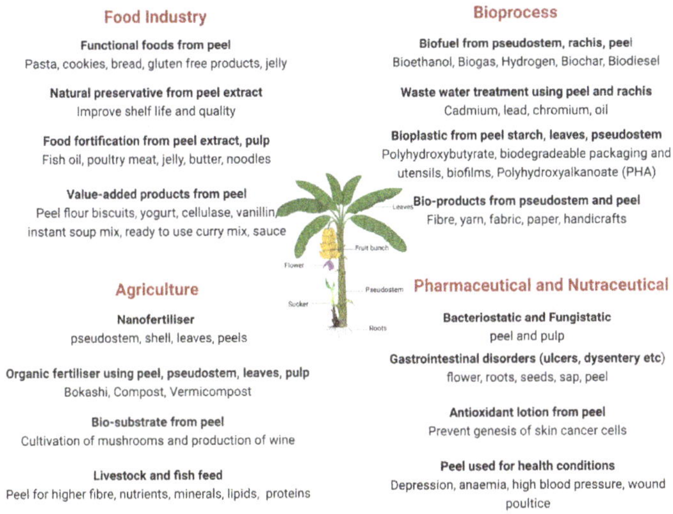

Figure 4. Use of banana waste for the circular economy.

7. Conclusions

Food waste and by-products have been recognised as sustainable sources of nutrients. Banana, as one of the important fruits consumed globally, provides banana peel as a source of many bioactive nutrients, such as carotenoids, phytosterols, carbohydrates and biogenic amines, along with many micronutrients and macronutrients. These components can serve as functional and nutritional ingredients in various industries, including food, pharmaceuticals, nutraceuticals, agriculture and bioprocessing. Food fortification with banana peels to provide nutrients can help with improving food security and improving health outcomes. Further, banana peels can also serve as the ingredients for biofuels and bioproducts that can help recycle waste from banana production, thus contributing to the circular economy. Overall, banana peel can serve as a raw material for value-added products in food, agriculture, bioprocess and nutraceutical industries, thus providing sustainable and environmentally friendly solutions for this waste.

Author Contributions: Conceptualisation, S.K.P.; methodology, M.Y., S.G. and S.K.P.; writing—original draft preparation, M.Y. and S.G.; writing—review and editing, S.K.P.; supervision, S.K.P. All authors have read and agreed to the published version of the manuscript.

Funding: This research received no external funding.

Institutional Review Board Statement: Not applicable.

Informed Consent Statement: Not applicable.

Data Availability Statement: Not applicable.

Conflicts of Interest: The authors declare no conflicts of interest.

References

1. Lucas Aguirre, J.C.; Quintero Castaño, V.D. Rheological characterization of starch and morphological evaluation of 20 varieties of musaceae (*Musa* sp.), from the Fedeplátano germplasm bank, Chinchiná—Caldas, Colombia. *Acta Agron.* **2016**, *65*, 218–225. [CrossRef]
2. Alzate Acevedo, S.; Díaz Carrillo, Á.J.; Flórez-López, E.; Grande-Tovar, C.D. Recovery of banana waste-loss from production and processing: A contribution to a circular economy. *Molecules* **2021**, *26*, 5282. [CrossRef]
3. Ploetz, R.C.; Kema, G.H.J.; Ma, L.-J. Impact of diseases on export and smallholder production of banana. *Annu. Rev. Phytopathol.* **2015**, *53*, 269–288. [CrossRef] [PubMed]
4. Mohd Zaini, H.; Roslan, J.; Saallah, S.; Munsu, E.; Sulaiman, N.S.; Pindi, W. Banana peels as a bioactive ingredient and its potential application in the food industry. *J. Funct. Foods* **2022**, *92*, 105054. [CrossRef]
5. Rattanapan, C.; Ounsaneha, W. The potential for value-added banana production waste under circular economy concept. *Int. J. Environ. Sci. Dev.* **2022**, *13*, 92–96. [CrossRef]
6. Panigrahi, N.; Thompson, A.J.; Zubelzu, S.; Knox, J.W. Identifying opportunities to improve management of water stress in banana production. *Sci. Hortic.* **2021**, *276*, 109735. [CrossRef]
7. World Population Review. Banana Consumption by Country 2024. 2024. Available online: https://worldpopulationreview.com/country-rankings/banana-consumption-by-country (accessed on 10 January 2025).
8. Amarasinghe, N.K.; Wickramasinghe, I.; Wijesekara, I.; Thilakarathna, G.; Deyalage, S.T. Functional, physicochemical, and antioxidant properties of flour and cookies from two different banana varieties (*Musa acuminata* cv. Pisang awak and Musa acuminata cv. Red dacca). *Int. J. Food Sci.* **2021**, *2021*, 6681687. [CrossRef]
9. Torres, L.C.; Zamora, L.C. Benefits in Latin America and the Caribbean about production of Cavendish AAA banana resistant to black Sigatoka. *Rev. Bionatura* **2018**, *3*, 740–744. [CrossRef]
10. Varma, V.; Bebber, D.P. Climate change impacts on banana yields around the world. *Nat. Clim. Chang.* **2019**, *9*, 752–757. [CrossRef]
11. Heslop-Harrison, J.S.; Schwarzacher, T. Domestication, genomics and the future for banana. *Ann. Bot.* **2007**, *100*, 1073–1084. [CrossRef]
12. Santosh, D.T.; Tiwari, K.N.; Reddy, R.G. Banana bunch covers for quality banana production—A review. *Int. J. Curr. Microbiol. Appl. Sci.* **2017**, *6*, 1275–1291. [CrossRef]
13. Provin, A.P.; Medeiros d'Alva, A.; de Aguiar Dutra, A.R.; Salgueirinho Osório de Andrade Guerra, J.B.; Leal Vieira Cubas, A. Closing the cycle: Circular economy strategies for the textile industry using banana farming waste. *J. Clean. Prod.* **2024**, *470*, 143352. [CrossRef]
14. Tewari, S.; Reshamwala, S.M.S.; Bhatt, L.; Kale, R.D. Vegan leather: A sustainable reality or a marketing gimmick? *Environ. Sci. Pollut. Res.* **2024**, *31*, 3361–3375. [CrossRef]
15. Afzal, M.F.; Khalid, W.; Akram, S.; Khalid, M.A.; Zubair, M.; Kauser, S.; Mohamedahmed, K.A.; Aziz, A.; Siddiqui, S.A. Bioactive profile and functional food applications of banana in food sectors and health: A review. *Int. J. Food Prop.* **2022**, *25*, 2286–2300. [CrossRef]
16. Anirudh, M.K.; Nandhu Lal, A.M.; Harikrishnan, M.P.; Jose, J.; Thasim, J.; Warrier, A.S.; Venkatesh, R.; Vaddevolu, U.B.P.; Kothakota, A. Sustainable seedling pots: Development and characterisation of banana waste and natural fibre-reinforced composites for horticultural applications. *Int. J. Biol. Macromol.* **2024**, *270*, 132070. [CrossRef] [PubMed]
17. Bishnoi, S.; Sharma, S.; Agrawal, H. Exploration of the potential application of banana peel for its effective valorization: A review. *Indian J. Microbiol.* **2023**, *63*, 398–409. [CrossRef]
18. Mohd Dom, Z.; Azhar, A.Z.; Masaudin, S.N.A. Utilization of banana peel as functional ingredient in product development. *Adv. Agric. Food Res. J.* **2023**, *4*, a0000209. [CrossRef]
19. Shaukat, N.; Farooq, U.; Akram, K.; Shafi, A.; Hayat, Z.; Naz, A.; Hakim, A.; Hayat, K.; Naseem, S.; Khan, M.Z. Antimicrobial potential of banana peel: A natural preservative to improve food safety. *Asian J. Agric. Biol.* **2023**, *3*, 1–6. [CrossRef]
20. Manzoor, A.; Ahmad, S. Banana peel: Characteristics and consideration of its extract for use in meat products preservation: A review. *ACS Food Sci. Technol.* **2021**, *1*, 1492–1506. [CrossRef]
21. Narwal, P.; Kapoor, B.; Negi, N.P. Decoding the chemical repertoire, antimicrobial synergy, and antioxidant mastery of banana pulp and peel extracts. *J. Plant Biochem. Biotechnol.* **2024**, *33*, 436–446. [CrossRef]
22. Roselli, V.; Pugliese, G.; Leuci, R.; Brunetti, L.; Gambacorta, L.; Tufarelli, V.; Piemontese, L. Green methods to recover bioactive compounds from food industry waste: A sustainable practice from the perspective of the circular economy. *Molecules* **2024**, *29*, 2682. [CrossRef]

23. Hendrawan, Y.; Larasati, R.; Wibisono, Y.; Umam, C.; Sutan, S.M.; Hawa, L.C. Extraction of phenol and antioxidant compounds from Kepok banana skin with PEF pre-treatment. *IOP Conf. Ser. Earth Environ. Sci.* **2019**, *305*, 012065. [CrossRef]
24. Aguiar Pascoalino, L.; Barros, L.; Barreira, J.C.M.; Oliveira, M.B.P.P.; Reis, F.S. Closing the loop: Exploring apple pomace as a source of bioactive compounds in the framework of circular economy. *Sustain. Food Technol.* **2025**, *3*, 81–95. [CrossRef]
25. García-González, A.G.; Rivas-García, P.; Escamilla-Alvarado, C.; Ramírez-Cabrera, M.A.; Paniagua-Vega, D.; Galván-Arzola, U.; Cano-Gómez, J.J.; Escárcega-González, C.E. Fruit and vegetable waste as a raw material for obtaining functional antioxidants and their applications: A review of a sustainable strategy. *Biofuels Bioprod. Biorefin.* **2025**, *19*, 231–249. [CrossRef]
26. Barreira, J.C.M.; Arraibi, A.A.; Ferreira, I.C.F.R. Bioactive and functional compounds in apple pomace from juice and cider manufacturing: Potential use in dermal formulations. *Trends Food Sci. Technol.* **2019**, *90*, 76–87. [CrossRef]
27. Heuzé, V.; Tran, G. Banana (General). 2016. Available online: https://www.feedipedia.org/node/4670 (accessed on 17 January 2025).
28. Huang, P.-H.; Cheng, Y.-T.; Lu, W.-C.; Chiang, P.-Y.; Yeh, J.-L.; Wang, C.-C.; Liang, Y.-S.; Li, P.-H. Changes in nutrient content and physicochemical properties of cavendish bananas var. Pei Chiao during ripening. *Horticulturae* **2024**, *10*, 384. [CrossRef]
29. Arias, P.; Dankers, C.; Liu, P.; Pilkauskas, P. *The World Banana Economy, 1985–2002*; Food and Agriculture Organization of the United Nations: Rome, Italy, 2003.
30. Hikal, W.M.; Said-Al Ahl, H.A.H.; Bratovcic, A.; Tkachenko, K.G.; Sharifi-Rad, J.; Kačániová, M.; Elhourri, M.; Atanassova, M. Banana peels: A waste treasure for human being. *Evid. Based Complement. Alternat. Med.* **2022**, *2022*, 7616452. [CrossRef]
31. Amah, D.; van Biljon, A.; Brown, A.; Perkins-Veazie, P.; Swennen, R.; Labuschagne, M. Recent advances in banana (*Musa* spp.) biofortification to alleviate vitamin A deficiency. *Crit. Rev. Food Sci. Nutr.* **2019**, *59*, 3498–3510. [CrossRef] [PubMed]
32. Häkkinen, M. Reappraisal of sectional taxonomy in *Musa* (Musaceae). *Taxon* **2013**, *62*, 809–813. [CrossRef]
33. Robinson, J.C.; Galán Saúco, V. Taxonomic classification, cultivars and breeding. In *Bananas and Plantains*; Robinson, J.C., Galán Saúco, V., Eds.; CABI: Wallingford, UK, 2010; pp. 21–49. [CrossRef]
34. Thangavelu, R.; Saraswathi, M.S.; Uma, S.; Loganathan, M.; Backiyarani, S.; Durai, P.; Edwin Raj, E.; Marimuthu, N.; Kannan, G.; Swennen, R. Identification of sources resistant to a virulent Fusarium wilt strain (VCG 0124) infecting Cavendish bananas. *Sci. Rep.* **2021**, *11*, 3183. [CrossRef]
35. Sidhu, J.S.; Zafar, T.A. Bioactive compounds in banana fruits and their health benefits. *Food Qual. Saf.* **2018**, *2*, 183–188. [CrossRef]
36. Castillo, M.; de Guzman, M.J.K.; Aberilla, J.M. Environmental sustainability assessment of banana waste utilization into food packaging and liquid fertilizer. *Sustain. Prod. Consum.* **2023**, *37*, 356–368. [CrossRef]
37. Gumisiriza, R.; Hawumba, J.F.; Okure, M.; Hensel, O. Biomass waste-to-energy valorisation technologies: A review case for banana processing in Uganda. *Biotechnol. Biofuels* **2017**, *10*, 11. [CrossRef]
38. Kagezi, G.H.; Kangire, A.; Tushemereirwe, W.; Bagamba, F.; Kikulwe, E.; Muhangi, J.; Gold, C.S.; Ragama, P. Banana bacterial wilt incidence in Uganda. *Afr. Crop Sci. J.* **2006**, *14*, 83–91. [CrossRef]
39. Sagar, N.A.; Pareek, S.; Sharma, S.; Yahia, E.M.; Lobo, M.G. Fruit and vegetable waste: Bioactive compounds, their extraction, and possible utilization. *Compr. Rev. Food Sci. Food Saf.* **2018**, *17*, 512–531. [CrossRef]
40. Almeyda, E.; Ipanaqué, W. Recent developments of artificial intelligence for banana: Application areas, learning algorithms, and future challenges. *Eng. Agríc.* **2022**, *42*, e20210144. [CrossRef]
41. Al-Dairi, M.; Pathare, P.B.; Al-Yahyai, R.; Jayasuriya, H.; Al-Attabi, Z. Evaluation of chemical quality attributes in bruised bananas during storage. *LWT* **2024**, *197*, 115904. [CrossRef]
42. Al-Dairi, M.; Pathare, P.B.; Al-Yahyai, R.; Jayasuriya, H.; Al-Attabi, Z. Postharvest quality, technologies, and strategies to reduce losses along the supply chain of banana: A review. *Trends Food Sci. Technol.* **2023**, *134*, 177–191. [CrossRef]
43. Sugianti, C.; Imaizumi, T.; Thammawong, M.; Nakano, K. Recent postharvest technologies in the banana supply chain. *Rev. Agric. Sci.* **2022**, *10*, 123–137. [CrossRef]
44. Khawas, P.; Deka, S.C. Comparative nutritional, functional, morphological, and diffractogram study on culinary banana (*Musa* ABB) peel at various stages of development. *Int. J. Food Prop.* **2016**, *19*, 2832–2853. [CrossRef]
45. Ounsaneha, W.; Buadit, T.; Rattanapan, C. Assessment of human health impact based on life cycle assessment: A case study of Thai retread tire. *IOP Conf. Ser. Mat. Sci. Eng.* **2019**, *773*, 012038. [CrossRef]
46. Roibás, L.; Elbehri, A.; Hospido, A. Evaluating the sustainability of Ecuadorian bananas: Carbon footprint, water usage and wealth distribution along the supply chain. *Sustain. Prod. Consum.* **2015**, *2*, 3–16. [CrossRef]
47. Veliz, K.; Chico-Santamarta, L.; Ramirez, A.D. The environmental profile of Ecuadorian export banana: A life cycle assessment. *Foods* **2022**, *11*, 3288. [CrossRef] [PubMed]
48. Housagul, S.; Sirisukpoka, U.; Boonyawanich, S.; Pisutpaisal, N. Biomethane production from co-digestion of banana peel and waste glycerol. *Energy Procedia* **2014**, *61*, 2219–2223. [CrossRef]
49. Vu, H.T.; Scarlett, C.J.; Vuong, Q.V. Optimization of ultrasound-assisted extraction conditions for recovery of phenolic compounds and antioxidant capacity from banana (*Musa cavendish*) peel. *J. Food Process. Preserv.* **2017**, *41*, e13148. [CrossRef]

50. Schieber, A.; Stintzing, F.C.; Carle, R. By-products of plant food processing as a source of functional compounds—Recent developments. *Trends Food Sci. Technol.* **2001**, *12*, 401–413. [CrossRef]
51. Zhang, P.; Whistler, R.L.; BeMiller, J.N.; Hamaker, B.R. Banana starch: Production, physicochemical properties, and digestibility—A review. *Carbohydr. Polym.* **2005**, *59*, 443–458. [CrossRef]
52. Vu, H.T.; Scarlett, C.J.; Vuong, Q.V. Phenolic compounds within banana peel and their potential uses: A review. *J. Funct. Foods* **2018**, *40*, 238–248. [CrossRef]
53. Sampath Kumar, K.P.; Bhowmik, D.; Duraivel, S.; Umadevi, M. Traditional and medicinal uses of banana. *J. Parmacogn. Phytochem.* **2012**, *1*, 51–63.
54. Youssef, M.A.; Eman, A.A.; Abeir, M.E. Evaluation the bioactive compounds extracted from dried banana (*Musa* sp.) peels which obtained by different drying methods. *Curr. Sci. Int.* **2018**, *7*, 135–148.
55. Sinanoglou, V.J.; Tsiaka, T.; Aouant, K.; Mouka, E.; Ladika, G.; Kritsi, E.; Konteles, S.J.; Ioannou, A.-G.; Zoumpoulakis, P.; Strati, I.F.; et al. Quality assessment of banana ripening stages by combining analytical methods and image analysis. *Appl. Sci.* **2023**, *13*, 3533. [CrossRef]
56. Marta, H.; Cahyana, Y.; Djali, M.; Pramafisi, G. The properties, modification, and application of banana starch. *Polymers* **2022**, *14*, 3092. [CrossRef]
57. Emaga, T.H.; Andrianaivo, R.H.; Wathelet, B.; Tchango, J.T.; Paquot, M. Effects of the stage of maturation and varieties on the chemical composition of banana and plantain peels. *Food Chem.* **2007**, *103*, 590–600. [CrossRef]
58. Oyeyinka, B.O.; Afolayan, A.J. Comparative evaluation of the nutritive, mineral, and antinutritive composition of *Musa sinensis* L. (banana) and *Musa paradisiaca* L.(plantain) fruit compartments. *Plants* **2019**, *8*, 598. [CrossRef]
59. Emaga, T.H.; Bindelle, J.; Agneesens, R.; Buldgen, A.; Wathelet, B.; Paquot, M. Ripening influences banana and plantain peels composition and energy content. *Trop. Anim. Health Prod.* **2011**, *43*, 171–177. [CrossRef] [PubMed]
60. Sial, T.A.; Khan, M.N.; Lan, Z.; Kumbhar, F.; Ying, Z.; Zhang, J.; Sun, D.; Li, X. Contrasting effects of banana peels waste and its biochar on greenhouse gas emissions and soil biochemical properties. *Process Saf. Environ. Prot.* **2019**, *122*, 366–377. [CrossRef]
61. Hassan, H.F.; Hassan, U.F.; Usher, O.A.; Ibrahim, A.B.; Tabe, N.N. Exploring the potentials of banana (*Musa sapietum*) peels in feed formulation. *Int. J. Adv. Res. Chem. Sci.* **2018**, *5*, 10–14. [CrossRef]
62. Nagarajaiah, S.B.; Jamuna Prakash, J.P. Chemical composition and antioxidant potential of peels from three varieties of banana. *Asian J. Food Agro-Ind.* **2011**, *4*, 31–46.
63. Tsamo, C.V.P.; Herent, M.-F.; Tomekpe, K.; Emaga, T.H.; Quetin-Leclercq, J.; Rogez, H.; Larondelle, Y.; Andre, C. Phenolic profiling in the pulp and peel of nine plantain cultivars (*Musa* sp.). *Food Chem.* **2015**, *167*, 197–204. [CrossRef]
64. Abiodun-Solanke, A.; Falade, K.O. A review of the uses and methods of processing banana and plantain (*Musa* spp.) into storable food products. *J. Agric. Res. Develop.* **2010**, *9*, 85–166. [CrossRef]
65. Pereira, A.; Maraschin, M. Banana (*Musa* spp.) from peel to pulp: Ethnopharmacology, source of bioactive compounds and its relevance for human health. *J. Ethnopharmacol.* **2015**, *160*, 149–163. [CrossRef]
66. Khaliq, A.; Amreen; Jameel, N.; Krauth, S.J. Knowledge and practices on the prevention and management of diarrhea in children under-2 years among women dwelling in urban slums of Karachi, Pakistan. *Matern. Child Health J.* **2022**, *26*, 1442–1452. [CrossRef] [PubMed]
67. García, O.P.; Martínez, M.; Romano, D.; Camacho, M.; de Moura, F.F.; Abrams, S.A.; Khanna, H.K.; Dale, J.L.; Rosado, J.L. Iron absorption in raw and cooked bananas: A field study using stable isotopes in women. *Food Nutr. Res.* **2015**, *59*, 25976. [CrossRef] [PubMed]
68. Tavakkoli-Kakhki, M.; Motavasselian, M.; Mosaddegh, M.; Esfahani, M.M.; Kamalinejad, M.; Nematy, M.; Eslami, S. Omega-3 and omega-6 content of medicinal foods for depressed patients: Implications from the Iranian Traditional Medicine. *Avicenna J. Phytomed.* **2014**, *4*, 225–230.
69. Laufenberg, G.; Kunz, B.; Nystroem, M. Transformation of vegetable waste into value added products: (A) the upgrading concept; (B) practical implementations. *Bioresour. Technol.* **2003**, *87*, 167–198. [CrossRef] [PubMed]
70. Kanazawa, K.; Sakakibara, H. High content of dopamine, a strong antioxidant, in cavendish banana. *J. Agric. Food Chem.* **2000**, *48*, 844–848. [CrossRef]
71. Septembre-Malaterre, A.; Stanislas, G.; Douraguia, E.; Gonthier, M.-P. Evaluation of nutritional and antioxidant properties of the tropical fruits banana, litchi, mango, papaya, passion fruit and pineapple cultivated in Réunion French Island. *Food Chem.* **2016**, *212*, 225–233. [CrossRef]
72. Alexandra Pazmiño-Durán, E.; Giusti, M.M.; Wrolstad, R.E.; Glória, M.B.A. Anthocyanins from banana bracts (*Musa* X *paradisiaca*) as potential food colorants. *Food Chem.* **2001**, *73*, 327–332. [CrossRef]
73. Rao, U.S.M.; Mohd, K.S.; Muhammad, A.; Ahmad, B.A.; Mohamad, M.; Ali, R.M. Taxonomical, phytochemical and pharmacological reviews of *Musa sapientum* var. paradisiaca. *Res. J. Pharm. Technol.* **2014**, *7*, 1356–1361.
74. Lewis, D.A.; Fields, W.N.; Shaw, G.P. A natural flavonoid present in unripe plantain banana pulp (*Musa sapientum* L. var. paradisiaca) protects the gastric mucosa from aspirin-induced erosions. *J. Ethnopharmacol.* **1999**, *65*, 283–288. [CrossRef]

75. Boots, A.W.; Haenen, G.R.M.M.; Bast, A. Health effects of quercetin: From antioxidant to nutraceutical. *Eur. J. Pharmacol.* **2008**, *585*, 325–337. [CrossRef]
76. Ullah, A.; Munir, S.; Badshah, S.L.; Khan, N.; Ghani, L.; Poulson, B.G.; Emwas, A.-H.; Jaremko, M. Important flavonoids and their role as a therapeutic agent. *Molecules* **2020**, *25*, 5243. [CrossRef] [PubMed]
77. Zheng, X.; Zhang, X.; Zeng, F. Biological functions and health benefits of flavonoids in fruits and vegetables: A contemporary review. *Foods* **2025**, *14*, 155. [CrossRef] [PubMed]
78. Someya, S.; Yashiki, Y.; Okubo, K. Antioxidant compounds from bananas (*Musa cavendish*). *Food Chem.* **2002**, *79*, 351–354. [CrossRef]
79. Tsamo, C.V.P.; Herent, M.-F.; Tomekpe, K.; Emaga, T.H.; Quetin-Leclercq, J.; Rogez, H.; Larondelle, Y.; Andre, C.M. Effect of boiling on phenolic profiles determined using HPLC/ESI-LTQ-Orbitrap-MS, physico-chemical parameters of six plantain banana cultivars (*Musa* sp). *J. Food Comp. Anal.* **2015**, *44*, 158–169. [CrossRef]
80. Van den Berg, H.; Faulks, R.; Granado, H.F.; Hirschberg, J.; Olmedilla, B.; Sandmann, G.; Southon, S.; Stahl, W. The potential for the improvement of carotenoid levels in foods and the likely systemic effects. *J. Sci. Food Agric.* **2000**, *80*, 880–912. [CrossRef]
81. Bufka, J.; Vaňková, L.; Sýkora, J.; Křížková, V. Exploring carotenoids: Metabolism, antioxidants, and impacts on human health. *J. Funct. Foods* **2024**, *118*, 106284. [CrossRef]
82. Aquino, C.F.; Salomão, L.C.C.; Pinheiro-Sant'Ana, H.M.; Ribeiro, S.M.R.; De Siqueira, D.L.; Cecon, P.R. Carotenoids in the pulp and peel of bananas from 15 cultivars in two ripening stages. *Rev. Ceres* **2018**, *65*, 217–226. [CrossRef]
83. Davey, M.W.; Keulemans, J.; Swennen, R. Methods for the efficient quantification of fruit provitamin A contents. *J. Chromatogr. A* **2006**, *1136*, 176–184. [CrossRef]
84. Wójcik, W.; Łukasiewicz, M.; Puppel, K. Biogenic amines: Formation, action and toxicity—A review. *J. Sci. Food Agric.* **2021**, *101*, 2634–2640. [CrossRef]
85. Shalaby, A.R. Significance of biogenic amines to food safety and human health. *Food Res. Int.* **1996**, *29*, 675–690. [CrossRef]
86. Ademakinwa, A.N.; Agunbiade, M.O. Banana peel wastes as a source of tyrosinase useful in the production of l-DOPA. *Sustain. Chem. Pharm.* **2022**, *30*, 100853. [CrossRef]
87. Marangoni, F.; Poli, A. Phytosterols and cardiovascular health. *Pharmacol. Res.* **2010**, *61*, 193–199. [CrossRef] [PubMed]
88. Ismail, S.; Dubey, P.K.; Mishra, A.A.; Ashka, F. Valorisation of banana peel and mango peel as functional ingredients in baked products: A review. *Int. J. Food Sci. Technol.* **2024**, *59*, 5938–5950. [CrossRef]
89. Aquino, C.F.; Salomão, L.C.C.; Ribeiro, S.M.R.; De Siqueira, D.L.; Cecon, P.R. Carbohydrates, phenolic compounds and antioxidant activity in pulp and peel of 15 banana cultivars. *Rev. Bras. Frutic.* **2016**, *38*, e-090. [CrossRef]
90. Aboul-Enein, A.M.; Salama, Z.A.; Gaafar, A.A.; Aly, H.F.; Bou-Elella, F.A.; Ahmed, H.A. Identification of phenolic compounds from banana peel (*Musa paradaisica* L.) as antioxidant and antimicrobial agents. *J. Chem. Pharmaceut. Res.* **2016**, *8*, 46–55.
91. El-Zawawy, N.A. Antioxidant, antitumor, antimicrobial studies and quantitative phytochemical estimation of ethanolic extracts of selected fruit peels. *Int. J. Curr. Microbiol. Appl. Sci.* **2015**, *4*, 298–309.
92. Waghmare, J.S.; Kurhade, A.H. GC-MS analysis of bioactive components from banana peel (*Musa sapientum* peel). *Eur. J. Exp. Biol.* **2014**, *4*, 10–15.
93. González-Montelongo, R.; Lobo, M.G.; González, M. Antioxidant activity in banana peel extracts: Testing extraction conditions and related bioactive compounds. *Food Chem.* **2010**, *119*, 1030–1039. [CrossRef]
94. Rasidek, N.A.M.; Nordin, M.F.M.; Shameli, K. Formulation and evaluation of semisolid jelly produced by *Musa acuminata* Colla (AAA Group) peels. *Asian Pac. J. Trop. Biomed.* **2016**, *6*, 55–59. [CrossRef]
95. Mokbel, M.S.; Hashinaga, F. Antibacterial and antioxidant activities of banana (*Musa*, AAA cv. Cavendish) fruits peel. *Am. J. Biochem. Biotechnol.* **2005**, *1*, 125–131. [CrossRef]
96. Kapadia, S.P.; Pudakalkatti, P.S.; Shivanaikar, S. Detection of antimicrobial activity of banana peel (*Musa paradisiaca* L.) on Porphyromonas gingivalis and Aggregatibacter actinomycetemcomitans: An in vitro study. *Contemp. Clin. Dent.* **2015**, *6*, 496–499. [CrossRef]
97. Kavitha, V.; Manonmani, G.; Devakumar, J. Antimicrobial efficacy of fresh and dried banana peel extracts: An in vitro study. *Drug Invent. Today* **2019**, *11*, 1237.
98. Semangoen, T.; Chotigawin, R.; Sangnim, T.; Chailerd, N.; Pahasup-anan, T.; Huanbutta, K. Evaluation of banana peel extract (*Musa sapientum* L.) as a natural antimicrobial for livestock farming. *J. Sustain. Agric. Environ.* **2024**, 3, e12118. [CrossRef]
99. Lino, P.; Corrêa, C.; Archondo, M.; Leite-Dellova, D. Evaluation of post-surgical healing in rats using a topical preparation based on extract of *Musa sapientum* epicarp. *Rev. Bras. De Farmacogn.* **2011**, *21*, 491–496. [CrossRef]
100. Saleem, M.; Saeed, M.T. Potential application of waste fruit peels (orange, yellow lemon and banana) as wide range natural antimicrobial agent. *J. King Saud Univ. Sci.* **2020**, *32*, 805–810. [CrossRef]
101. Okorondu, S.I.; Akujobi, C.O.; Nwachukwu, I.N. Antifungal properties of *Musa paradisiaca* (plantain) peel and stalk extracts. *Int. J. Biol. Chem. Sci.* **2012**, *6*, 1527–1534. [CrossRef]

applied sciences

MDPI

Review

Harnessing Agri-Food Waste as a Source of Biopolymers for Agriculture

Carlotta Valle [1], Monica Voss [1,2], Emanuela Calcio Gaudino [1], Claudio Forte [2], Giancarlo Cravotto [1] and Silvia Tabasso [1,*]

[1] Dipartimento di Scienza e Tecnologia del Farmaco, University of Turin, Via P. Giuria 9, 10125 Turin, TO, Italy; carlotta.valle@unito.it (C.V.); monica.voss@unito.it (M.V.); emanuela.calcio@unito.it (E.C.G.); giancarlo.cravotto@unito.it (G.C.)

[2] Dipartimento di Scienze Veterinarie, University of Turin, Largo Paolo Braccini 2, 10095 Grugliasco, TO, Italy; claudio.forte@unito.it

* Correspondence: silvia.tabasso@unito.it; Tel.: +39-011-6703686

Abstract: Nowadays, the world is facing a general problem of resource overconsumption and waste overproduction: to address these two issues, the United Nations delivered the 12th Sustainable Development Goal (SDG), which has the objective of ensuring sustainable consumption and production patterns. Currently, polymers are present in every aspect of our lives and have the disadvantage of mostly coming from fossil sources and causing pollution when disposed of the wrong way. Agriculture plays a key role in the overall world environmental issues, being responsible for the creation of between 13 and 21% of global greenhouse gas (GHG) emissions. Moreover, it represents a continuously growing field, producing large amounts of waste. These residues can cause serious environmental concerns and high costs when disposed. However, agri-food waste (AFW) is a natural source of natural biopolymers, such as lignin, cellulose, pectin, and starch, but can also be used as a substrate to produce other non-toxic and biodegradable biopolymers, such as chitosan, polyhydroxyalkanoates (PHAs), and polylactic acid (PLA) through microbial fermentation. These polymers find applications in agricultural practices such as mulching films, soil stabilizers, hydrogels, nanocarriers, and coating for seeds, fruits, and vegetables. The employment of AFW in the production of non-toxic, sustainable, and biodegradable biopolymers for their agricultural utilization is an example of a virtuous circular economy approach that could help agriculture to be more sustainable.

Keywords: agri-food waste; biopolymers; agriculture; biomass; lignin; cellulose; pectin; chitosan; PLA; PHA

1. Introduction

In 2023, EU Bioplastics reported that biopolymer global production capacity is expected to reach 7.43 million tons by 2028; however, the market demand corresponds to 400 million tons per year. The large difference between biopolymer supply and demand shows the need for raw material alternatives at sustainable costs to increase production [1]. This is also the reason why plastics of fossil origin are still widely used. Therefore, reducing the consumption of fossil fuels and using renewable and environmentally safe resources will probably be the main advantage of bioplastics, along with the possibility of biodegradability at the end of their life [2]. While biopolymers offer an alternative to fossil-based materials, their production still demands substantial energy input, largely sourced from fossil fuels. One approach involving renewable energy usage relies on the implementation of an integrated system, using AFW-derived energy to power the process (e.g., through AFW incineration to produce steam and electricity) [3].

Worldwide, around 1.3 million tons per year of food produced for human consumption is lost or wasted and this is equivalent to a third of all food produced [4]; for this

128. Akpomie, K.G.; Conradie, J. Banana peel as a biosorbent for the decontamination of water pollutants. A review. *Environ. Chem. Lett.* **2020**, *18*, 1085–1112. [CrossRef]
129. Ashraf, M.A.; Hussain, M.; Mahmood, K.; Wajid, A.; Yusof, M.; Alias, Y.; Yusoff, I. Removal of acid yellow-17 dye from aqueous solution using eco-friendly biosorbent. *Desalin. Water Treat.* **2013**, *51*, 4530–4545. [CrossRef]
130. Al-Shaalan, N.H.; Ali, I.; Alothman, Z.A.; Al-Wahaibi, L.H.; Alabdulmonem, H. High performance removal and simulation studies of diuron pesticide in water on MWCNTs. *J. Mol. Liq.* **2019**, *289*, 111039. [CrossRef]
131. Beni, A.A.; Esmaeili, A. Biosorption, an efficient method for removing heavy metals from industrial effluents: A review. *Environ. Technol. Innov.* **2020**, *17*, 100503. [CrossRef]
132. Singh, S.; Kumar, V.; Datta, S.; Dhanjal, D.S.; Sharma, K.; Samuel, J.; Singh, J. Current advancement and future prospect of biosorbents for bioremediation. *Sci. Total Environ.* **2020**, *709*, 135895. [CrossRef]
133. Vilardi, G.; Di Palma, L.; Verdone, N. Heavy metals adsorption by banana peels micro-powder: Equilibrium modeling by non-linear models. *Chin. J. Chem. Eng.* **2018**, *26*, 455–464. [CrossRef]
134. Ayawei, N.; Ebelegi, A.N.; Wankasi, D. Modelling and interpretation of adsorption isotherms. *J. Chem.* **2017**, *2017*, 3039817. [CrossRef]
135. Iftekhar, S.; Ramasamy, D.L.; Srivastava, V.; Asif, M.B.; Sillanpää, M. Understanding the factors affecting the adsorption of Lanthanum using different adsorbents: A critical review. *Chemosphere* **2018**, *204*, 413–430. [CrossRef]
136. Mahindrakar, K.V.; Rathod, V.K. Utilization of banana peels for removal of strontium (II) from water. *Environ. Technol. Innov.* **2018**, *11*, 371–383. [CrossRef]
137. El-Din, G.A.; Amer, A.A.; Malsh, G.; Hussein, M. Study on the use of banana peels for oil spill removal. *Alex. Eng. J.* **2018**, *57*, 2061–2068. [CrossRef]
138. Agama-Acevedo, E.; Islas-Hernandez, J.J.; Osorio-Díaz, P.; Rendón-Villalobos, R.; Utrilla-Coello, R.G.; Angulo, O.; Bello-Pérez, L.A. Pasta with unripe banana flour: Physical, texture, and preference study. *J. Food Sci.* **2009**, *74*, S263–S267. [CrossRef]
139. Segundo, C.; Román, L.; Lobo, M.; Martinez, M.M.; Gómez, M. Ripe banana flour as a source of antioxidants in layer and sponge cakes. *Plant Foods Hum. Nutr.* **2017**, *72*, 365–371. [CrossRef]
140. Ho, L.-H.; Abdul Aziz, N.A.; Azahari, B. Physico-chemical characteristics and sensory evaluation of wheat bread partially substituted with banana (*Musa acuminata* × *balbisiana* cv. Awak) pseudo-stem flour. *Food Chem.* **2013**, *139*, 532–539. [CrossRef] [PubMed]
141. Eshak, N.S. Sensory evaluation and nutritional value of balady flat bread supplemented with banana peels as a natural source of dietary fiber. *Ann. Agric. Sci.* **2016**, *61*, 229–235. [CrossRef]
142. Yancy, W.S., Jr.; Foy, M.; Chalecki, A.M.; Vernon, M.C.; Westman, E.C. A low-carbohydrate, ketogenic diet to treat type 2 diabetes. *Nutr. Metab.* **2005**, *2*, 34. [CrossRef] [PubMed]
143. Arun, K.B.; Persia, F.; Aswathy, P.S.; Chandran, J.; Sajeev, M.S.; Jayamurthy, P.; Nisha, P. Plantain peel—A potential source of antioxidant dietary fibre for developing functional cookies. *J. Food Sci. Technol.* **2015**, *52*, 6355–6364. [CrossRef]
144. Rehman, S.-u.; Nadeem, M.; Ahmad, F.; Mushtaq, Z. Biotechnological production of xylitol from banana peel and its impact on physicochemical properties of rusks. *J. Agric. Sci. Technol.* **2013**, *15*, 747–756.
145. Ahmed, Z.; El-Sharnouby, G.A.; El-Waseif, M.A. Use of banana peel as a by-product to increase the nutritive value of the cake. *J. Food Dairy Sci.* **2021**, *12*, 87–97. [CrossRef]
146. Borges, A.d.M.; Pereira, J.; Júnior, A.S.; de Lucena, E.M.P.; de Sales, J.C. Stability of cake pre-mixture made with 60% of green banana flour. *Ciênc. Agrotecnol.* **2010**, *34*, 173–181. [CrossRef]
147. Amini Khoozani, A.; Birch, J.; Bekhit, A.E.-D.A. Production, application and health effects of banana pulp and peel flour in the food industry. *J. Food Sci. Technol.* **2019**, *56*, 548–559. [CrossRef] [PubMed]
148. Mohd Zaini, H.B.; Bin Sintang, M.D.; Pindi, W. The roles of banana peel powders to alter technological functionality, sensory and nutritional quality of chicken sausage. *Food Sci. Nutr.* **2020**, *8*, 5497–5507. [CrossRef] [PubMed]
149. Devatkal, S.K.; Kumboj, R.; Paul, D. Comparative antioxidant effect of BHT and water extracts of banana and sapodilla peels in raw poultry meat. *J. Food Sci. Technol.* **2014**, *51*, 387–391. [CrossRef] [PubMed]
150. Santhoskumar, A.U.; Vaishnavi, R.; Karunakaran, T.; Chitra, N.J. Studies on mechanical properties and biodegradation of edible food wrapper from banana peel. *Asian J. Adv. Basic Sci.* **2019**, *7*, 1–4. [CrossRef]
151. Sarkar, D.; Prajapati, S.; Poddar, K.; Sarkar, A. Ethanol production by *Klebsiella* sp. SWET4 using banana peel as feasible substrate. *Biomass Conv. Bioref.* **2022**, *12*, 1479–1491. [CrossRef]

102. Kamal, A.M.; Taha, M.S.; Mousa, A.M. The radioprotective and anticancer effects of banana peels extract on male mice. *J. Food Nutr. Res.* **2019**, *7*, 827–835. [CrossRef]
103. Dahham, S.S.; Agha, M.T.; Tabana, Y.M.; Abdul Majid, A.M.S. Antioxidant activities and anticancer screening of extracts from banana fruit (*Musa sapientum*). *Acad. J. Cancer Res.* **2015**, *8*, 28–34.
104. Krinsky, N.I.; Johnson, E.J. Carotenoid actions and their relation to health and disease. *Mol. Asp. Med.* **2005**, *26*, 459–516. [CrossRef]
105. Mondal, A.; Banerjee, S.; Bose, S.; Das, P.P.; Sandberg, E.N.; Atanasov, A.G.; Bishayee, A. Cancer preventive and therapeutic potential of banana and its bioactive constituents: A systematic, comprehensive, and mechanistic review. *Front. Oncol.* **2021**, *11*, 697143. [CrossRef] [PubMed]
106. Navghare, V.V.; Dhawale, S.C. *In vitro* antioxidant, hypoglycemic and oral glucose tolerance test of banana peels. *Alex. J. Med.* **2017**, *53*, 237–243. [CrossRef]
107. Vasu, P.; Khan, N.D.; Khan, Z.H.; Mular, S.M. *In vitro* antidiabetic activity of methanolic extract of *Citrus limon*, *Punica granatum*, *Musa acuminata* peel. *Int. J. Appl. Res.* **2017**, *3*, 804–806.
108. Rao, U.S.M.; Ahmad, B.A.; Mohd, K.S.; Zin, T. Antiulcer activity of *Musa paradisiaca* (banana) tepal and skin extracts in ulcer induced Albino mice. *Malay. J. Anal. Sci.* **2016**, *20*, 1203–1216. [CrossRef]
109. Bansal, J.; Malviya, R.; Malaviya, T.; Bhardwaj, V.; Sharma, P.K. Evaluation of banana peel pectin as excipient in solid oral dosage form. *Glob. J. Pharmacol.* **2014**, *8*, 275–278. [CrossRef]
110. Khamsucharit, P.; Laohaphatanalert, K.; Gavinlertvatana, P.; Sriroth, K.; Sangseethong, K. Characterization of pectin extracted from banana peels of different varieties. *Food Sci. Biotechnol.* **2017**, *27*, 623–629. [CrossRef]
111. Venkatanagaraju, E.; Bharathi, N.; Sindhuja, R.H.; Chowdhury, R.R.; Sreelekha, Y. Extraction and purification of pectin from agro-industrial wastes. In *Pectins-Extraction, Purification, Characterization and Applications*; Masuelli, M., Ed.; IntechOpen: London, UK, 2020. [CrossRef]
112. Ibrahim, H.M.M. Green synthesis and characterization of silver nanoparticles using banana peel extract and their antimicrobial activity against representative microorganisms. *J. Rad. Res. Appl. Sci.* **2015**, *8*, 265–275. [CrossRef]
113. Navaladian, S.; Viswanathan, B.; Viswanath, R.P.; Varadarajan, T.K. Thermal decomposition as route for silver nanoparticles. *Nanoscale Res. Lett.* **2006**, *2*, 44–48. [CrossRef] [PubMed]
114. Solano-Ruiz, E.; Berrú, R.S.; Ocotlán-Flores, J.; Saniger, J.M. Synthesis of silver nanoparticles by sonochemical induced reduction application in SERS. *J. Nano Res.* **2010**, *9*, 77–81. [CrossRef]
115. El-Faham, A.; Elzatahry, A.A.; Al-Othman, Z.A.; Elsayed, E.A. Facile method for the synthesis of silver nanoparticles using 3-hydrazino-isatin derivatives in aqueous methanol and their antibacterial activity. *Int. J. Nanomed.* **2014**, *9*, 1167–1174. [CrossRef]
116. Sreeram, K.J.; Nidhin, M.; Nair, B.U. Microwave assisted template synthesis of silver nanoparticles. *Bull. Mater. Sci.* **2008**, *31*, 937–942. [CrossRef]
117. Kokila, T.; Ramesh, P.S.; Geetha, D. Biosynthesis of silver nanoparticles from Cavendish banana peel extract and its antibacterial and free radical scavenging assay: A novel biological approach. *Appl. Nanosci.* **2015**, *5*, 911–920. [CrossRef]
118. Narayanamma, A. Natural synthesis of silver nanoparticles by banana peel extract and as an antibacterial agent. *IOSR J. Polymer Textile Eng.* **2016**, *3*, 17–25. [CrossRef]
119. Food and Agriculture Organization of the United Nations. *Crops and Livestock Products*; Food and Agriculture Organization of the United Nations: Rome, Italy, 2024. Available online: https://www.fao.org/faostat/en/#data/QCL/metadata (accessed on 28 December 2024).
120. Guerrero, A.B.; Aguado, P.L.; Sánchez, J.; Curt, M.D. GIS-based assessment of banana residual biomass potential for ethanol production and power generation: A case study. *Waste Biomass Valor.* **2016**, *7*, 405–415. [CrossRef]
121. Oji, C.; Okoro, I.; Nnaji, J. Optimization of bioethanol production from banana peels: An alternative energy source. *J. Chem. Soc. Nigeria* **2024**, *49*, 487–499. [CrossRef]
122. Guerrero, A.B.; Ballesteros, I.; Ballesteros, M. The potential of agricultural banana waste for bioethanol production. *Fuel* **2018**, *213*, 176–185. [CrossRef]
123. Li, D.; Wang, M.-Q.; Lee, C. The waste treatment and recycling efficiency of industrial waste processing based on two-stage data envelopment analysis with undesirable inputs. *J. Clean. Prod.* **2020**, *242*, 118279. [CrossRef]
124. Alharbi, O.M.L.; Basheer, A.A.; Khattab, R.A.; Ali, I. Health and environmental effects of persistent organic pollutants. *J. Mol. Liq.* **2018**, *263*, 442–453. [CrossRef]
125. Ali, I.; Alothman, Z.A.; Alwarthan, A. Supra molecular mechanism of the removal of 17-β-estradiol endocrine disturbing pollutant from water on functionalized iron nano particles. *J. Mol. Liq.* **2017**, *241*, 123–129. [CrossRef]
126. Zhou, L.; Yu, Q.; Cui, Y.; Xie, F.; Li, W.; Li, Y.; Chen, M. Adsorption properties of activated carbon from reed with a high adsorption capacity. *Ecol. Eng.* **2017**, *102*, 443–450. [CrossRef]
127. Kyzas, G.Z.; Bikiaris, D.N. Recent modifications of chitosan for adsorption applications: A critical and systematic review. *Mar. Drugs* **2015**, *13*, 312–337. [CrossRef] [PubMed]

75. Boots, A.W.; Haenen, G.R.M.M.; Bast, A. Health effects of quercetin: From antioxidant to nutraceutical. *Eur. J. Pharmacol.* **2008**, *585*, 325–337. [CrossRef]
76. Ullah, A.; Munir, S.; Badshah, S.L.; Khan, N.; Ghani, L.; Poulson, B.G.; Emwas, A.-H.; Jaremko, M. Important flavonoids and their role as a therapeutic agent. *Molecules* **2020**, *25*, 5243. [CrossRef] [PubMed]
77. Zheng, X.; Zhang, X.; Zeng, F. Biological functions and health benefits of flavonoids in fruits and vegetables: A contemporary review. *Foods* **2025**, *14*, 155. [CrossRef] [PubMed]
78. Someya, S.; Yashiki, Y.; Okubo, K. Antioxidant compounds from bananas (*Musa cavendish*). *Food Chem.* **2002**, *79*, 351–354. [CrossRef]
79. Tsamo, C.V.P.; Herent, M.-F.; Tomekpe, K.; Emaga, T.H.; Quetin-Leclercq, J.; Rogez, H.; Larondelle, Y.; Andre, C.M. Effect of boiling on phenolic profiles determined using HPLC/ESI-LTQ-Orbitrap-MS, physico-chemical parameters of six plantain banana cultivars (*Musa* sp). *J. Food Comp. Anal.* **2015**, *44*, 158–169. [CrossRef]
80. Van den Berg, H.; Faulks, R.; Granado, H.F.; Hirschberg, J.; Olmedilla, B.; Sandmann, G.; Southon, S.; Stahl, W. The potential for the improvement of carotenoid levels in foods and the likely systemic effects. *J. Sci. Food Agric.* **2000**, *80*, 880–912. [CrossRef]
81. Bufka, J.; Vaňková, L.; Sýkora, J.; Křížková, V. Exploring carotenoids: Metabolism, antioxidants, and impacts on human health. *J. Funct. Foods* **2024**, *118*, 106284. [CrossRef]
82. Aquino, C.F.; Salomão, L.C.C.; Pinheiro-Sant'Ana, H.M.; Ribeiro, S.M.R.; De Siqueira, D.L.; Cecon, P.R. Carotenoids in the pulp and peel of bananas from 15 cultivars in two ripening stages. *Rev. Ceres* **2018**, *65*, 217–226. [CrossRef]
83. Davey, M.W.; Keulemans, J.; Swennen, R. Methods for the efficient quantification of fruit provitamin A contents. *J. Chromatogr. A* **2006**, *1136*, 176–184. [CrossRef]
84. Wójcik, W.; Łukasiewicz, M.; Puppel, K. Biogenic amines: Formation, action and toxicity—A review. *J. Sci. Food Agric.* **2021**, *101*, 2634–2640. [CrossRef]
85. Shalaby, A.R. Significance of biogenic amines to food safety and human health. *Food Res. Int.* **1996**, *29*, 675–690. [CrossRef]
86. Ademakinwa, A.N.; Agunbiade, M.O. Banana peel wastes as a source of tyrosinase useful in the production of l-DOPA. *Sustain. Chem. Pharm.* **2022**, *30*, 100853. [CrossRef]
87. Marangoni, F.; Poli, A. Phytosterols and cardiovascular health. *Pharmacol. Res.* **2010**, *61*, 193–199. [CrossRef] [PubMed]
88. Ismail, S.; Dubey, P.K.; Mishra, A.A.; Ashka, F. Valorisation of banana peel and mango peel as functional ingredients in baked products: A review. *Int. J. Food Sci. Technol.* **2024**, *59*, 5938–5950. [CrossRef]
89. Aquino, C.F.; Salomão, L.C.C.; Ribeiro, S.M.R.; De Siqueira, D.L.; Cecon, P.R. Carbohydrates, phenolic compounds and antioxidant activity in pulp and peel of 15 banana cultivars. *Rev. Bras. Frutic.* **2016**, *38*, e-090. [CrossRef]
90. Aboul-Enein, A.M.; Salama, Z.A.; Gaafar, A.A.; Aly, H.F.; Bou-Elella, F.A.; Ahmed, H.A. Identification of phenolic compounds from banana peel (*Musa paradaisica* L.) as antioxidant and antimicrobial agents. *J. Chem. Pharmaceut. Res.* **2016**, *8*, 46–55.
91. El-Zawawy, N.A. Antioxidant, antitumor, antimicrobial studies and quantitative phytochemical estimation of ethanolic extracts of selected fruit peels. *Int. J. Curr. Microbiol. Appl. Sci.* **2015**, *4*, 298–309.
92. Waghmare, J.S.; Kurhade, A.H. GC-MS analysis of bioactive components from banana peel (*Musa sapientum* peel). *Eur. J. Exp. Biol.* **2014**, *4*, 10–15.
93. González-Montelongo, R.; Lobo, M.G.; González, M. Antioxidant activity in banana peel extracts: Testing extraction conditions and related bioactive compounds. *Food Chem.* **2010**, *119*, 1030–1039. [CrossRef]
94. Rasidek, N.A.M.; Nordin, M.F.M.; Shameli, K. Formulation and evaluation of semisolid jelly produced by *Musa acuminata* Colla (AAA Group) peels. *Asian Pac. J. Trop. Biomed.* **2016**, *6*, 55–59. [CrossRef]
95. Mokbel, M.S.; Hashinaga, F. Antibacterial and antioxidant activities of banana (*Musa*, AAA cv. Cavendish) fruits peel. *Am. J. Biochem. Biotechnol.* **2005**, *1*, 125–131. [CrossRef]
96. Kapadia, S.P.; Pudakalkatti, P.S.; Shivanaikar, S. Detection of antimicrobial activity of banana peel (*Musa paradisiaca* L.) on Porphyromonas gingivalis and Aggregatibacter actinomycetemcomitans: An in vitro study. *Contemp. Clin. Dent.* **2015**, *6*, 496–499. [CrossRef]
97. Kavitha, V.; Manonmani, G.; Devakumar, J. Antimicrobial efficacy of fresh and dried banana peel extracts: An in vitro study. *Drug Invent. Today* **2019**, *11*, 1237.
98. Semangoen, T.; Chotigawin, R.; Sangnim, T.; Chailerd, N.; Pahasup-anan, T.; Huanbutta, K. Evaluation of banana peel extract (*Musa sapientum* L.) as a natural antimicrobial for livestock farming. *J. Sustain. Agric. Environ.* **2024**, *3*, e12118. [CrossRef]
99. Lino, P.; Corrêa, C.; Archondo, M.; Leite-Dellova, D. Evaluation of post-surgical healing in rats using a topical preparation based on extract of *Musa sapientum* epicarp. *Rev. Bras. De Farmacogn.* **2011**, *21*, 491–496. [CrossRef]
100. Saleem, M.; Saeed, M.T. Potential application of waste fruit peels (orange, yellow lemon and banana) as wide range natural antimicrobial agent. *J. King Saud Univ. Sci.* **2020**, *32*, 805–810. [CrossRef]
101. Okorondu, S.I.; Akujobi, C.O.; Nwachukwu, I.N. Antifungal properties of *Musa paradisiaca* (plantain) peel and stalk extracts. *Int. J. Biol. Chem. Sci.* **2012**, *6*, 1527–1534. [CrossRef]

reason, the United Nations has delivered the 12th sustainable development goal (SDG), which aims to ensure sustainable consumption and production patterns [5]. These losses occur during the processing of fruits and vegetables, as substantial amounts of pulp, stalks, peels, or seeds are generated and are destined for disposal [6]. These undervalued materials are rich sources of biopolymers, such as polysaccharides, proteins, and dietary fiber, as well as other valuable functional molecules such as tannins, flavonoids, polyphenols, and fatty acids [7]. Recently, the reuse of agri-food waste (AFW) as a raw material to obtain biopolymers for application in agriculture has gained increasing interest. Biopolymers with varying purity degrees and properties can be obtained through microbiological fermentation or biotechnological processes, as well as by treatment of biomass waste with enabling technologies to shorten the extraction/conversion times. The employment of sustainable, promising processes for sustainable agricultural operations may help to solve the dual challenges of waste management and natural resource conservation. Utilizing renewable resources, such as AFW, for biopolymer production offers the advantages of lower dependence on fossil fuels, resource recovery, and waste management [8]. The use of AFW mitigates the emission of greenhouse gases (GHGs) [9], for example, the GHG emissions from synthetic plastics production range between 1.8 and 3.55 (t CO_2 eq./t plastic). Comparatively, for polylactic acid (PLA), the GHG emissions range between 0.4 and 1.3 (t CO_2 eq./t plastic) [10]. Furthermore, the production of biopolymers from waste aligns with the principles of the circular economy, as this process transforms AFW into value-added products, minimizing the environmental impact and promoting efficient use of natural resources. Biopolymers deriving from AFW are indeed a renewable, often biodegradable, non-toxic, and environmentally friendly alternative to polymers obtained from non-renewable sources, such as petroleum [11,12]. However, the production cost of PLA is higher compared to conventional plastics: it costs between 1.31–2.02 € to manufacture petroleum-based plastics, meanwhile polyhydroxyalkanoates (PHA) and PLA cost between 4.00–12.02 € and 0–2.0 €, respectively [13].

In the agricultural sector, biopolymers derived from AFW have several applications, including mulching films, hydrogels (for water, biocides and/or biostimulants accumulation, and controlled release), soil stabilizers, and, lastly, coating for seeds, fruits, and vegetables [14,15]. The potential of using biopolymers to address the urgent challenges of both waste management and agricultural sustainability promises to promote a more resilient and environmentally conscious agricultural system. The sourcing of new, sustainable raw materials for biopolymer production and the examination of the practical implications of using them in agriculture are essential to advance sustainable agricultural practices in the 21st century [2].

The purpose of this review is to define and investigate the potential of AFW to produce biopolymers for agricultural applications; thus, for the sake of this review, only polymers directly deriving from agri-food waste or from microorganisms fed with agri-food waste will be explored. The proposed approach includes: (i) the identification of sustainable strategies to mitigate the environmental issues related to waste disposal, (ii) the process and technologies used to recover higher yields, and (iii) how the biopolymers can lead to higher crop yields and better preservation of fruits and vegetables, among other benefits.

2. Sources of Biopolymers

2.1. Lignin: Extraction Methods and Applications in Agriculture

Lignin is a high molecular weight aromatic biopolymer (Figure 1) that can be found mainly in the woody stems of plants, but it is also a component of plant cell walls [16]; this characteristic makes lignin one of the most abundant polymers on earth. This polymer is mainly composed of the monomeric moieties of sinapyl alcohol, coniferyl alcohol, and *p*-coumaryl alcohol.

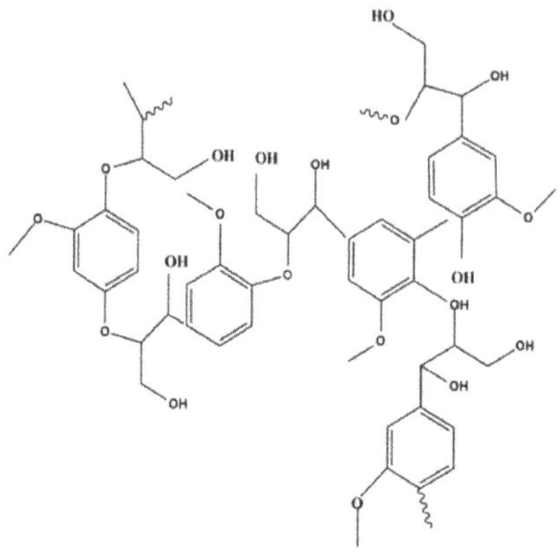

Figure 1. Lignin chemical structure.

Valorization of lignin, a complex organic polymer found in the cell walls of plants, refers to the processes of converting it into higher-value products. Traditionally, lignin has been considered a by-product of the paper and pulp industry [17], often burnt for energy or used as a low-value filler in various applications. Nonetheless, with increasing interest in sustainable and renewable resources (pruning waste, straw, stalks, and bagasse), there has been growing attention towards finding more valuable uses for lignin.

Overall, lignin valorization holds great promise for reducing the environmental impact of industries relying on lignocellulosic biomass, creating new revenue streams and contributing to the development of a more sustainable bioeconomy. Biorefineries mainly using food industry waste are gaining importance in lignin production. Indeed, the only way to make a biorefinery economically viable encompasses the valorization of all the waste components, including lignin.

However, developing cost-effective and scalable processes aimed at the creation of a market for lignin-derived products is still challenging. Lignin has been explored for different applications in the medical, pharmaceutical, and electrochemical fields and as a plasticizer [18]. Moreover, depolymerized lignin is a platform chemical that can potentially substitute aromatic chemicals deriving from the fossil-fuel industry [18,19]. Delignification usually aims to remove lignin from a lignocellulosic biomass, obtaining a relatively pure holocellulose fraction for further purposes and lignin as a by-product. The literature displays a lot of different delignification methods, such as alkaline, acid delignification, organosolv process, autohydrolysis, or enzymatic methods [20]. The lignin pyrolysis method has also been explored to obtain lignin biochar. Although biochar may have agricultural applications, it is important to note that lignin biochar is usually exploited as a filler/additive for biocomposites, rather than biopolymer itself. Additionally, the roasting method is not effective in reducing GHG emissions [21]. The possibility to further valorize the lignin for different applications paved the way for research aiming to improve delignification processes through combination with green technologies such as microwave, ultrasound (US), or steam explosion, affording higher lignin yields with reduced extraction time. Recently, solvents such as ionic liquids, natural deep eutectic solvents (NaDESs), and dioxane have also been investigated for lignin extraction [20].

Delignification techniques employed in recent studies are shown in Table 1, where the best yields were found coupling alkaline delignification with other different pretreatments.

Table 1. Delignification processes.

Entry	Biomass	Delignification Method	Yield on Total Lignin (%)	Yield on Biomass (%)	Ref.
1	Tobacco (*Nicotiana tabacum*) stalks	Hydrothermal prehydrolysis + Alkaline delignification	65.0	N/A	[22]
		Hydrothermal prehydrolysis + Dioxane delignification	37.2	N/A	
		Phosphotungstic acid prehydrolysis + Alkaline delignification	67.4	N/A	
		Phosphotungstic acid prehydrolysis + Dioxane delignification	45.5	N/A	
		Alkaline delignification	60.6	N/A	
		Dioxane delignification	35.6	N/A	
2	Grape (*Vitis vinifera*) stalks	US ChLA Assisted (60–120 min)	N/A	40–50	[23]
		US ChLAGly Assisted (60–120 min)	N/A	27.96–33.38	
		US ChLevA Assisted (60–120 min)	N/A.	73.99–72.31	
		US Alkaline Assisted (60–120 min)	N/A	30.05–25.05	
3	Corn (*Zea mays*) cobs	Alkaline extraction	73.76	N/A	[24]
4		Autohydrolysis + Organosolv Delignification	76.0	N/A	[25]
		Autohydrolysis + Alkaline Delignification	93.0	N/A	
5	Waste of pineapple (*Ananas comosus*) leaves	Laccase from *Pleurotus djamor*	78.5	N/A	[26]

N/A: not available.

Indeed, Liu et al. (Entry 1) [22] observed that hydrothermal prehydrolysis promoted the depolymerization of lignin macromolecules, facilitating the delignification process. It was also observed that lignin obtained by alkaline delignification has a lower molecular weight and a higher content of hydroxyl groups, making it an interesting source of value-added functional bio-based materials. In addition, Fialho et al. (Entry 4) [25] successfully removed 93% of lignin from corncobs by coupling autohydrolysis with alkaline delignification, which was also more selective in lignin removal compared to dioxane delignification. Another advantage of a two-step delignification process is the fractionation of lignocellulosic components, as the first step of both the reported studies successfully removed and fractionated the hemicellulose fraction from the biomass. Conversely, corncobs treated solely with alkaline delignification afforded a lower delignification yield (Entry 3) [24].

The enzymatic treatment of pineapple leaf waste was reported in Entry 5 [26], presenting an effective removal of lignin from the biomass with high selectivity (78.5%); unfortunately, the lignin-rich fraction obtained was not characterized.

Ultrasound-assisted (UA) delignification using NaDES as green solvents is also an interesting route to explore; in Entry 2 [23], different NaDES were tested with different extraction times (60 to 120 min). When US technology was used in combination with NaDES, a higher delignification time was required; however, higher yields were achieved. This phenomenon probably occurred because NaDES are viscous and need more time for an optimal mass transfer. Overall, chlorine chloride-levulinic acid (ChLevA) afforded the highest yield of delignification and had a similar extraction yield at 60 and 120 min. Unfortunately, the recovery of lignin extracted with NaDES was very low (30 and 10 mg) for ChLA (choline chloride-lactic acid) and ChLevA, respectively, starting from 2.5 g of grape stalks, while for ChLAGly (choline chloride-lactic acid-glycerol), no lignin was recovered. This is probably caused by the depolymerization of lignin promoted by the treatment; moreover, NaDES could act as strong solubilizers towards lignin, preventing its precipitation, because deep eutectic solvents (DES) possess a very high polarity and can form H-bonds with the -OH moieties of lignin. This makes DES very efficient in lignin solubilization, however lignin isolation from the extract can be difficult, owing to the strong interaction with the solvent. However, the lignin precipitated from NaDES extraction was found to be less modified and presented a more cross-linked structure when compared to lignin obtained with alkaline treatment.

Lignin derived from food waste or agri-food by-products serves various purposes, categorizable into agricultural and food industry realms. In the agricultural field, lignin is a widely explored biopolymer, particularly for its application in soil enrichment.

This is because lignin's intrinsic resistance to biodegradation makes it a durable contributor to soil organic matter [27]. Additionally, it integrates into a diverse range of polymers utilized in various materials, including those in outdoor applications like greenhouse construction and irrigation systems [28]. Moreover, lignin serves as a versatile agent, functioning as a dispersant, surfactant, insecticide [29], fungicide [30], antibacterial agent [31], or fertilizer [32]. Moreover, in the food industry, lignin finds utility as an additive in food packaging materials, enhancing their properties [20].

Poor mechanical properties of lignin make it unsuitable for direct plastic applications, however lignin added to other biopolymers can also improve the characteristics of the polymeric compound, since the chromophores naturally present in lignin absorb UV light and provide UV resistance to the composite [33–36]. Moreover, the addition of lignin has been proved to improve the thermal stability [37] and the tensile strength of other biopolymers, such as polylactic acid (PLA) [38]. The addition of lignin nanoparticles to a bioplastic derived from *Kappaphycus alvarezii* has been found capable of enhancing surface roughness, water barrier, hydrophobicity, and antimicrobial properties of the polymer, making it a perfect candidate for fruit packaging [39]. Overall, the incorporation of lignin in plastic biopolymers provides better performances, making them perform well as mulching films. Many works present in the literature investigated the functionalization of polymers with lignin, mainly for their strong antimicrobial characteristics [40,41]. All the characteristics mentioned above make lignin a perfect additive to polymers used for mulching films, food packaging, and fruit and vegetable coatings.

2.2. Pectin: Importance and Applicability as Biopolymer

Pectin is a heteropolysaccharide mainly composed of methylated D-galacturonic acid units (Figure 2); this biopolymer can have a different composition, length, degree of esterification, and ramifications. The different characteristics that pectin can undertake are directly linked to the natural variability of the polymer. Pectin is present in all plant tissues, but it is particularly abundant in fruit peels and pomace; for this reason, peels are the part used more often for pectin extraction, with being citrus peel the most common AFW source for commercial manufacturing of pectin [42].

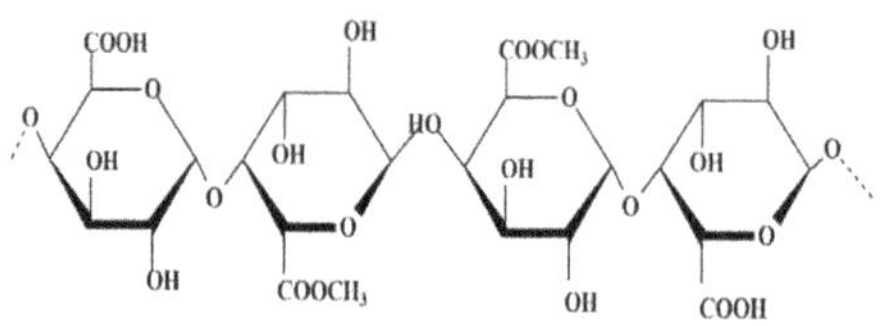

Figure 2. Pectin chemical structure.

Pectin can be extracted using different methods: the most used one is the extraction under conventional heating (70–100 °C) and acidic conditions (pH 1.5–2.5) [43]. After the extraction step, pectin is usually isolated through ethanol precipitation. However, in conventional processes, pectin yield is lower compared to novel techniques [44].

In Table 2, some recent works on pectin extraction from biomass are reported. The yields and quality of pectin, measured by the degree of esterification, vary depending on the plant source, cultivar, growth conditions, and part of the plant being extracted.

Table 2. Pectin extraction methods and yields.

Entry	Biomass	Extraction Method	Yield on Biomass d.w. (%)	Ref.
1	Apple (*Malus* sp.) pomace *Citrus reticulata* peel	SWE	16.68 21.95	[45]
2	Pomegranate (*Punica granatum*) peel	UAE + citric acid	31.89	[46]
3	Grape (*Vitis vinifera*) pomace	UAE + citric acid	25.46	[47]
4	*Citrus maxima* peel (albedo and flavedo)	MAE + HCl	Albedo: 5.57 Flavedo: 3.09	[48]
5	*Ananas comosus* (L.) peel	MAE + H_2SO_4 Hot bath extraction + H_2SO_4	2.43 2.00	[49]
6	Orange (*Citrus* × *sinensis*) peel (albedo)	Hot bath extraction + HCl	18.73	[50]
7	Onion (*Allium cepa*) waste	PUAE + HAE HAE PUAE	20.32 16.22 9.83	[51]

d.w.: dry weight.

For instance, subcritical water extraction (SWE) reveals different optimal extraction temperatures for apple pomace and citrus peel. The yield for apple pomace is 16.68% and for citrus peel is 21.95%, achieved through extractions at 150 °C and 120 °C, respectively. (Entry 1) [45]. In Entries 2 [46] and 3 [47] pectin was extracted from grape pomace and pomegranate peels using similar protocols, but the yield on pomegranate peel is 1.25 times higher in comparison to grape pomace. Furthermore, it can be noticed in Entry 4 [48] that the yield obtained from *Citrus maxima* albedo is significantly higher compared to the one of flavedo. The extraction method also influenced the amount of pectin recovered: for example, in Entry 5 [49] the yield of pectin on ananas peel was higher by 21.50% using microwave-assisted extraction (MAE) compared to conventional conditions (Table 2, Entry 1).

The usage of different methods and technologies to valorise AFW as a source of pectin represents a promising path, since pectin has different agricultural applications, such as an antimicrobial agent [52] that will help in the biotic control of the soil and will promote growth of crops; used as a hydrogel to retain water, thus helping to maintain soil moisture; and it is also used as a biopolymer to cover fruits, increasing its shelf-life. Therefore, AFW represents a low-priced source of this polysaccharide. For instance, pectin can be applied directly in the agricultural field as a hydrogel agent for drought stress management. A common virulence factor of plant pathogenic bacteria and fungi is the ability to produce pectinases, so plant pathogenic microbes can degrade pectins [53]. However, it has been found that the addition of agricultural waste polysaccharides improved plant growth and increased plant resistance against soil-borne pathogens [54]; additionally, pectin was found to be beneficial to *Flavobacterium*, a common plant holobiont [55].

Sayed et al. [56], for example, fabricated a guar gum-pectin/polyacrylamide/ZnO super absorbent hydrogel with satisfactory water retention and adequate biodegradation in soil, and Sharma et al. [57] created pectin nanoparticles which showed good performance in the retention of soil moisture. Applying a cascade treatment to orange peel, Figueira et al. (Table 2, Entry 6) [50] were able to obtain pectin with high anhydruyronic acid content and hesperidin, a high added-value compound. Şen et al. [51] (Table 2, Entry 7) observed that the combination of pulsed ultrasound assisted extraction (PUAE) and hot acid extraction (HAE) brought the highest yield on onion waste biomass compared to the two methods used separately.

Pectin is also a very important polysaccharide for fruit coating; Pholsin et al. [58] used cocoa shell powder as a raw material to recover the pectin, that was then used to coat tomatoes and was found to be more successful in maintaining the fruit characteristics compared to both the control and edible coating without pectin. Edible pectin films for fruit protection can also be functionalized with antimicrobial agents to enhance the protection against fruit pathogens, because the pectin does have bacteriostatic properties; in plants they serve as a barrier that protects them from biotic and abiotic stress [59] and pectin-derived saccharides have been found to exert antimicrobial activity [60]. Thus, pectin coating functionalized with *trans*-cinnamaldehyde was able to reduce weight loss, keep the firmness of the fruit, and reduce pericarp browning while maintaining the taste of rambutan [61].

2.3. Starch: A Promising Biopolymer for Agricultural Applications

Starch is a promising polysaccharide deriving from abundant renewable sources; for certain applications, it can be a valid replacement of traditional plastics. Starch is synthesized by most plants as an energy reservoir, stored in granular form within the cytosol of plant cells. This biopolymer is mainly composed of amylose and amylopectin; this latter forms the main crystalline parts, while the amorphous regions are formed by branches of both. The starch chemical structure can be seen in Figure 3 [62].

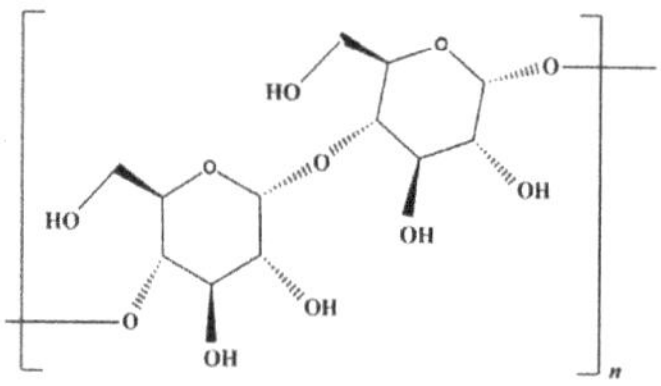

Figure 3. Starch chemical structure.

Most AFW could pose as a valuable source of starch; indeed, there are several works in the literature that report its extraction from various residual biomasses (Table 3).

Table 3. Starch extraction techniques and biomass sources.

Entry	Biomass	Extraction Method	Yield on Biomass dw (%)	Ref.
1	Plantain (*Musa* × *paradisiaca*) peel waste	Ascorbic acid solution (1 to 5%)	23.8–32.5	[63]
2	Litchi (*Litchi chinensis*) seed	0.16% sodium bisulfite solution	21.4	[64]
3	Cassava (*Manihot esculenta*) peel Cassava (*Manihot esculenta*) bagasse	Sodium acetate buffer Sodium acetate buffer	30 7.5	[65]
4	Pineapple (*Ananas comosus*) stem	Wet milling	30	[66]
5	Mango (*Mangifera indica*) kernel	UAE in 1% sodium bisulfite solution	54	[67]

Hernández-Carmona et al. [63], for example, observed that increasing the concentration of ascorbic acid from 1% to 5% *w/v* increased the extraction of starch from 23.8% to 32.5%. Kaur et al. (Table 3, Entry 2) [64] found that the highest starch recovery from biomass was obtained using a temperature of 30 °C and an extraction time of 11 h; moreover, the modification of extracted starch with citric acid caused a depletion of swelling power, solubility, and amylose content, coupled with an increase in water absorption capacity. Cassava peel yielded more starch compared to bagasse (Table 3, Entry 3) [65], and this starch was also more thermostable. Starch extracted from pineapple stem was compared to a commercial one, and was found to have a higher amylose content, gelatinization temperature, gelatinization enthalpy, and pasting temperature, making it thermoplastic (Table 3, Entry 4) [66].

Mieles-Gómez et al. (Table 3, Entry 5) [67] extracted starch rich in phenolic compounds with UAE; the employment of this technology enhanced the water-holding capacity, solubility, oil-holding capacity, and swelling power of starch, when compared to the polysaccharide obtained with conventional extraction.

Starch can be employed in agriculture for the creation of mulch films showing beneficial characteristics, such as maintaining soil moisture, preventing soil erosion, limiting the growth of weed species, and regulating solar radiation [62–68]. Additionally, it can be used for the encapsulation of biocontrol bacteria [69]; this encapsulation protects helpful microorganisms that can prevent the cultivated plant from biotic and abiotic stresses. Starch can also be employed in the construction of superabsorbent polymers for irrigation optimization and controlled release of biocides and biofertilizers [70].

2.4. Cellulose: A Resourceful Biopolymer in All Its Forms

Cellulose is the most common polysaccharide in the world. The chemical structure can be seen in Figure 4: traditionally obtained from plants, it is formed by β-D-Glucopyranose molecules [71]. Nowadays, there is a growing interest in the use of bacterial cellulose (BC), which is characterized by the absence of lignin and hemicellulose, allowing for the avoidance of the downstream purification process. Another advantage is that BC is generally endowed with a high crystallinity and water holding capacity. Furthermore, bacteria are a more sustainable source compared to plants because they do not require extensive land use, which disrupts natural habitats. Additionally, bacteria can feed on waste, making them an environmentally friendly option [72]; however, its high production costs are a strong discouragement for its employment [73].

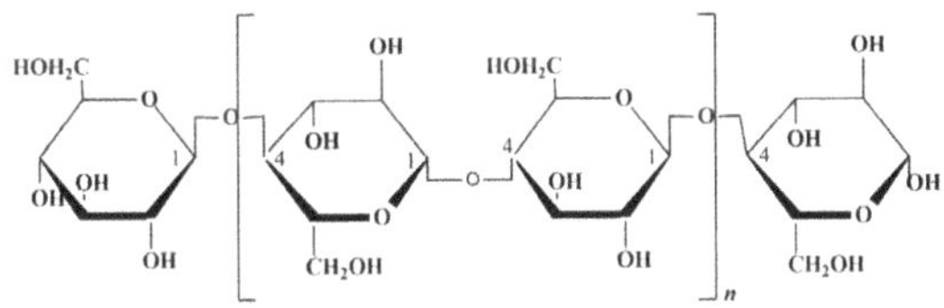

Figure 4. Cellulose chemical structure.

Cellulose is present in the materials market in various derivative forms, such as methyl and ethyl cellulose, carboxymethyl cellulose (CMC), cellulose acetate, cellulose sulphates, amino celluloses, and hydroxypropyl methylcellulose. Cellulose can be present in crystalline, semicrystalline, or amorphous form, in nano or microparticles and in nanofibrils: each of these types of cellulose has its peculiar characteristics and can be used in the production of different materials [74].

As is evident in Table 4, cellulose purification is usually preceded by a delignification step and is often achieved by bleaching the sample or treating it under mild acid/alkaline conditions.

Palm leaves (23.5% of lignin) treated with alkaline hemicellulose removal followed by bleaching yielded a final cellulosic solid with a relatively high degree of purity (87.12%) and crystallinity (69.9%) (Table 4, Entry 1) [75]. By carrying out the cascade removal of hemicellulose and lignin, Fialho et al. (Table 4, Entry 2) [25] obtained a final solid with a glucan content of 77% after both organosolv and alkaline delignification.

The production of CMC from young and mature coconut coir is reported in Entry 3, reporting an improvement in the characteristics directly proportional to the number of bleaching times; it was also documented that mature coconut coir yields more cellulose and more CMC compared to the young ones [76].

Table 4. Cellulose purification from agri-food waste.

Entry	Biomass	Purification Method			Cellulose (%)	Characteristics	Ref.
1	Palm (*Arecaceae* sp.) leaves	Alkaline hemicellulose removal		H_2O_2 Delignification	N/A	87.12% purity, crystallinity of 69.9%	[75]
2	Corn (*Zea mays*) cobs	Hemicellulose removal by autohydrolysis		Organosolv Delignification	77	90% enzymatic hydrolysis yield	[25]
		Hemicellulose removal by autohydrolysis		Alkaline Delignification	77	89.2% enzymatic hydrolysis yield	
3	Mature Coconut (*Cocos nucifera* L.) coir	Alkaline pretreatment	$NaClO_2$/acetic acid hemicellulose and lignin removal	H_2O_2 1 to 3 times	60.76–45.78	79.88–80.43% CMC yield from cellulose	[76]
	Young Coconut (*Cocos nucifera* L.) coir				50.34–40.76	70.67–74.76% CMC yield from cellulose	
4	Bamboo (*Bambusa vulgaris*) culms	Ligninolytic enzymes producing fungi (*Paracremonium* sp. LCB1 + *Clonostachys compactiuscula* LCN1)			N/A	Selectivity of delignification >10	[77]
5	*Melia dubia* woody biomass	Enzolv method			52	20% increase in crystallinity	[78]

N/A: not available.

Cellulose purification can also be done with biological methods involving, for example, the use of a consortium of two ligninolytic fungal strains with very low levels of cellulase activity (Table 4, Entry 4) [64], or with thermosolvent laccase enzymes combined with steam treatment and ethanol (Table 4, Entry 5) [78].

As mentioned before, cellulose can be also obtained in the form of BC, considering that BC-forming microorganisms can grow on AFW. Recent works focusing on BC production from waste are detailed in Table 5, revealing relatively modest yields and final titers. Future research efforts should concentrate on optimizing fermentation conditions to increase BC yield, as feeding waste biomass to bacteria is not enough to make the process economically feasible.

Table 5. BC from waste biomass.

Entry	Biomass	Pretreatment	Microorganism	Yield % (g Cellulose/g Biomass)	Final Titer (g/L)	Ref.
1	Orange (*Citrus × sinensis*) processing waste	UAE diluted acid hydrolysis	*Komagataeibacter sucrofermentans* DSM 15973	5.82	N/A	[79]
2	Mango (*Mangifera indica*) peel waste	Nitric acid hydrolysis	*Achromobacter* S3	N/A	1.22	[80]
3	Olive (*Olea europaea*) pomace	Mild acid treatment + enzymatic hydrolysis	*Novacetimonas hansenii* 53582	9.60	0.68	[73]

N/A: not available.

In this context, another cost-cutting strategy is the approach adopted by Karanicola et al. (Table 5, Entry 1) [79], as an alternative to fermentation optimization. This approach was aimed at the full utilization of the biomass, which entailed the extraction of valuable compounds such as essential oils and pectin before the fermentation; using this method, the cost of the entire process is supported by the obtainment of multiple value-added molecules from biomass.

Cellulose and its derivatives possess numerous applications in agriculture, as they have been employed for the controlled release of fertilizers [81] and essential micronutrients [82]. Another use of cellulose is as a fruit coating agent for ripening delay [83] or as a hydrogel for drought management [84,85]. Moreover, mulch films obtained with a

CMC–PVA (polyvinyl alcohol) blend were able to degrade within two months [86], while those formed with CMC and soybean protein, compared with high density polyethylene (HDPE), led to an increased germination of cabbage seeds [87]. The production of mulch films was also found to be feasible in combination with antimicrobial elements such as silver nanowires for an enhanced crop protection [88].

2.5. Chitosan, an Important Adjuvant for Plant, Fruit, and Seed Protection

Chitin is the most abundant amino polysaccharide present in the world; it is formed by N-acetylglucosamine (GlcNAc) residues [89], and it is present in nature in crystalline forms of α, β, and γ-chitin. Traditionally, the main source of chitin is the crustacean waste deriving from the fishing industry [90], but it is not as sustainable as fungal chitin. Chitosan is a direct derivative of chitin obtained usually by alkaline deacetylation [91]; the result is a copolymer formed by β (1–4) linked D-glucosamine (GlcN) units. Chitosan (Figure 5) is characterized by a variable percentage of GlcNAc residues, as deacetylation is never complete. Chitin, and consequently chitosan, can be extracted from fungal waste biomass as a secondary product; in addition, fungi can also successfully grow on agri-food waste.

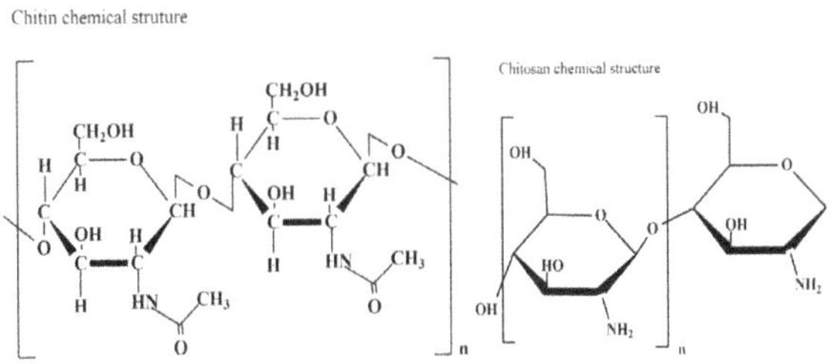

Figure 5. Chitin and chitosan chemical structure.

Tayel et al. [92] successfully extracted chitosan from *Aspergillus niger* fungal waste following the production of citric acid, a common food additive. *Mucor circinelloides* was able to grow on sugar bagasse hydrolysate and corn steep solids, and a total of 1.36 g/L of chitosan was extracted from the fungal biomass [93]. Habibi et al. [94] obtained chitosan from *Aspergillus terrerus* grown using apple waste as the sole carbon source. Corn steep liquor and Cassava wastewater were used as a carbon and nitrogen source by *Mucor subtilissimus* UCP 1262 and *Lichtheimia hyalospora* UCP 1266; the chitosan extracted from the deriving fungal biomass exhibited a deacetylation degree >80%, comparable to commercial chitosan [95].

In each of the papers mentioned, the fungal biomass first underwent an alkaline treatment, then chitosan was extracted from the alkali-insoluble fraction using a mild acid treatment.

The antimicrobial properties of chitosan are useful in agriculture, as they can be employed to increase the shelf life of fresh fruit [93,96]. Moreover, chitosan can also be used in seed protection, since it facilitates the attachment of the biological priming agent *Trichoderma asperellum* to the seed [97], therefore significantly increasing seedling germination and inhibiting diseases [98]. Chitosan nanoparticles can also be used for the controlled release of herbicides, as they have been successfully used to encapsulate imazapic and imazapyr to reduce their toxicity and increase the efficacy of the treatment [99].

2.6. Polyhydroxyalkanoates (PHAs): A Versatile and Biodegradable Bioplastic

Chemically, PHAs are polyesters consisting of hydroxyalkanoate monomers linked by ester bonds. The general chemical structure of PHAs can be seen in Figure 6. PHAs are biopolymers naturally produced by several bacteria and archaea and stored under the form of granules inside the cell as a reserve carbon source. The specific chemical structure of PHAs can vary widely depending on the microbial species producing them, the carbon source provided during cultivation, and the environmental conditions. To synthetize PHAs, these microorganisms usually use glucose or fatty acids as a substrate: AFW can be a good and cheap source of these compounds, thus representing a good strategy for cutting costs, given that PHA is 15 times more expensive than petroleum-based plastics [100].

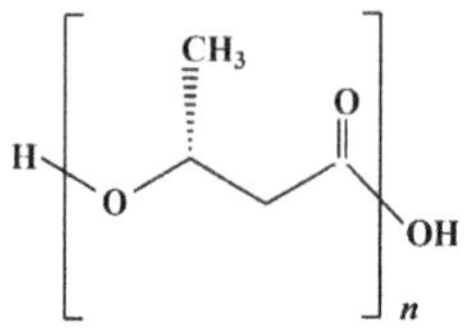

Figure 6. Polyhydroxyalkanoates chemical structure.

Some recent works are reported in Table 6.

Table 6. PHA production from agri-food waste.

Entry	Biomass	Pretreatment	Culture	Comments	Ref.
1	Olive (*Olea europaea*) mill wastewater	Dephenolization and microbial pretreatment (acidogenic fermentation) for conversion of COD in VFA	Mixed culture—activated sludge	22% of COD converted in PHA	[101]
2	Corn (*Zea mays*) straw	Microwave-assisted acid saccharification	Mixed microbial culture from dairy industry waste	76.3% PHA dw/Biomass dw in 3 L bioreactor—420 mg/L final concentration	[102]
3	Mixture of molasses and olive oil	Acid pretreatment	*Cupriavidus necator* DSM 428	2.03 g/L final concentration of PHA	[103]
4	Lignin-rich black liquor from sugarcane (*Saccharum officinarum* L.) bagasse	Alkaline pretreatment	*Pseudomonas monteilii* BCC19149	237.9 mg/L of PHA final titer under optimized conditions	[104]
5	Fruit waste	Acidogenic fermentation	Mixed culture—activated sludge	Y SFP/FW 0.79 $gCOD \cdot gCOD^{-1}$ Y PHA/SFP 0.98 $gCOD \cdot gCOD^{-1}$ Y PHA/FW 0.45 $gCOD \cdot gCOD^{-1}$	[105]
6	Potato (*Solanum tuberosum*) peels waste	Acid hydrolysis	*Bacillus circulans* MTCC 8167	232 mg/L of PHA final titer	[106]

Campanari et al. [101] (Table 6, Entry 1) utilized a multi-stage process in which they selected PHA-storing microorganisms under feast and famine (FF) conditions in a sequencing batch reactor (SBR) and found out that the PHA production is widely influenced by the selection of PHA producers, which is in turn influenced by the organic loading rate (ORL). Verdini et al. (Table 6, Entry 2) [102] used corn straw saccharified with the aid of microwave technology and obtained a greater yield during scaling up in a stirred tank 3L bioreactor, compared to a shake flask. Moreover, the authors were also able to determine the specific PHA that was being produced, which was poly(3-hydroxybutyrate-*co*-3-hydroxyvalerate) (PHBV). Similarly, Kag et al. (Table 6, Entry 6) [106] saccharified a waste biomass to accumulate PHA, however they obtained a lower final titer.

Razzaq et al. [103] (Table 6, Entry 3) determined that the type of PHA monomer is influenced by the selection of the substrate, as *C.necator* usually produces short-chain length PHAs, but with the mixture that they used as a substrate (without the addition of a nitrogen source), they were able to produce medium-chain length PHAs. In Entry 4 [104], *P. monteilii* was grown on sugarcane bagasse black liquor through a two-phase fermentation: (i) optimization of cell growth in batch configuration, (ii) pulse-feeding in fed-batch configuration, with the C:N ratio finely tuned to maximize the production of PHA. During the growth phase, the percentage of PHA in the total cell concentration was less than 1%; however, in the accumulation phase this percentage was increased to 12.8%.

Matos et al. [63] (Table 6, Entry 5) proposed a three-step valorization of fruit waste on a pilot scale: the first step aimed to convert the biomass into soluble fermentation products (SFP) through fermentation, then the SPF media, rich in butyrate, was fed to the second and third steps. The second step was aimed at microbial culture selection to improve PHA yield, and the third step involved PHA accumulation. Overall, the process was able to afford a remarkable maximum PHA content of 80.5% and a storage yield of 0.98 gCOD·gCOD^{-1}; the overall process yielded 0.45 gCOD·gCOD^{-1}. A more complex fermentation set-up, comprising the separation of SPF production step and the PHA creation, and the accumulation step can dramatically improve the PHA yield.

The usage of PHAs in agriculture has been explored mainly as mulch films [107,108]; for this application, PHA is usually blended with other biodegradable components to improve its processability. For example, the addition of medium-chain length PHA increased the processability of PLA [109] and slowed down its biodegradation rate [108]. It was found that the addition of a small percentage of UV stabilizer drastically slowed down the degradation rate of polybutylene adipate terephthalate (PBAT)/PHA mulching films when subjected to accelerated aging test (AAT). On the other hand, the addition of a hydrolysis resistant additive was found to be ineffective in protecting the mulching film from degradation [110]. Othman et al. [111] tested different combinations of PHA/polycaprolactone (PCL) films for active mulching purposes and found them to be a good and biodegradable alternative to traditional ones, as they improved rice seed germination.

2.7. Polylactic Acid (PLA): A Non-Toxic Alternative for Mulching Films and Food Packaging

PLA is a biopolymer with properties like those of conventional plastics deriving from fossil fuels; the PLA chemical structure can be seen in Figure 7. Since the properties of PLA are better when the polymer consists exclusively of the stereoisomer L- (L-LA) [112], the cheaper way is trough fermentation. Lactic acid (LA) can derive from the fermentation catalysed by a wide range of microorganisms that can, consequently, grow on and exploit a wide range of substrates; for this reason, a broad spectrum of AFW finds its usage in this stream of production. The optimization of the use of AFW as fermentation substrates to produce LA is crucial for both a waste valorization approach and for the cost reduction of the process [113]. Before the fermentation step, the biomass typically undergoes a pre-treatment, which may be physical, biological, chemical, or a combination of these methods: this is because agricultural by-products, which are usually lignocellulosic, need to be delignified and saccharified for the microorganism to use them as substrates [114]. Alternatively, the process needs to employ a microorganism with both the capacity to degrade complex lignocellulosic biomasses and to produce LA; this approach is called consolidated bioprocessing (CBP) [115].

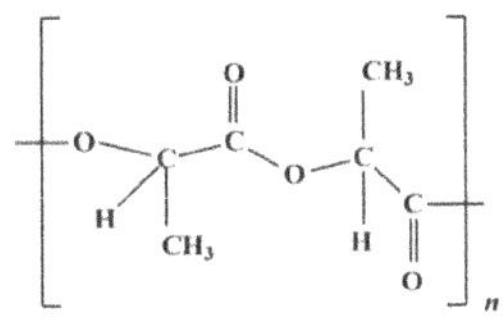

Figure 7. Polylactic acid chemical structure.

In Table 7, some of the agricultural waste biomasses that have been investigated as substrates to produce LA are reported. In this case, different microorganisms can afford different yields of LA.

Table 7. Biotransformation of biomass in LA.

Entry	Biomass	Biomass Pretreatment	Saccharification	LA Producer	Yield	LA Titer (g/L)	Ref.
1	Grape (*Vitis vinifera*) stalks	Steam explosion	Enzymatic saccharification	*Lactobacillus rhamnosus* IMC501	0.98 g/g Glu	20	[116]
2	Olive (*Olea europaea*) pits	Diluted acid (H_2SO_4)		*Bacillus coagulans* DSM2314	0.71 g/g OPH	61.3	[117]
3	Sugarcane (*Saccharum officinarum* L.) bagasse	Hydrothermal hydrolysation		*Bacillus coagulans* DSM2314	0.66 g/g SBH	52.4	
4	Potato (*Solanum tuberosum*) peel waste		*Rhizopus oryzae* (CBP)		0.039 g/g PPW	3.14	[118]
5	Corncob (*Zea mays*) Residue	Alkali pretreatment	*Clostridium thermocellum* ΔpyrF:p2638-BGL	*Geobacillus stearothermophilus* 2H-3	0.74 g/g CR	51.36	[119]

An example of this was the study elaborated by D'Ambrosio et al. [116] (Table 7, Entry 1) where a maximum theoretical yield of LA from glucose was obtained; however, the LA yield calculated on the starting biomass was affected by a lower yield in the previous step (38.1%, enzymatic saccharification). Entries 2 and 3 [117] report a non-sterile *B. coagulans* DSM2314 fermentation that led to satisfying yields when fermenting xylose-rich hydrolysates; in fact, they were very similar to the one obtained when fermenting pure xylose (0.77 g LA/g Xyl); moreover, the lack of sterility is an advantage as it greatly reduces operation costs.

Liu et al. [119] (Table 7, Entry 5) followed a different path when treating corncob residue, as they applied an alkali delignification followed by a whole-cell enzymatic saccharification with *C. thermocellum* ΔpyrF: p2638-BGL, an engineered strain with enhanced β-glucosidase activity [120]. The saccharified fraction was then fermented by *G. stearothermophilus* 2H-3, a contaminant strain found in corncob residue that can produce high quantities of LA with very little substrate competition by other fermentation products. Another advantage of this fermentation method is that saccharification and fermentation can be carried out in a one-pot successive fermentation.

Despite the advantage of not requiring additional pre-treatments, CBP with *R. oryzae* [118] (Table 7, Entry 4) afforded a much lower LA final titer and yield on biomass, but it is worthy to note that this bioprocess mainly aimed to produce bioethanol, which was obtained with a final titer of 18.83 g/L.

Overall, to enhance the competitiveness of PLA compared to fossil-derived plastics, a thorough optimization process is required, encompassing biomass pre-treatment, fermentation parameters, and, ultimately, the selection of bacterial and fungal strains [121].

PLA, being both biodegradable and non-toxic, is a particularly useful biopolymer in agricultural applications. To be specific, PLA has been widely investigated in its role in the formation of mulch films [122] and food packaging [123] and has been proposed for the construction of shade nets and greenhouse covers [124].

The use of lignocellulosic waste such as wood flour [125,126], cork [125], rubber wood sawdust [127], bark [128], spinach stems, tomato pomace, cocoa shells [122], and orange peels [129] as fillers in PLA films successfully improved its mechanical characteristics.

Poly (lactic acid) (PLA) is a biodegradable, bio-based polymer that has attracted a lot of attention as an alternative to conventional petroleum-based plastics [112]. However, it is not the only sustainable alternative. Others include polybutylene succinate (PBS) and polybutylene adipate terephthalate (PBAT), which have several advantages and limitations compared to PLA [130,131].

PBS is a biodegradable polyester that has properties similar to polypropylene. It is derived from succinic acid and 1,4-butanediol, which can be produced from renewable resources. PBS offers good thermal and mechanical properties, making it suitable for various applications, such as packaging films, bags, and disposable items. One advantage of PBS over PLA is its greater flexibility and strength, which can be an advantage in certain applications. However, PBS biodegrades at a slower rate than PLA and can be more expensive to produce [130].

PBAT, on the other hand, is another biodegradable polyester that is often used in blends with PLA to improve its flexibility and strength [132]. PBAT is synthesized from adipic acid, 1,4-butanediol, and terephthalic acid, all of which can be obtained from renewable resources. PBAT offers good flexibility and elongation properties [132], making it suitable for applications that require elasticity [131], such as compostable bags and films [130]. Compared to PLA, PBAT generally has better elongation at break and toughness. However, depending on the specific formulation, it may have lower stiffness and heat resistance than PLA [130].

Although both PBS and PBAT offer advantages over PLA in terms of certain mechanical properties and flexibility, they can also present challenges, such as slower biodegradation rates, higher production costs, and potential compatibility issues in blends [130]. In addition, the choice between these polymers often depends on application-specific requirements, including mechanical performance, biodegradability, and cost considerations. However, PLA is currently most used in agricultural applications due to its high biodegradability and biocompatibility, making it more suitable for sustainable agricultural applications [130].

3. Sustainable Strategies and Future Challenges for AFW Disposal

The sustainability of agricultural supply chains has become a priority concern in recent years, driven by growing awareness of environmental issues, resource scarcity, and the need to feed a growing world population [133].

The challenge at hand involves managing waste generated throughout the supply chain, from farm to fork. Innovative waste management models are emerging as important tools to address this issue and improve the sustainability of food supply chains [31]. AFW recovery holds great potential for economic, social, and environmental benefits. Many countries around the world have implemented food and waste recovery strategies. The reuse of AFW encompasses the conversion into useful products, such as biofuels and chemicals [134]. To achieve this goal, various methods have been exploited to add value to AFW, including pyrolysis, anaerobic digestion, decomposition, chemical and thermal treatment, solid catalysis, extraction, and fermentation [31,135].

Biorefineries are an innovative approach to waste management in the agricultural food supply chain, aiming for the complete utilization of all the components of AFW. This goal represents an environmentally friendly solution, as it encompasses advanced biochemical and biotechnological processes to convert organic waste streams into high-value products such as biofuels, biochemicals, and bioplastics. By utilizing by-products that would otherwise be discarded, biorefineries not only reduce waste, but also contribute to the development of a more sustainable and resource-efficient supply chain. Governments play a critical role in facilitating the transition to innovative waste management models in agri-food supply chains. By implementing supportive policies, regulations, and financial incentives, policy makers can encourage investment in sustainable waste management infrastructure, drive research and development of waste reduction technologies, and create a favourable environment for innovation and collaboration [136].

The development toward the sustainability of food supply chains depends on the adoption of innovative waste management models that prioritize resource efficiency, minimize waste generation, and create value from waste streams. Converting agricultural waste into potential valuable products and closing the loop aligns with the principles of the circular economy in agriculture and requires collaboration among all actors in the food supply chain.

Adopting a circular economy approach, harnessing technological advances, promoting biorefineries and by-product recovery, fostering collaborative partnerships, engaging consumers, and providing policy support and incentives is crucial to build resilient, sustainable, and environmentally friendly agri-food supply chains.

4. Conclusions

Fossil-based plastics represent one of the greatest problems of this last century; being fabricated from non-renewable resources and being (mostly) non-biodegradable, they contribute to GHG emissions and pollution both at the beginning and at the end of their life [137].

This review explores the feasibility of converting AFW into biopolymers and the possible employment of this in agriculture as mulch films, hydrogels, biocides, biostimulants, soil stabilizers, and for seed, fruit, and vegetable coatings [14,15]. Biopolymers can represent a good alternative to fossil-derived plastic, as they are renewable and often biodegradable. Replacing conventional plastics with biopolymers is desirable, as it represents a significant step towards sustainability, given the ubiquitous use of polymers across various industries.

The agricultural sector, and, more broadly, the food industry, would be one of the first fields to benefit from the use of biopolymers, as they can help mitigate the environmental footprint of an industry already heavily impacting the environment. Moreover, the employment of biopolymers would improve the safety of the food produced, as microplastics are a potential health hazard that is becoming more and more alarming [138]. Furthermore, the capacity of biopolymers to biodegrade helps to simplify the agricultural process (e.g., mulching films composed by biopolymers do not have to be removed anymore since they naturally biodegrade). However, some gaps still need to be filled for biopolymers to be considered a valid commercial alternative to fossil-based plastic; further research should be done on the improvement of mechanical properties of biopolymers. Indeed, since they are often too fragile for their intended use, mixtures of biopolymers pose as a feasible alternative to modulate the polymer features, making them suitable for the intended application.

According to an economic point of view, the employment of AFW helps with cutting the cost of operations, but not enough to compete with traditional plastic biopolymers. An overall process optimization is necessary to improve employment and popularity of biopolymers in the agricultural field. The employment of a cascade process where more than one compound is produced from AFW can help to cut the cost of production, by producing value-added products. In processes where microbial production is concerned, optimization of the fermentation parameters and genetic strain improvement are crucial for the economic feasibility of the process.

Future actions should focus on researching ways to optimize processes, in order to make them more economically viable and more environmentally friendly. Each region should focus on local AFW production to reduce costs and greenhouse gas emissions caused by transportation. Furthermore, investing in renewable energy sources, such as solar or wind energy to enhance AFW, can further strengthen sustainability efforts, ensuring that our actions can pave the way for a greener and more sustainable future.

When the process of converting AFW into biopolymers is optimized and scaled at the industrial level, the costs of the final product will be reduced, and biopolymers can become more competitive than traditional plastic, thus making the agro-industrial sector more sustainable.

Author Contributions: Writing—original draft preparation, C.V. and M.V.; writing—review and editing, S.T., E.C.G. and G.C.; supervision, C.F. and S.T. All authors have read and agreed to the published version of the manuscript.

Funding: This work was supported by the University of Turin (Ricerca Locale 2023), and the Agritech National Research Center and received funding from the European Union Next-GenerationEU (Piano Nazionale di Ripresa e Resilienza (PNRR)—missione 4 componente 2, investimento 1.4—D.D. 1032 17/06/2022, CN00000022).

Conflicts of Interest: The authors declare no conflicts of interest.

Abbreviations

AFW	Agri-food waste
NaDES	Natural deep eutectic solvents
UA	Ultrasound assisted
US	Ultrasounds
ChLA	Choline chloride-lactic acid
ChLevA	Choline chloride-levulinic acid
ChLAGly	Choline chloride-lactic acid-glycine
PLA	Polylactic acid
SWE	Subcritical water extraction
MAE	Microwave assisted extraction
PUAE	Pulsed ultrasound assisted extraction
HAE	Hot acid extraction
BC	Bacterial cellulose
CMC	Carboxymethyl cellulose
PVA	Polyvinyl alcohol
HDPE	High density polyethylene
GlcNAC	N-Acetylglucosamine
GlcN	Glucosamine
FF	Feast and famine
SBR	Sequencing batch reactor
ORL	Organic loading rate
PHBV	Poly(3-hydroxybutyrate-*co*-3-hydroxyvalerate)
SFP	Soluble fermentation products
PBAT	Polybutylene adipate terephthalate
AAT	Accelerated aging test
PCV	Polycaprolactone
LA	Lactic acid
GHG	Greenhouse gases

References

1. Bioplastics Market Development Update 2023. Available online: https://www.european-bioplastics.org/market/ (accessed on 19 February 2024).
2. Visco, A.; Scolaro, C.; Facchin, M.; Brahimi, S.; Belhamdi, H.; Gatto, V.; Beghetto, V. Agri-Food Wastes for Bioplastics: European Prospective on Possible Applications in Their Second Life for a Circular Economy. *Polymers* **2022**, *14*, 2752. [CrossRef] [PubMed]
3. Yates, M.R.; Barlow, C.Y. Life Cycle Assessments of Biodegradable, Commercial Biopolymers—A Critical Review. *Resour. Conserv. Recycl.* **2013**, *78*, 54–66. [CrossRef]
4. Jorge, A.M.S.; Gaspar, M.C.; Henriques, M.H.F.; Braga, M.E.M. Edible Films Produced from Agrifood By-Products and Wastes. *Innov. Food Sci. Emerg. Technol.* **2023**, *88*, 103442. [CrossRef]
5. Ensure Sustainable Consumption and Production Patterns. Available online: https://sdgs.un.org/goals/goal12 (accessed on 20 February 2024).
6. Voss, M.; Gaudino, E.C.; Tabasso, S.; Forte, C.; Cravotto, G. Current Emerging Green Technologies for the Valorization of Grape and Cherry Wastes. *Curr. Food Sci. Technol. Rep.* **2023**, *1*, 47–61. [CrossRef]
7. Valdés, A.; Mellinas, A.C.; Ramos, M.; Burgos, N.; Jiménez, A.; Garrigós, M.C. Use of Herbs, Spices and Their Bioactive Compounds in Active Food Packaging. *RSC Adv.* **2015**, *5*, 40324–40335. [CrossRef]
8. Bolaji, I.; Nejad, B.; Billham, M.; Mehta, N.; Smyth, B.; Cunningham, E. Multi-Criteria Decision Analysis of Agri-Food Waste as a Feedstock for Biopolymer Production. *Resour. Conserv. Recycl.* **2021**, *172*, 105671. [CrossRef]
9. Nabuurs, G.-J.; Aoki, L.; Humpenöder, F.; Boone Kauffman, J.; Ayala-Niño, F.; Emmet-Booth, J.P.; Mrabet, R.; Abu Hatab, A.; Bustamante, M.; Clark, H.; et al. 7—Agriculture, Forestry and Other Land Uses (AFOLU). In *Climate Change 2022—Mitigation of Climate Change: Working Group III Contribution to the Sixth Assessment Report of the Intergovernmental Panel on Climate Change*; Cambridge University Press: Cambridge, UK, 2023. [CrossRef]
10. Fahim, I.S.; Chbib, H.; Mahmoud, H.M. The Synthesis, Production & Economic Feasibility of Manufacturing PLA from Agricultural Waste. *Sustain. Chem. Pharm.* **2019**, *12*, 100142. [CrossRef]
11. Puglia, D.; Pezzolla, D.; Gigliotti, G.; Torre, L.; Bartucca, M.L.; Del Buono, D. The Opportunity of Valorizing Agricultural Waste, through Its Conversion into Biostimulants, Biofertilizers, and Biopolymers. *Sustainability* **2021**, *13*, 2710. [CrossRef]

12. Pu, W.; Shen, C.; Wei, B.; Yang, Y.; Li, Y. A Comprehensive Review of Polysaccharide Biopolymers for Enhanced Oil Recovery (EOR) from Flask to Field. *J. Ind. Eng. Chem.* **2018**, *61*, 1–11. [CrossRef]
13. Maraveas, C. Environmental Sustainability of Greenhouse Covering Materials. *Sustainability* **2019**, *11*, 6129. [CrossRef]
14. Kalia, S.; Avérous, L. *Biopolymers: Biomedical and Environmental Applications*; Scrivener Publishing: Austin, TX, USA, 2011; ISBN 978-0-470-63923-8.
15. Klein, M.; Poverenov, E. Natural Biopolymer-Based Hydrogels for Use in Food and Agriculture. *J. Sci. Food Agric.* **2020**, *100*, 2337–2347. [CrossRef]
16. Brunow, G. Lignin Chemistry and Its Role in Biomass Conversion. In *Biorefineries-Industrial Processes and Products: Status Quo and Future Directions*; Wiley-VCH Verlag GmbH: Weinheim, Germany, 2008; Volume 2, pp. 151–163. ISBN 3527310274.
17. Collins, M.N.; Nechifor, M.; Tanasă, F.; Zănoagă, M.; McLoughlin, A.; Stróżyk, M.A.; Culebras, M.; Teacă, C.A. Valorization of Lignin in Polymer and Composite Systems for Advanced Engineering Applications—A Review. *Int. J. Biol. Macromol.* **2019**, *131*, 828–849. [CrossRef]
18. Yu, O.; Kim, K.H. Lignin to Materials: A Focused Review on Recent Novel Lignin Applications. *Appl. Sci.* **2020**, *10*, 4626. [CrossRef]
19. Aro, T.; Fatehi, P. Production and Application of Lignosulfonates and Sulfonated Lignin. *ChemSusChem* **2017**, *10*, 1861–1877. [CrossRef]
20. Cassoni, A.C.; Costa, P.; Vasconcelos, M.W.; Pintado, M. Systematic Review on Lignin Valorization in the Agro-Food System: From Sources to Applications. *J. Environ. Manag.* **2022**, *317*, 115258. [CrossRef]
21. Lu, X.; Gu, X. A Review on Lignin Pyrolysis: Pyrolytic Behavior, Mechanism, and Relevant Upgrading for Improving Process Efficiency. *Biotechnol. Biofuels Bioprod.* **2022**, *15*, 106. [CrossRef] [PubMed]
22. Liu, Z.; Peng, X.; Xu, Y.; Tang, T.; Wei, S.; Liu, B.; Wang, D.; Wang, H.M.; Jiang, J. Macromolecular Structural Characteristics and Functional Potential of Tobacco Stalk Lignin from the Phosphotungstic Acid-Assisted Delignification Process. *Biomass Bioenergy* **2023**, *170*, 106706. [CrossRef]
23. Salgado-Ramos, M.; Tabasso, S.; Calcio Gaudino, E.; Moreno, A.; Mariatti, F.; Cravotto, G. An Innovative, Green Cascade Protocol for Grape Stalk Valorization with Process Intensification Technologies. *Appl. Sci.* **2022**, *12*, 7417. [CrossRef]
24. Yang, X.; Li, X.; Zhu, L.; Liang, J.; Zhu, J. Production of Hemicellulose Sugars Combined with the Alkaline Extraction Lignin Increased the Hydro-Depolymerization of Cellulose from Corn Cob. *Sustainability* **2023**, *15*, 9041. [CrossRef]
25. Fialho, J.; Moniz, P.; Duarte, L.C.; Carvalheiro, F. Green Fractionation Approaches for the Integrated Upgrade of Corn Cobs. *ChemEngineering* **2023**, *7*, 35. [CrossRef]
26. Banerjee, R.; Chintagunta, A.D.; Ray, S. Laccase Mediated Delignification of Pineapple Leaf Waste: An Ecofriendly Sustainable Attempt towards Valorization. *BMC Chem.* **2019**, *13*, 58. [CrossRef] [PubMed]
27. Hernes, P.J.; Kaiser, K.; Dyda, R.Y.; Cerli, C. Molecular Trickery in Soil Organic Matter: Hidden Lignin. *Environ. Sci. Technol.* **2013**, *47*, 9077–9085. [CrossRef] [PubMed]
28. Yiamsawas, D.; Kangwansupamonkon, W.; Kiatkamjornwong, S. Lignin-Based Nanogels for the Release of Payloads in Alkaline Conditions. *Eur. Polym. J.* **2021**, *145*, 110241. [CrossRef]
29. Jiang, Y.; Chen, Y.; Tian, D.; Shen, F.; Wan, X.; Xu, L.; Chen, Y.; Zhang, H.; Hu, J.; Shen, F. Fabrication and Characterization of Lignin–Xylan Hybrid Nanospheres as Pesticide Carriers with Enzyme-Mediated Release Property. *Soft Matter* **2020**, *16*, 9083–9093. [CrossRef] [PubMed]
30. Fischer, J.; Beckers, S.J.; Yiamsawas, D.; Thines, E.; Landfester, K.; Wurm, F.R. Targeted Drug Delivery in Plants: Enzyme-Responsive Lignin Nanocarriers for the Curative Treatment of the Worldwide Grapevine Trunk Disease Esca. *Adv. Sci.* **2019**, *6*, 1802315. [CrossRef] [PubMed]
31. Voss, M.; Valle, C.; Gaudino, E.C.; Tabasso, S.; Forte, C.; Cravotto, G. Unlocking the Potential of Agrifood Waste for Sustainable Innovation in Agriculture. *Recycling* **2024**, *9*, 25. [CrossRef]
32. Zhang, S.; Fu, X.; Tong, Z.; Liu, G.; Meng, S.; Yang, Y.; Helal, M.I.D.; Li, Y.C. Lignin-Clay Nanohybrid Biocomposite-Based Double-Layer Coating Materials for Controllable-Release Fertilizer. *ACS Sustain. Chem. Eng.* **2020**, *8*, 18957–18965. [CrossRef]
33. Muhammed, A.P.; Thangarasu, S.; Oh, T.H. Green Interconnected Network Structure of Chitosan-Microcrystalline Cellulose-Lignin Biopolymer Film for Active Packaging Applications. *Int. J. Biol. Macromol.* **2023**, *253*, 127471. [CrossRef]
34. Rodrigues, J.S.; de Freitas, A.D.S.; Lopes, H.S.M.; Pires, A.A.F.; Lemes, A.P.; Ferreira, M.; Botaro, V.R. Improvement of UV Stability of Thermoplastic Starch Matrix by Addition of Selected Lignin Fraction—Photooxidative Degradation. *Int. J. Biol. Macromol.* **2023**, *230*, 123142. [CrossRef] [PubMed]
35. Li, B.; Zhou, M.; Huo, W.; Cai, D.; Qin, P.; Cao, H.; Tan, T. Fractionation and Oxypropylation of Corn-Stover Lignin for the Production of Biobased Rigid Polyurethane Foam. *Ind. Crops Prod.* **2020**, *143*, 111887. [CrossRef]
36. Xie, J.; Hse, C.Y.; Shupe, T.F.; Hu, T. Physicochemical Characterization of Lignin Recovered from Microwave-Assisted Delignified Lignocellulosic Biomass for Use in Biobased Materials. *J. Appl. Polym. Sci.* **2015**, *132*, 42635. [CrossRef]
37. Morandim-Giannetti, A.A.; Agnelli, J.A.M.; Lanças, B.Z.; Magnabosco, R.; Casarin, S.A.; Bettini, S.H.P. Lignin as Additive in Polypropylene/Coir Composites: Thermal, Mechanical and Morphological Properties. *Carbohydr. Polym.* **2012**, *87*, 2563–2568. [CrossRef]
38. Gao, Y.; Qu, W.; Liu, Y.; Hu, H.; Cochran, E.; Bai, X. Agricultural Residue-Derived Lignin as the Filler of Polylactic Acid Composites and the Effect of Lignin Purity on the Composite Performance. *J. Appl. Polym. Sci.* **2019**, *136*, 47915. [CrossRef]

39. Rizal, S.; Alfatah, T.; Abdul Khalil, H.P.S.; Yahya, E.B.; Abdullah, C.K.; Mistar, E.M.; Ikramullah, I.; Kurniawan, R.; Bairwan, R.D. Enhanced Functional Properties of Bioplastic Films Using Lignin Nanoparticles from Oil Palm-Processing Residue. *Polymers* **2022**, *14*, 5126. [CrossRef]
40. Yang, W.; Owczarek, J.S.; Fortunati, E.; Kozanecki, M.; Mazzaglia, A.; Balestra, G.M.; Kenny, J.M.; Torre, L.; Puglia, D. Antioxidant and Antibacterial Lignin Nanoparticles in Polyvinyl Alcohol/Chitosan Films for Active Packaging. *Ind. Crops Prod.* **2016**, *94*, 800–811. [CrossRef]
41. Rai, S.; Dutta, P.K.; Mehrotra, G.K. Natural Antioxidant and Antimicrobial Agents from Agrowastes: An Emergent Need to Food Packaging. *Waste Biomass Valorization* **2020**, *11*, 1905–1916. [CrossRef]
42. Zdunek, A.; Pieczywek, P.M.; Cybulska, J. The Primary, Secondary, and Structures of Higher Levels of Pectin Polysaccharides. *Compr. Rev. Food Sci. Food Saf.* **2021**, *20*, 1101–1117. [CrossRef] [PubMed]
43. Bagherian, H.; Zokaee Ashtiani, F.; Fouladitajar, A.; Mohtashamy, M. Comparisons between Conventional, Microwave- and Ultrasound-Assisted Methods for Extraction of Pectin from Grapefruit. *Chem. Eng. Process. Process Intensif.* **2011**, *50*, 1237–1243. [CrossRef]
44. Kumar, S.; Reddy, A.R.L.; Basumatary, I.B.; Nayak, A.; Dutta, D.; Konwar, J.; Purkayastha, M.D.; Mukherjee, A. Recent Progress in Pectin Extraction and Their Applications in Developing Films and Coatings for Sustainable Food Packaging: A Review. *Int. J. Biol. Macromol.* **2023**, *239*, 124281. [CrossRef] [PubMed]
45. Wang, X.; Chen, Q.; Lü, X. Pectin Extracted from Apple Pomace and Citrus Peel by Subcritical Water. *Food Hydrocoll.* **2014**, *38*, 129–137. [CrossRef]
46. Podetti, C.; Riveros-Gomez, M.; Román, M.C.; Zalazar-García, D.; Fabani, M.P.; Mazza, G.; Rodríguez, R. Polyphenol-Enriched Pectin from Pomegranate Peel: Multi-Objective Optimization of the Eco-Friendly Extraction Process. *Molecules* **2023**, *28*, 7656. [CrossRef] [PubMed]
47. Vakilian, K.; Nateghi, L.; Javadi, A.; Anarjan, N. Optimization of Conventional and Ultrasound-Assisted Extraction of Pectin from Unripe Grape Pomace: Extraction Yield, Degree of Esterification, and Galacturonic Acid Content. *J. Food Meas. Charact.* **2023**, *17*, 5777–5793. [CrossRef]
48. Arora, S.; Kataria, P.; Ahmad, W.; Mishra, R.; Upadhyay, S.; Dobhal, A.; Bisht, B.; Hussain, A.; Kumar, V.; Kumar, S. Microwave Assisted Green Extraction of Pectin from Citrus Maxima Albedo and Flavedo, Process Optimization, Characterisation and Comparison with Commercial Pectin. *Food Anal. Methods* **2023**, *17*, 105–118. [CrossRef]
49. Zakaria, N.A.; Abd Rahman, N.H.; Rahman, R.A.; Zaidel, D.N.A.; Hasham, R.; Illias, R.M.; Mohamed, R.; Ahmad, R.A. Extraction Optimization and Physicochemical Properties of High Methoxyl Pectin from Ananas Comosus Peel Using Microwave-Assisted Approach. *J. Food Meas. Charact.* **2023**, *17*, 3354–3367. [CrossRef]
50. Figueira, O.; Pereira, V.; Castilho, P.C. A Two-Step Approach to Orange Peel Waste Valorization: Consecutive Extraction of Pectin and Hesperidin. *Foods* **2023**, *12*, 3834. [CrossRef]
51. Şen, E.; Göktürk, E.; Hajiyev, V.; Uğuzdoğan, E. Comparisons of Pulsed Ultrasound-Assisted and Hot-Acid Extraction Methods for Pectin Extraction under Dual Acid Mixtures from Onion (*Allium Cepa* L.) Waste. *Food Sci. Nutr.* **2023**, *11*, 7320–7329. [CrossRef] [PubMed]
52. Minzanova, S.T.; Mironov, V.F.; Arkhipova, D.M.; Khabibullina, A.V.; Mironova, L.G.; Zakirova, Y.M.; Milyukov, V.A. Biological Activity and Pharmacological Application of Pectic Polysaccharides: A Review. *Polymers* **2018**, *10*, 1407. [CrossRef] [PubMed]
53. Lionetti, V.; Cervone, F.; Bellincampi, D. Methyl Esterification of Pectin Plays a Role during Plant–Pathogen Interactions and Affects Plant Resistance to Diseases. *J. Plant Physiol.* **2012**, *169*, 1623–1630. [CrossRef] [PubMed]
54. Al-Hazmi, N.E.; Naguib, D.M. Agricultural Wastes Polysaccharides Promising Soil Fertilizer Improves Plant Growth and Resistance against Soil-Borne Pathogens. *Plant Soil.* **2024**, *495*, 675–697. [CrossRef]
55. Kraut-Cohen, J.; Shapiro, O.H.; Dror, B.; Cytryn, E. Pectin Induced Colony Expansion of Soil-Derived Flavobacterium Strains. *Front. Microbiol.* **2021**, *12*, 651891. [CrossRef]
56. Sayed, A.; Mohamed, M.M.; Abdel-raouf, M.E.S.; Mahmoud, G.A. Radiation Synthesis of Green Nanoarchitectonics of Guar Gum-Pectin/Polyacrylamide/Zinc Oxide Superabsorbent Hydrogel for Sustainable Agriculture. *J. Inorg. Organomet. Polym. Mater.* **2022**, *32*, 4589–4600. [CrossRef]
57. Sharma, R.; Bajpai, J.; Bajpai, A.K.; Acharya, S.; Kumar, B.; Singh, R.K. Assessment of Water Retention Performance of Pectin-Based Nanocarriers for Controlled Irrigation in Agriculture. *Agric. Res.* **2017**, *6*, 139–149. [CrossRef]
58. Pholsin, R.; Shiekh, K.A.; Jafari, S.; Kijpatanasilp, I.; Na Nan, T.; Suppavorasatit, I.; Assatarakul, K. Impact of Pectin Edible Coating Extracted from Cacao Shell Powder on Postharvest Quality Attributes of Tomato (*Lycopersicon esculentum* Mill.) Fruit during Storage. *Food Control* **2024**, *155*, 110023. [CrossRef]
59. Shin, Y.; Chane, A.; Jung, M.; Lee, Y. Recent Advances in Understanding the Roles of Pectin as an Active Participant in Plant Signaling Networks. *Plants* **2021**, *10*, 1712. [CrossRef]
60. Wu, M.C.; Li, H.C.; Wu, P.H.; Huang, P.H.; Wang, Y.T. Assessment of Oligogalacturonide from Citrus Pectin as a Potential Antibacterial Agent against Foodborne Pathogens. *J. Food Sci.* **2014**, *79*, M1541–M1544. [CrossRef]
61. Sun, X.; Wall, M.; Follett, P.; Liang, P.; Xu, S.; Zhong, T. Effect of Pectin Coatings Containing Trans-Cinnamaldehyde on the Postharvest Quality of Rambutan. *HortScience* **2023**, *58*, 11–15. [CrossRef]
62. Gamage, A.; Liyanapathiranage, A.; Manamperi, A.; Gunathilake, C.; Mani, S.; Merah, O.; Madhujith, T. Applications of Starch Biopolymers for a Sustainable Modern Agriculture. *Sustainability* **2022**, *14*, 6085. [CrossRef]

63. Hernández-Carmona, F.; Morales-Matos, Y.; Lambis-Miranda, H.; Pasqualino, J. Starch Extraction Potential from Plantain Peel Wastes. *J. Environ. Chem. Eng.* **2017**, *5*, 4980–4985. [CrossRef]
64. Kaur, J.; Borah, A.; Chutia, H.; Gupta, P. Extraction, Modification, and Characterization of Native Litchi Seed (*Litchi chinesis* Sonn.) Starch. *J. Sci. Food Agric.* **2024**, *104*, 215–224. [CrossRef]
65. Weligama Thuppahige, V.T.; Moghaddam, L.; Welsh, Z.G.; Wang, T.; Xiao, H.W.; Karim, A. Extraction and Characterisation of Starch from Cassava (*Manihot esculenta*) Agro-Industrial Wastes. *LWT* **2023**, *182*, 114787. [CrossRef]
66. Nakthong, N.; Wongsagonsup, R.; Amornsakchai, T. Characteristics and Potential Utilizations of Starch from Pineapple Stem Waste. *Ind. Crops Prod.* **2017**, *105*, 74–82. [CrossRef]
67. Mieles-Gómez, L.; Quintana, S.E.; García-Zapateiro, L.A. Ultrasound-Assisted Extraction of Mango (*Mangifera indica*) Kernel Starch: Chemical, Techno-Functional, and Pasting Properties. *Gels* **2023**, *9*, 136. [CrossRef]
68. Rosseto, M.; Krein, D.D.C.; Balbé, N.P.; Dettmer, A. Starch–Gelatin Film as an Alternative to the Use of Plastics in Agriculture: A Review. *J. Sci. Food Agric.* **2019**, *99*, 6671–6679. [CrossRef]
69. Saberi Riseh, R.; Hassanisaadi, M.; Vatankhah, M.; Kennedy, J.F. Encapsulating Biocontrol Bacteria with Starch as a Safe and Edible Biopolymer to Alleviate Plant Diseases: A Review. *Carbohydr. Polym.* **2023**, *302*, 120384. [CrossRef]
70. Supare, K.; Mahanwar, P.A. Starch-Derived Superabsorbent Polymers in Agriculture Applications: An Overview. *Polym. Bull.* **2022**, *79*, 5795–5824. [CrossRef]
71. Klemm, D.; Heublein, B.; Fink, H.P.; Bohn, A. Cellulose: Fascinating Biopolymer and Sustainable Raw Material. *Angew. Chem.—Int. Ed.* **2005**, *44*, 3358–3393. [CrossRef]
72. El-Bestawy, E.; Eltaweil, A.S.; Khallaf, N.S. Effective Production of Bacterial Cellulose Using Acidic Dairy Industry By-Products and Agro Wastes. *Sustain. Chem. Pharm.* **2023**, *33*, 101064. [CrossRef]
73. Sagdic-Oztan, C.; Koschella, A.; Heinze, T.; Karaguler, N.G.; Tuter, M. Preparation of Bacterial Cellulose Using Enzymatic Hydrolysate of Olive Pomace as Carbon Source. *Bioresources* **2023**, *18*, 4168–4181. [CrossRef]
74. Yekta, R.; Abedi-Firoozjah, R.; Azimi Salim, S.; Khezerlou, A.; Abdolmaleki, K. Application of Cellulose and Cellulose Derivatives in Smart/Intelligent Bio-Based Food Packaging. *Cellulose* **2023**, *30*, 9925–9953. [CrossRef]
75. Chen, R.; Hou, Y.; Zhang, J.; Cui, J.; Li, G. Feasibility for the Preparation of Aerogels with Celluloses Extracted Mildly from Waste Palm Leaves. *Nord. Pulp Pap. Res. J.* **2023**, *38*, 197–207. [CrossRef]
76. Klunklin, W.; Hinmo, S.; Thipchai, P.; Rachtanapun, P. Effect of Bleaching Processes on Physicochemical and Functional Properties of Cellulose and Carboxymethyl Cellulose from Young and Mature Coconut Coir. *Polymers* **2023**, *15*, 3376. [CrossRef] [PubMed]
77. Zhao, B.; Liu, R.; Guo, Q.; Xu, G.; Zhang, L.; Sun, P.; Cao, Y.; Hu, S. The Use of Newly Isolated Fungal Cultures for the Selective Delignification of Bamboo Culms. *Front. Bioeng. Biotechnol.* **2023**, *11*, 1265420. [CrossRef] [PubMed]
78. Muniraj, I.K.; Anbu, P.V.; Parthiban, K.T.; Uthandi, S. A New Enzolv Process for Simultaneous Delignification and Lignin-Derived Chemical Production from the Woody Biomass of Melia Dubia. *Biomass Convers. Biorefin.* **2023**, *13*, 14557–14571. [CrossRef]
79. Karanicola, P.; Patsalou, M.; Stergiou, P.Y.; Kavallieratou, A.; Evripidou, N.; Christou, P.; Panagiotou, G.; Damianou, C.; Papamichael, E.M.; Koutinas, M. Ultrasound-Assisted Dilute Acid Hydrolysis for Production of Essential Oils, Pectin and Bacterial Cellulose via a Citrus Processing Waste Biorefinery. *Bioresour. Technol.* **2021**, *342*, 126010. [CrossRef] [PubMed]
80. Hasanin, M.S.; Abdelraof, M.; Hashem, A.H.; El Saied, H. Sustainable Bacterial Cellulose Production by Achromobacter Using Mango Peel Waste. *Microb. Cell Fact.* **2023**, *22*, 24. [CrossRef]
81. Ahmad, D.F.B.A.; Wasli, M.E.; Tan, C.S.Y.; Musa, Z.; Chin, S.F. Eco-Friendly Cellulose-Based Hydrogels Derived from Wastepapers as a Controlled-Release Fertilizer. *Chem. Biol. Technol. Agric.* **2023**, *10*, 36. [CrossRef]
82. Callaghan, C.; Califano, D.; Feresin Gomes, M.H.; Pereira de Carvalho, H.W.; Edler, K.J.; Mattia, D. Cellulose Acetate Microbeads for Controlled Delivery of Essential Micronutrients. *ACS Sustain. Chem. Eng.* **2023**, *11*, 4749–4758. [CrossRef]
83. Ali, S.; Akbar Anjum, M.; Sattar Khan, A.; Nawaz, A.; Ejaz, S.; Khaliq, G.; Iqbal, S.; Ullah, S.; Naveed Ur Rehman, R.; Moaaz Ali, M.; et al. Carboxymethyl Cellulose Coating Delays Ripening of Harvested Mango Fruits by Regulating Softening Enzymes Activities. *Food Chem.* **2022**, *380*, 131804. [CrossRef] [PubMed]
84. Das, D.; Bhattacharjee, S.; Bhaladhare, S. Preparation of Cellulose Hydrogels and Hydrogel Nanocomposites Reinforced by Crystalline Cellulose Nanofibers (CNFs) as a Water Reservoir for Agriculture Use. *ACS Appl. Polym. Mater.* **2023**, *5*, 2895–2904. [CrossRef]
85. Das, D.; Prakash, P.; Rout, P.K.; Bhaladhare, S. Synthesis and Characterization of Superabsorbent Cellulose-Based Hydrogel for Agriculture Application. *Starch/Staerke* **2021**, *73*, 1900284. [CrossRef]
86. Xu, Y.; Li, Q.; Man, L. Bamboo-Derived Carboxymethyl Cellulose for Liquid Film as Renewable and Biodegradable Agriculture Mulching. *Int. J. Biol. Macromol.* **2021**, *192*, 611–617. [CrossRef] [PubMed]
87. Wang, Q.; Duan, Y.; Huang, Y.; Teng, Y.; Li, C.; Tao, Y.; Lu, J.; Du, J.; Wang, H. Multifunctional Soybean Protein Isolate-Graft-Carboxymethyl Cellulose Composite as All-Biodegradable and Mechanically Robust Mulch Film for "Green" Agriculture. *Carbohydr. Polym.* **2024**, *323*, 121410. [CrossRef]
88. Chen, Y.; Yang, S.; Qiu, Z.; Li, Y.; Qiu, F.; Zhang, T. Fabrication of Flexible AgNW/Cellulose Hybrid Film with Heat Preservation and Antibacterial Properties for Agriculture Application. *Cellulose* **2021**, *28*, 8693–8704. [CrossRef]
89. Islam, S.; Bhuiyan, M.A.R.; Islam, M.N. Chitin and Chitosan: Structure, Properties and Applications in Biomedical Engineering. *J. Polym. Environ.* **2017**, *25*, 854–866. [CrossRef]

90. Peniche Covas, C.A.; Argüelles-Monal, W.; Goycoolea, F.M. Chitin and Chitosan: Major Sources, Properties and Applications. In *Monomers, Polymers and Composites from Renewable Resources*; Elsevier: Amsterdam, The Netherlands, 2008; pp. 517–542. [CrossRef]
91. De Queiroz Antonino, R.S.C.M.; Fook, B.R.P.L.; Lima, V.A.D.O.; Rached, R.Í.D.F.; Lima, E.P.N.; Lima, R.J.D.S.; Covas, C.A.P.; Fook, M.V.L. Preparation and Characterization of Chitosan Obtained from Shells of Shrimp (*Litopenaeus vannamei* Boone). *Mar. Drugs* **2017**, *15*, 141. [CrossRef] [PubMed]
92. Tayel, A.A.; Moussa, S.H.; El-Tras, W.F.; Elguindy, N.M.; Opwis, K. Antimicrobial Textile Treated with Chitosan from Aspergillus Niger Mycelial Waste. *Int. J. Biol. Macromol.* **2011**, *49*, 241–245. [CrossRef]
93. Zininga, J.T.; Puri, A.K.; Dlangamandla, N.; Wang, Z.; Singh, S.; Permaul, K. Integrated Biorefinery of Mucor Circinelloides Biomass and Sugarcane Bagasse for Application of High-Value Biopolymers. *Biomass Convers. Biorefin.* **2023**. [CrossRef]
94. Habibi, A.; Karami, S.; Varmira, K.; Hadadi, M. Key Parameters Optimization of Chitosan Production from Aspergillus Terreus Using Apple Waste Extract as Sole Carbon Source. *Bioprocess. Biosyst. Eng.* **2021**, *44*, 283–295. [CrossRef] [PubMed]
95. de Souza, A.F.; Galindo, H.M.; de Lima, M.A.B.; Ribeaux, D.R.; Rodríguez, D.M.; da Silva Andrade, R.F.; Gusmão, N.B.; de Campos-Takaki, G.M. Biotechnological Strategies for Chitosan Production by Mucoralean Strains and Dimorphism Using Renewable Substrates. *Int. J. Mol. Sci.* **2020**, *21*, 4286. [CrossRef]
96. Romanazzi, G.; Moumni, M. Chitosan and Other Edible Coatings to Extend Shelf Life, Manage Postharvest Decay, and Reduce Loss and Waste of Fresh Fruits and Vegetables. *Curr. Opin. Biotechnol.* **2022**, *78*, 102834. [CrossRef]
97. Tiwari, M.; Singh, R.; Jha, R.; Singh, P. Heritable Priming by Trichoderma: A Sustainable Approach for Wheat Protection against Bipolaris Sorokiniana. *Front. Plant Sci.* **2022**, *13*, 1050765. [CrossRef] [PubMed]
98. Prasad, R.D.; Chandrika, K.S.V.P.; Godbole, V. A Novel Chitosan Biopolymer Based Trichoderma Delivery System: Storage Stability, Persistence and Bio Efficacy against Seed and Soil Borne Diseases of Oilseed Crops. *Microbiol. Res.* **2020**, *237*, 126487. [CrossRef] [PubMed]
99. Maruyama, C.R.; Guilger, M.; Pascoli, M.; Bileshy-José, N.; Abhilash, P.C.; Fraceto, L.F.; De Lima, R. Nanoparticles Based on Chitosan as Carriers for the Combined Herbicides Imazapic and Imazapyr. *Sci. Rep.* **2016**, *6*, 19768. [CrossRef] [PubMed]
100. Możejko-Ciesielska, J.; Kiewisz, R. Bacterial Polyhydroxyalkanoates: Still Fabulous? *Microbiol. Res.* **2016**, *192*, 271–282. [CrossRef] [PubMed]
101. Campanari, S.; E Silva, F.A.; Bertin, L.; Villano, M.; Majone, M. Effect of the Organic Loading Rate on the Production of Polyhydroxyalkanoates in a Multi-Stage Process Aimed at the Valorization of Olive Oil Mill Wastewater. *Int. J. Biol. Macromol.* **2014**, *71*, 34–41. [CrossRef] [PubMed]
102. Verdini, F.; Tabasso, S.; Mariatti, F.; Bosco, F.; Mollea, C.; Calcio Gaudino, E.; Cirio, A.; Cravotto, G. From Agri-Food Wastes to Polyhydroxyalkanoates through a Sustainable Process. *Fermentation* **2022**, *8*, 556. [CrossRef]
103. Razzaq, S.; Shahid, S.; Farooq, R.; Noreen, S.; Perveen, S.; Bilal, M. Sustainable Bioconversion of Agricultural Waste Substrates into Poly (3-Hydroxyhexanoate) (Mcl-PHA) by Cupriavidus Necator DSM 428. *Biomass Convers. Biorefin.* **2022**, *14*, 9429–9439. [CrossRef]
104. Unrean, P.; Napathorn, S.C.; Tee, K.L.; Wong, T.S.; Champreda, V. Lignin to Polyhydroxyalkanoate Bioprocessing by Novel Strain of Pseudomonas Monteilii. *Biomass Convers. Biorefin.* **2023**, *12*, 4651–4657. [CrossRef]
105. Matos, M.; Cruz, R.A.P.; Cardoso, P.; Silva, F.; Freitas, E.B.; Carvalho, G.; Reis, M.A.M. Combined Strategies to Boost Polyhydroxyalkanoate Production from Fruit Waste in a Three-Stage Pilot Plant. *ACS Sustain. Chem. Eng.* **2021**, *9*, 8270–8279. [CrossRef]
106. Kag, S.; Kumar, P.; Kataria, R. Potato Peel Waste as an Economic Feedstock for PHA Production by Bacillus Circulans. *Appl. Biochem. Biotechnol.* **2023**. [CrossRef]
107. Cascone, G.; D'Emilio, A.; Buccellato, E.; Mazzarella, R. New Biodegradable Materials for Greenhouse Soil Mulching. *Acta Hortic.* **2008**, *801 Pt 1*, 283–290. [CrossRef]
108. Tian, Y.; Wang, J.H. Polyhydroxyalkanoates for Biodegradable Mulch Films Applications. *ACS Symp. Ser.* **2020**, *1373*, 145–160. [CrossRef]
109. Botta, L.; Mistretta, M.C.; Palermo, S.; Fragalà, M.; Pappalardo, F. Characterization and Processability of Blends of Polylactide Acid with a New Biodegradable Medium-Chain-Length Polyhydroxyalkanoate. *J. Polym. Environ.* **2015**, *23*, 478–486. [CrossRef]
110. Wang, J.H.; Tian, Y.; Zhou, B. Degradation and Stabilization of Poly(Butylene Adipate-Co-Terephthalate)/Polyhydroxyalkanoate Biodegradable Mulch Films Under Different Aging Tests. *J. Polym. Environ.* **2022**, *30*, 1366–1379. [CrossRef]
111. Othman, N.A.F.; Selambakkannu, S.; Seko, N. Biodegradable Dual-Layer Polyhydroxyalkanoate (Pha)/Polycaprolactone (Pcl) Mulch Film for Agriculture: Preparation and Characterization. *Energy Nexus* **2022**, *8*, 100137. [CrossRef]
112. Mehta, R.; Kumar, V.; Bhunia, H.; Upadhyay, S.N. Synthesis of Poly(Lactic Acid): A Review. *J. Macromol. Sci. Part C Polym. Rev.* **2005**, *45*, 325–349. [CrossRef]
113. Ahmad, A.; Banat, F.; Taher, H. A Review on the Lactic Acid Fermentation from Low-Cost Renewable Materials: Recent Developments and Challenges. *Environ. Technol. Innov.* **2020**, *20*, 101138. [CrossRef]
114. Huang, J.; Wang, J.; Liu, S. Advanced Fermentation Techniques for Lactic Acid Production from Agricultural Waste. *Fermentation* **2023**, *9*, 765. [CrossRef]
115. Mazzoli, R. Current Progress in Production of Building-Block Organic Acids by Consolidated Bioprocessing of Lignocellulose. *Fermentation* **2021**, *7*, 248. [CrossRef]

116. D'ambrosio, S.; Zaccariello, L.; Sadiq, S.; D'Albore, M.; Battipaglia, G.; D'Agostino, M.; Battaglia, D.; Schiraldi, C.; Cimini, D. Grape Stalk Valorization: An Efficient Re-Use of Lignocellulosic Biomass through Hydrolysis and Fermentation to Produce Lactic Acid from Lactobacillus Rhamnosus IMC501. *Fermentation* **2023**, *9*, 616. [CrossRef]
117. Cox, R.; Narisetty, V.; Castro, E.; Agrawal, D.; Jacob, S.; Kumar, G.; Kumar, D.; Kumar, V. Fermentative Valorisation of Xylose-Rich Hemicellulosic Hydrolysates from Agricultural Waste Residues for Lactic Acid Production under Non-Sterile Conditions. *Waste Manag.* **2023**, *166*, 336–345. [CrossRef]
118. Ozer Uyar, G.E.; Uyar, B. Potato Peel Waste Fermentation by Rhizopus Oryzae to Produce Lactic Acid and Ethanol. *Food Sci. Nutr.* **2023**, *11*, 5908–5917. [CrossRef]
119. Liu, Y.J.; Zhang, Y.; Chi, F.; Chen, C.; Wan, W.; Feng, Y.; Song, X.; Cui, Q. Integrated Lactic Acid Production from Lignocellulosic Agricultural Wastes under Thermal Conditions. *J. Environ. Manag.* **2023**, *342*, 118281. [CrossRef]
120. Qi, K.; Chen, C.; Yan, F.; Feng, Y.; Bayer, E.A.; Kosugi, A.; Cui, Q.; Liu, Y.J. Coordinated β-Glucosidase Activity with the Cellulosome Is Effective for Enhanced Lignocellulose Saccharification. *Bioresour. Technol.* **2021**, *337*, 125441. [CrossRef] [PubMed]
121. Wan-Mothar, W.A.Q.I.; Khalid, N.I.; Rahim, M.H.A.; Luthfi, A.A.I.; Zaini, N.S.M.; Din, N.A.S.; Zaini, N.A.M. Underutilized Malaysian Agro-Industrial Wastes as Sustainable Carbon Sources for Lactic Acid Production. *Fermentation* **2023**, *9*, 905. [CrossRef]
122. Merino, D.; Zych, A.; Athanassiou, A. Biodegradable and Biobased Mulch Films: Highly Stretchable PLA Composites with Different Industrial Vegetable Waste. *ACS Appl. Mater. Interfaces* **2022**, *14*, 46920–46931. [CrossRef] [PubMed]
123. Swetha, T.A.; Bora, A.; Mohanrasu, K.; Balaji, P.; Raja, R.; Ponnuchamy, K.; Muthusamy, G.; Arun, A. A Comprehensive Review on Polylactic Acid (PLA)—Synthesis, Processing and Application in Food Packaging. *Int. J. Biol. Macromol.* **2023**, *234*, 123715. [CrossRef]
124. Maraveas, C.; Bayer, I.S.; Bartzanas, T. 4D Printing: Perspectives for the Production of Sustainable Plastics for Agriculture. *Biotechnol. Adv.* **2022**, *54*, 107785. [CrossRef] [PubMed]
125. Andrzejewski, J.; Szostak, M.; Barczewski, M.; Łuczak, P. Cork-Wood Hybrid Filler System for Polypropylene and Poly(Lactic Acid) Based Injection Molded Composites. Structure Evaluation and Mechanical Performance. *Compos. B Eng.* **2019**, *163*, 655–668. [CrossRef]
126. Dalu, M.; Temiz, A.; Altuntaş, E.; Demirel, G.K.; Aslan, M. Characterization of Tanalith E Treated Wood Flour Filled Polylactic Acid Composites. *Polym. Test.* **2019**, *76*, 376–384. [CrossRef]
127. Petchwattana, N.; Covavisaruch, S. Mechanical and Morphological Properties of Wood Plastic Biocomposites Prepared from Toughened Poly(Lactic Acid) and Rubber Wood Sawdust (*Hevea brasiliensis*). *J. Bionic Eng.* **2014**, *11*, 630–637. [CrossRef]
128. Borysiuk, P.; Boruszewski, P.; Auriga, R.; Danecki, L.; Auriga, A.; Rybak, K.; Nowacka, M. Influence of a Bark-Filler on the Properties of PLA Biocomposites. *J. Mater. Sci.* **2021**, *56*, 9196–9208. [CrossRef]
129. Sambudi, N.S.; Lin, W.Y.; Harun, N.Y.; Mutiari, D. Modification of Poly(Lactic Acid) with Orange Peel Powder as Biodegradable Composite. *Polymers* **2022**, *14*, 4126. [CrossRef] [PubMed]
130. Coiai, S.; Di Lorenzo, M.L.; Cinelli, P.; Righetti, M.C.; Passaglia, E. Binary Green Blends of Poly(Lactic Acid) with Poly(Butylene Adipate-Co-Butylene Terephthalate) and Poly(Butylene Succinate-Co-Butylene Adipate) and Their Nanocomposites. *Polymers* **2021**, *13*, 2489. [CrossRef] [PubMed]
131. Zhang, J.; Hirschberg, V.; Rodrigue, D. Mechanical Fatigue of Biodegradable Polymers: A Study on Polylactic Acid (PLA), Polybutylene Succinate (PBS) and Polybutylene Adipate Terephthalate (PBAT). *Int. J. Fatigue* **2022**, *159*, 106798. [CrossRef]
132. Hamad, K.; Kaseem, M.; Ayyoob, M.; Joo, J.; Deri, F. Polylactic Acid Blends: The Future of Green, Light and Tough. *Prog. Polym. Sci.* **2018**, *85*, 83–127. [CrossRef]
133. Duque-Acevedo, M.; Belmonte-Ureña, L.J.; Cortés-García, F.J.; Camacho-Ferre, F. Agricultural Waste: Review of the Evolution, Approaches and Perspectives on Alternative Uses. *Glob. Ecol. Conserv.* **2020**, *22*, e00902. [CrossRef]
134. Lin, C.S.K.; Pfaltzgraff, L.A.; Herrero-Davila, L.; Mubofu, E.B.; Abderrahim, S.; Clark, J.H.; Koutinas, A.A.; Kopsahelis, N.; Stamatelatou, K.; Dickson, F.; et al. Food Waste as a Valuable Resource for the Production of Chemicals, Materials and Fuels. Current Situation and Global Perspective. *Energy Environ. Sci.* **2013**, *6*, 426–464. [CrossRef]
135. O'Connor, J.; Hoang, S.A.; Bradney, L.; Dutta, S.; Xiong, X.; Tsang, D.C.W.; Ramadass, K.; Vinu, A.; Kirkham, M.B.; Bolan, N.S. A Review on the Valorisation of Food Waste as a Nutrient Source and Soil Amendment. *Environ. Pollut.* **2021**, *272*, 115985. [CrossRef]
136. Ong, K.L.; Kaur, G.; Pensupa, N.; Uisan, K.; Lin, C.S.K. Trends in Food Waste Valorization for the Production of Chemicals, Materials and Fuels: Case Study South and Southeast Asia. *Bioresour. Technol.* **2018**, *248*, 100–112. [CrossRef]
137. Biobased, Biodegradable and Compostable Plastics. Available online: https://environment.ec.europa.eu/topics/plastics/biobased-biodegradable-and-compostable-plastics_en#:~:text=Biobased%20plastics%20are%20fully%20or,not%20necessarily%20biodegradable%20or%20compostable (accessed on 9 February 2024).
138. Zhao, B.; Rehati, P.; Yang, Z.; Cai, Z.; Guo, C.; Li, Y. The Potential Toxicity of Microplastics on Human Health. *Sci. Total Environ.* **2024**, *912*, 168946. [CrossRef] [PubMed]

MDPI AG
Grosspeteranlage 5
4052 Basel
Switzerland
Tel.: +41 61 683 77 34

Applied Sciences Editorial Office
E-mail: applsci@mdpi.com
www.mdpi.com/journal/applsci

www.ingramcontent.com/pod-product-compliance
Lightning Source LLC
Chambersburg PA
CBHW041950260926
51850CB00019B/2285
* 9 7 8 3 7 2 5 8 6 3 0 4 4 *